CONTESTING CASTRO

Today they stand as enemies, but in the 1950s, few countries were as closely intertwined as Cuba and the United States. Thousands of Americans (including Ernest Hemingway and Errol Flynn) lived on the island, and, in the United States, dancehalls swayed to the mambo beat. The strong-arm Batista regime depended on Washington's support, and it invited American gangsters like Meyer Lansky to build fancy casinos for U.S. tourists. Major league scouts searched for Cuban talent: The New York Giants even offered a contract to a young pitcher named Fidel Castro. In 1955, Castro did come to the United States, but not for baseball: He toured the country to raise money for a revolution.

Thomas Paterson tells the fascinating story of Castro's insurrection, from that early fund-raising trip to Batista's fall and the flowering of the Cuban Revolution that has bedeviled the United States for more than three decades. With evocative prose and a swift-moving narrative, Paterson recreates the love-hate relationship between the two nations, then traces the intrigue of the insurgency, the unfolding revolution, and the sources of the Bay of Pigs invasion, CIA assassination plots, and the missile crisis. The drama ranges from Havana blackjack tables to Miami streets; from the crowded deck of the *Granma*, the frail boat that carried the *Fidelistas* to Cuba from Mexico, to the Eisenhower and Kennedy White Houses; from Batista's fortified palace to mountain hideouts where Raúl Castro held Americans hostage. Drawing upon impressive international research, including recently declassified documents and interviews, Paterson reveals how Washington, fixed on the issue of Communism, failed to grasp the widespread disaffection from Batista. The Eisenhower administration alienated Cubans by supplying arms to a hated regime, by sustaining Cuba's economic dependency, and by publicly backing Batista. As Batista self-destructed, U.S. officials launched third-force conspiracies in a vain attempt to block

CONTESTING CASTRO

The United States and the Triumph of the Cuban Revolution

THOMAS G. PATERSON

New York Oxford
OXFORD UNIVERSITY PRESS
1994

Oxford University Press

Oxford New York Toronto
Delhi Bombay Calcutta Madras Karachi
Kuala Lumpur Singapore Hong Kong Tokyo
Nairobi Dar es Salaam Cape Town
Melbourne Aukland

and associated companies in
Berlin Ibadan

Published by Oxford University Press, Inc.,
200 Madison Avenue, New York, New York 10016

Library of Congress Cataloging-in-Publication Data
Paterson, Thomas G., 1941–
Contesting Castro : the United States and the triumph
of the Cuban Revolution / Thomas G. Paterson.
p. cm. Includes bibliographical references and index.
ISBN 0-19-508630-9
1. United States—Foreign relations—Cuba.
2. Cuba—Foreign relations—United States.
3. Castro, Fidel, 1927– .
4. Cuba—History—Revolution, 1959. I. Title.
E183.8.C9P36 1994
327.7307291—dc20 93-24260

1 3 5 7 9 8 6 4 2

Printed in the United States of America
on acid-free paper

For Elizabeth

PREFACE

This book seeks not only to explain the origins of the U.S. collision with revolutionary Cuba and the reasons for the U.S. failure to block Fidel Castro's rise to power, but also to tell the dramatic tale of the Cuban insurrection against the Batista dictatorship in 1956–59. The many documents which undergird *Contesting Castro*—Cuban, American, British, Canadian, and more—permitted an unusual opportunity to move back and forth between high-level policymaking, as in the National Security Council and the State Department, and the ground level of clandestine meetings, bombings, assassinations, mountain skirmishes, gun-smuggling, kidnappings, urban terrorism, political rivalries, consumer protests against electric and telephone rates, journalists' scoops, FBI stake-outs, CIA conspiracies, casino gambling, baseball, diplomats' debates, and much more.

The British Ambassador to Cuba once remarked that "the picture changed so rapidly from day to day—and even at times from hour to hour."[1] The location of the story shifts again and again, from Havana to Santiago de Cuba, from Mexico to the Sierra Maestra Mountains, from New York to Key West, from Texas to California, from Washington to Miami, from Britain to Venezuela, and many places in between.

The introduction sets the questions and themes for this book. Part I establishes the Cuban-American relationship at mid-century—the extensive political, economic, cultural, and military links that generated both Cuba's attraction to the United States and its rejection of dependency. The book turns next, in Parts II and III, to the hair-trigger events that led to the surprising success of Fidel Castro's 26th of July Movement. The myriad ways that North Americans intersected with the rebellion from 1956 to 1959 fed the anti-Americanism that became ascendant among many anti-Batista forces. The interaction also solidified an anti-Castro U.S. policy based upon the conclusion that the Cuban uprising threatened U.S. interests. The fourth part of the book details several U.S. "third-force" plots hatched to dump an unwilling Batista in order to block an uncooperative Castro. The concluding section addresses the fundamental question—How did the United States let Cuba get away?—and explores the aftermath of Castro's victory and the lessons that Cubans and Americans drew from their clash. The embittering experiences of the late 1950s undermined accommodation for decades after.

My multinational, multiarchival research and exercising of the Freedom of Information Act (FOIA) to open classified documents from the Central Intelligence Agency (CIA), Federal Bureau of Investigation (FBI), and other institutions, and the sometimes confused, overlapping records-keeping and release policies of the U.S. government, have necessitated some decisions on citation form that the reader should know about. When a U.S. government document that I first read in the National Archives or a presidential library later appeared in a volume of the Department of State's series *Foreign Relations of the United States* (*FRUS*), I have usually cited the published document in *FRUS.* In some cases, I received State Department records under FOIA that were later filed and opened in the National Archives; for these materials, I have cited my original source—FOIA. When Spanish-language Cuban materials have also appeared in translation, I have tended to cite the English version.

Because Cubans frequently referred to Americans—the people of the United States—as North Americans, I have often used this expression in the book.

My research and writing have always benefited from the generous assistance of others. My heartiest thanks to them for their encouragement, thoughtfulness, and wisdom. Several friends and colleagues read drafts of the manuscript, and they will see the fruits of their recommendations and editing throughout: J. Garry Clifford (University of Connecticut), Frank Costigliola (University of Rhode Island), Javier Figueroa (Universidad de Puerto Rico), Walter LaFeber (Cornell University), Elizabeth Mahan (University of Connecticut), Robert J. McMahon (University of Florida), Dennis Merrill (University of Missouri at Kansas City), Alfred L. Padula (University of Maine at Portland), Louis A. Pérez, Jr. (University of South Florida), and Stephen G. Rabe (University of Texas at Dallas). I am pleased to acknowledge their constructive role in shaping this book, and I am very proud that Professors Figueroa, McMahon, Merrill, and Rabe are former students of mine who have become highly regarded scholars of U.S. relations with the Third World and/or Latin America.

Fellowships and research grants make research and writing possible for university professors. I received essential financial assistance for this project from the National Endowment for the Humanities, John Simon Guggenheim Memorial Foundation, Institute for the Study of World Politics, Hoover Presidential Library Association, American Philosophical Society, and Lyndon Baines Johnson Foundation. The Research Foundation of the University of Connecticut provided funds for travel and photocopying. This institution has long and eminently invigorated the intellectual life of the university.

Many people helped me find sources, answered queries, offered suggestions, provided documents, wrote letters of recommendation, or opened their homes during my research trips. I fear that some people who gave me a helping hand in this intellectual journey may not find their names here because of my oversight. They are no less appreciated. Special thanks, then, to Richard Baker, Chip Beadle, Jules Benjamin, Barton J. Bernstein, Richard D. Brown, Dan Caldwell, William Chafe, Thomas B. Curtis, Richard Cushing, Jorge Domínguez, Theodore Draper, Sylvia Figueroa, Miriam Gilbert, Paul Goodwin, Kenneth J. Hagan, Hugh Hamill, George Herring, Carl Hiller, John Jenswold, Constantine N. Kangles, Ronald D. Landa, Francisco López Segrera, Victor Lord, James T. Patterson, Anita Pawley, John Plank, Ronald Pruessen, Donald Ritchie, Riordan Roett, John Rourke, R. Roy Rubottom, Jr., Jeffrey Safford, Francisco Scarano, William Slany, George Smathers, Wayne Smith, Robert A. Stevenson, Mark Stoler, Stephen Streeter, Athan Theoharis, Seth Tillman, Nancy B. Tucker, Lucien S. Vandenbroucke, Park F. Wollam, Clarence R. Wyatt, and Thomas Zoumaras.

At the University of Connecticut, Elizabeth French, Laura Grant, Bradley Hale, Ellen Kerley, Shane Maddock, Paul Manning, Carl Murdock, Brian Murphy, Barney J. Rickman III, Noah Schwartz, and Lori Van Pembroke served well as research assistants over the years of this project. On numerous occasions Stephen Streeter patiently walked me through word-processing procedures and worked me out of computer-jams. The Interlibrary Loan Department of the Homer Babbidge Library proved indispensible. Robert Vrecenak and his staff always facilitated my search for sources with enviable good spirit and professionalism. Darlene Waller of Special Collections and Patrick McGlamery of the Map Library in the Babbidge Library also located sources for me. The Graphic and Photographic Division of the university's Center for Instructional Media and Technology helped prepare illustrations for this book.

My department chairs Bruce M. Stave and Edmund S. Wehrle and my deans Frank Vasington, Julius Elias, and Thomas Giolas facilitated leaves and the administration of fellowship awards. My departmental secretaries, Lisa Ferriere, Diedra Gosline, and Roberta Lusa, saw me through innumerable tasks. The Department of History has been a stimulating intellectual home, and I thank colleagues, students, and staff for making it such a lively and rewarding environment.

For interviews, I thank Juan Antonio Blanco, John H. Crimmins, Kelley V. Holbert, Francisco López Segrera, Hugo Pons, G. Harvey Summ, and Carlos Ciaño Zanetti.

Among the archivists and librarians who expertly guided me through documentary and manuscript collections were David Haight and Rod Soubers (Dwight D. Eisenhower Library), Will Johnson and Suzanne Forbes (John F. Kennedy Library), David Humphrey and Nancy Smith (Lyndon B. Johnson Library), Milton Gustafson, Kathy Nicastro, David Langbart, and Wilbert Mahoney (National Archives), Paul Marsden (National Archives of Canada), James H. Hutson (Library of Congress), Jerry Brovery (Federal Bureau of Investigation), Dean Allard (Naval Historical Center), Elena Danielson (Hoover Institution), Judith A. Schiff and William R. Massa, Jr. (Yale University Library), Nancy Bressler and Nancy Young (Princeton University), Martin Schmitt (University of Oregon Library), Fred Diers and Michael Swenson (Boise Cascade), Teresa Hickey (Bank of America), Tina L. Lavato (Natonal Defense University), Ronald Grele (Columbia University Oral History Project), Cary Conn (National Archives, Suitland, Maryland), Charles S. Kennedy (Foreign Affairs Oral History Program, Georgetown University), Susan Naulty (Richard Nixon Library and Birthplace), John Rothman (New York Times Archives), and Connell Gallagher (University of Vermont). The Geography and Map Division of the Library of Congress provided essential maps of Cuba in the late 1950s. My thanks also to the many other librarians and archivists who helped; their depositories are identified in the notes and bibliography.

To all of the people in federal agencies who processed my Freedom of Information Act requests: You cannot be blamed for the government's restrictive guidelines governing censorship and declassification. You can be praised for handling my many requests as expeditiously as staff and regulations allowed.

Among the editors at Oxford University Press who shepherded the book to publication were Sheldon Meyer, Leona Capeless, Joellyn Ausanka, Karen Wolny, and Stephanie Sakson, and I thank them for their excellent advice and editing.

This book is dedicated to M. Elizabeth Mahan, Director of the Center for Latin American and Caribbean Studies at the University of Connecticut, intellectual partner, and loving friend.

Storrs, Connecticut Thomas G. Paterson
Fall 1993

CONTENTS

CONTESTING
CASTRO

INTRODUCTION

A Puzzling Matter

THIS BOOK EXAMINES the sources of the bitter anti-Americanism of Castroism and the acrid American anti-Castroism that have coursed through U.S.-Cuba relations from the the 1950s to the 1990s. Why did these once close neighbors and allies become such inveterate enemies? President Dwight D. Eisenhower wondered aloud about the question at a 1959 news conference: "Here is a country that you would believe, on the basis of our history, would be one of our real friends." After all, he said, the United States had helped Cuba gain its independence from Spain in 1898 and then strove to put the island "on a sound basis." The president thought it "a puzzling matter to figure out just exactly why the Cubans and the Cuban [Castro] Government would be so unhappy when, after all, their principal market is right here, their best market. You would think they would want good relationships. I don't know exactly what the difficulty is."[1] This book plumbs "the difficulty."

When I traveled to Havana in 1985 to discuss this subject with Cuban specialists on U.S. foreign relations, conversation always turned to what it would take for the United States and Cuba to heal the wounds of a quarter-century's making. Diplomatic relations had been broken in 1961 and never resumed, although each nation had established Interests Sections in the other's capital under an agreement hammered out by the Jimmy Carter and Fidel Castro governments in the late 1970s. U.S. economic sanctions, first imposed in 1960, continued to pinch the island. The causes and legacy of the missile crisis of 1962 could still be seen in the large presence on the island of the Soviet Union, Cuba's economic and military lifeline. At the time of my visit, moreover, Fidel Castro was tauntingly reaffirming his support for Third World revolutions, including those in Central America, while right-wing ideologues in the Reagan Administration were clamoring for a crusade to prevent "the proliferation of Cuba-modeled states."[2] The inflammatory rhetoric rocketing between Washington and Havana impeded accommodation, as did the large, vocal Cuban exile community which had carved out a base of political power in Florida and whose ranks were constantly swelled by Cubans who fled their island homes by boat or plane.

Most of these characteristics of Cuba's relationship with the world and the United States have persisted, with one major exception: The political and economic disintegration of the Soviet Union and the end of the Cold War in the early 1990s pulled back the Soviets' subsidizing hand, deprived Cuba of an ideological partner, and sent the island into deeper economic crisis. The Castro regime entered a new time of troubles that many North Americans hoped would finally dispose of Fidel Castro and the Cuban Revolution.

Before these turbulent events, Cuba and the United States had long differed on bedrock issues. Since the 1950s, shaped by a staunch nationalism fueled after the turn of the century by anti-imperialist resentments against the United States, Cubans and their new hero Fidel Castro have insisted on absolute sovereignty and U.S. acceptance of Cuba's determination to follow an independent, revolutionary path. Shaped by a traditional hegemonic presumption toward the Caribbean and insistent upon protecting its strategic and economic interests, the United States has demanded the expulsion of the Soviets, an end to Cuba's endorsement of revolution in the Third World, and the removal of Castro's Communist government. The United States has sought Cuban compliance with U.S. Cold War and anti-revolutionary goals; Cuba has sought to demonstrate its unshackled freedom from the United States by continuing to challenge U.S. predominance in the Western Hemisphere. From the late 1950s onward, Cubans proclaimed a wrenching revolution that depreciated U.S. interests and influence, while North Americans strove to reclaim lost power on the island.

At root, the contest has been a classic one of a major power, driven by regional and global interests, at odds with a weaker state, driven by a sense of nationhood that rejected status as an obedient client. For several decades, the unflinching Castro has stuck to a revolutionary, Communist course and the unrelenting Uncle Sam has schemed to recover a hegemonic past. Castro has ruled through a repressive state that has provided a basic if meager level of welfare for the Cuban people. He has also survived everything the United States has hurled at him. The United States has largely blunted Castro's aspirations to spread the Cuban example abroad, and it has choked the Cuban economy. But several U.S. administrations have failed to depose him.

The Cuban-American schism, of course, dates from the turn of the century when the United States declared its wishes fiat in the Western Hemisphere. In 1898, President William McKinley claimed that the United States had special interests in Cuba because "it is right at our door."[3] The Platt amendment, attached to the Cuban constitution in a 1903 treaty, granted the United States the right to intervene on the island, but the provision had to be imposed on reluctant Cubans, who knew imperialism when they saw it. "The trouble about Cuba," Secretary of War Elihu Root said at the time, "is that, although it is technically a foreign country, practically and morally it occupies an intermediate position, since we have required it to become part of our political and military system."[4]

A few years later, when President Theodore Roosevelt grew impatient with unstable island politics and Cuban defiance of the U.S.-imposed order, he remarked that he was "so angry with that infernal little Cuban republic that I would like to wipe its people off the face of the earth."[5] When he ordered U.S. troops to Cuba in 1906, Roosevelt summarized the prevalent U.S. attitude toward "those ridiculous dagos": All he wanted, he said, was that Cubans "should behave themselves."[6] Cuban behavior usually fell short of U.S. expectations, even during the 1920s dictatorship of Gerardo Machado.

Nationalists led by Ramón Grau San Martín, rallying under the banner "Cuba for Cubans" in the revolution of 1933, unilaterally abrogated the Platt amendment, launched a host of reforms that diminished U.S. interests, and then refused to bend to U.S. demands that the changes be moderated. U.S. Ambassador Sumner Welles concluded that "the social revolution which is under way" was producing "confiscatory" decrees directed against American properties and interests. The Grau government, Welles feared, harbored an "intention of minimizing any form of American influence in Cuba."[7] The ambassador soon conspired with Sergeant Fulgencio Batista y Zaldívar to smash the independent-minded Cuban government. In early 1934, Batista overthrew the revolutionary regime, inaugurating the Batista era in Cuban history. The events of the 1930s and after led a later ambassador, Earl E. T. Smith, to the blunt conclusion that "the American Ambassador was the second most important man in Cuba, sometimes even more important than the President."[8]

Throughout the twentieth century, then, U.S. leaders expressed open disdain for Cuban sovereignty while they unabashedly claimed North American pre-eminence under the venerable Monroe Doctrine. CIA analysts claimed that although "popular suspicions of US motives make it necessary for governments to avoid the appearance of subservience to the US," Caribbean nations "must accommodate their policies to US security interests."[9] Latin American peoples would have to learn that "their dependence on the United States is greater than that of the United States on them," added the U.S. Commander-in-chief of the Caribbean Command.[10]

Such supremacist thinking fed on negative stereotypes of Latinos. Cartoonists of the early twentieth century, putting into visual image the derogatory utterances of U.S. officials, had sketched childlike people of color who required discipline from a stern Uncle Sam. The cartoons pictured Latin males debilitated by laziness, dishonesty, and cruelty, and seductive Latina maidens awaiting rescue from superior, masculine North Americans.[11] The image of passive, submissive, and occasionally unruly Cubans incapable of self-mastery and needing U.S. guardianship persisted into the 1950s. U.S. diplomats spoke of Cubans—especially those in rebellion against the U.S.-backed Batista regime—as an "emotional," "romantic," and "childish" people suffering from "excessive pride."[12] Cubans, North Americans believed, suffered from an "exaggerated nationalism."[13] "One nice thing about Cuba," however, *Houston Post* editors smugly wrote in 1956, "is that she can have a revolution any time hotbloods get in the mood without adding to world tension."[14]

Cubans bristled against such pompous North American condescension. But as one scholar has ably remarked about U.S. relations with Third World nations, "Americans are brilliant communicators but bad listeners."[15] A young Cuban who had grown critical of the contradiction in the United States between the practice of racial segregation and the rhetoric of equality and democracy wondered why North Americans understood so little about Cuba. "I will tell you why," he lectured. "It's because you think you are so superior to us that you don't *have* to know about us."[16] In contrast, Cubans *had* to know about the United States. The Canadian Ambassador to Cuba once observed that "the only country with which most Cubans are seriously concerned outside of their own is the United States and in few places is the influence of 'Tío Sam' more obvious . . . than in Cuba."[17]

Cubans generally liked Americans, and the attractions of the United States were many, especially for the wealthy and middle class—educational opportunities for their children, modern appliances for their homes, and a protective alliance with the world's superpower. But U.S. relations with Cuba bred a closeness that sometimes stimulated Cuban imitation of North American ways and at other times rejection and rebellion.[18] As a U.S. National Security Council report once observed, Latin Americans were "ambivalent" toward the United States. They admired advanced U.S. technology and material prosperity, "but they also express envy by disparaging US materialism," and they feared "the bogey of US economic imperialism." They admired U.S. democracy and representative government, "but they are also keenly aware of imperfections in US democracy and highly sensitive to any supposed suggestion of Anglo-American superiority over Latin or colored peoples."[19]

Prominent and ultimately triumphant Cuban dissidents who aligned with Fidel Castro shared this sensitivity. They believed, moreover, that many of their nation's woes stemmed from U.S. intrusions. It is understandable why they thought so, for U.S. decisions and activities and the interlocking of the Cuban and American economies and cultures had long directed and often dominated the life of the island. During the twentieth century, North Americans built "internal structures of hegemony" in Cuba through military interventions, occupations, threats, economic penetration, transfers of culture, and political manipulation.[20] "Hegemony" means the dominance or preponderant influence that permitted U.S. decisions to condition Cuba's politics, economy, culture, society, and military. U.S. hegemony empowered North Americans to set and maintain most of the rules by which Cubans lived and by which the Cuban-American relationship was governed.[21] Sometimes the United States boldly manhandled Cuba; at other times the U.S. grip was light, nearly invisible. The result was the same in any case: dependency. And the Cuban counterpoint to this foreign influence was a fervent nationalism laced with an anti-Americanism that Fidel Castro and many other opponents of the Batista regime imbibed, nurtured, and exploited.[22] "Cuban nationalism," the Cuban writer Carlos Alberto Montaner has explained, "like a bow being readied for combat, became menacingly taut at the sight of only one target: the United States."[23]

During the Batista fifties, both U.S. government and private interests developed ties with the dictatorship that sustained a structure of power that was often repressive, violent, and corrupt. Although North Americans praised the relationship for protecting their stakes, many Cubans resented their dependency—especially the unremitting curse of dependency on sugar that stemmed from the necessity to sell in the U.S. marketplace and from capricious U.S. decisions on import quotas. Cubans also shook accusing fingers at the U.S.-owned, monopolistic utility companies that they believed subverted the public interest. Some Cubans noted with disapproval the concentration of land and wealth, including U.S.-owned plantations, ranches, and mills, in the hands of an economic elite. Others charged that Batista and his officials were in league with U.S. companies, enriching themselves on U.S. business ventures. Criticism mounted: The United States had reduced Cuba to an economic satellite where corruption flourished.

Other aspects of Cuban-American relations spawned discontent with both Batista and the United States, especially among young students and professionals. The showy presence of North American mobsters who ran the legal casinos drew Cubans' ire, for these unsavory types collaborated with Batista to turn the island into a tourist playground of gambling, sex, and booze. Some Cubans regretted, moreover, that their country had become subsumed in the North American consumer culture, losing its independent identity. Many middle-class Cubans became anxious and restless, perceiving that their standard of living was declining compared with that of the United States. The active roles of the U.S. military, United States Information Service, Federal Bureau of Investigation, and Central Intelligence Agency in shoring up Batista's illegitimate regime also contributed to the popular view that Cubans were not masters of their own house.

Not all nationalistic Cubans embraced anti-Americanism, but even those who did not—U.S. officials admired them as "moderates"—sought to peel back the heavy overlay of North American power on the island. Cuba and the United States essentially shared a border, and borders are places where nationalism finds expression in contrasting ways. Some Cuban nationalists, like the *fidelistas,* drew sharp distinctions and built fences to emphasize separation. Others no less nationalistic, like the moderates, operated in the borderland differently, seeking opportunities for accommodation and reconciliation.[24] Castro disparaged the moderates as no better than the lackeys who had bent to U.S. dictates in the past and who had become subservient to a mentality of dependency. They could never defend Cuban interests, he declared, because they were always looking for U.S. answers to Cuban problems. Moderate nationalists retorted that Cuba's vital economic links with the United States necessitated a close and friendly relationship with the powerful northern neighbor, and that dependency could be reduced through agreement rather than through confrontation.

For the *fidelistas* who ultimately dominated Cuban politics, the proposition that "the friend [the United States] of my enemy [Batista] is my enemy" became prominent. They held responsible for Cuba's plight both *batistianos* and *norteamericanos,* which they grouped as the "selfish interests" and the "international oligarchy."[25] To oppose one was to oppose both, so integrated had they become. And the daily stories about the activities of U.S. customs agents, military advisers, businesses, CIA and FBI operatives, and notables like the casino boss Meyer Lansky ensured that a revolution to oust the Batista dictatorship also became a revolution to expel the United States from the island.

The anti-Americanism that blared from the Cuban Revolution and the anti-Castroism that sounded harshly from Washington for more than three decades solidified during the events of 1956–59, from the landing of Castro's small rebel force in Oriente Province to the rebel army's triumphant entrance into Havana. The U.S.-rebel interaction during the insurrection shaped Cuban and North American attitudes and policies, which have endured. It is to the tumultuous late 1950s that we must turn to understand what went wrong in the Cuban-American relationship after Fidel Castro's stunning victory in early 1959. We must examine the years of the insurrection if we are to fathom the familiar confrontations that came later—Castro's calls for hemispheric rebellion against U.S.-backed governments, inflammatory anti-American speeches that shocked even his sympathizers in North America, expropriation of U.S. economic interests, U.S trade embargo, Cuba's movement toward Communism and military alliance with the Soviet Union, severance of diplomatic relations, Bay of Pigs invasion, CIA covert plots and the spoiling schemes of Operation Mongoose, missile crisis, large migrations of Cubans to the United States, and repeated war scares.

This study explores why a superpower with extensive interests on the line failed to prevent the loss of a long dependent, weaker state just a hundred miles away. How did the United States let this one get away? The 1950s clash between Cuban nationalism and U.S. hegemony, between ragtag mountain rebels and hegemonic managers, is central to the answer. As North Americans discovered during the years of the rebellion, Cubans regularly made independent choices and Batista did not always follow U.S. advice. U.S. officials could not run the day-to-day affairs of the island. Seldom was U.S. manipulative power absolute. As Batista, Castro, and Eisenhower learned, powerful nations like the United States which attempt to determine the outcomes of civil wars must act within individual, national, regional, and international contexts that limit the exercise of power.

Influential individuals—Batista, Castro, the rebel commander Che Guevara, President Eisenhower, Secretary of State John Foster Dulles, Assistant Secretary of State for Inter-American Affairs R. Roy Rubottom, Jr., Ambassadors Arthur Gardner and Earl E. T. Smith, among them—decided whether to negotiate, conciliate, tolerate, resist, or fight; they floundered or persevered at key moments. Their words defined the relationship. This book especially assesses the stubborn Batista as an imperfect instrument of U.S. policy, and the charismatic Castro as a revolutionary leader, identifying the strengths and weaknesses of both. The first self-destructed; the second survived like Houdini. As for President Eisenhower, he was regularly briefed, but he seldom closely analyzed the Cuban condition. His management style delegated authority to subordinates who tended to bring a problem to him only when it had reached a critical, urgent stage. His advisers believed that they had Cuban matters under adequate control until Batista fell much faster than anyone had predicted; only then—in late 1958—did Cuba policy become a priority for Eisenhower. But by then it was too late to stop Castro.

To explain why and how the Cuban insurgents won against tremendous adversity and why and how the United States stumbled despite working diligently to block Castro, the book emphasizes the timing of events, the difficulties in trying to contain guerrilla war in inhospitable terrain, and the ardent Cuban nationalism that fueled rebellion. *Contesting Castro* also follows the backfiring machinations of the corrupt Batista government and its military establishment and the twists and turns of decisionmaking in the Eisenhower Administration. The narrative reveals a deadly combination of U.S. ignorance and arrogance and, U.S. claims to the contrary, the absence of even-handed neutrality during the civil war.

American officials at first strove to tame the rebellion against their ally Batista; then they tried to ease him out, hoping to install a moderate, pro-U.S. government. In the end the United States launched several covert "third-force" plots against Castro on the eve of his victory. U.S. officials frequently miscalculated and bungled clandestine operations. Spotlighted here are the poor quality, shallow analysis, and distorting hegemonic assumptions of U.S. leaders. Even as Castro neared the victory the United States had struggled to deny him, American officials assumed that no Cuban government could survive long without U.S. endorsement and that any Cuban leader would have to cut a deal with his northern neighbor.

Public opinion in the United States and elsewhere made U.S. officials hesitate to use force to prevent Castro's triumph. Segments of the U.S. press and Congress examined the Cuban rebellion and criticized Eisenhower's policy toward dictators. At the same time, no clearcut majoritarian opinion on the Castro insurrection took form among the American people. Washington nonetheless grew sensitive to mounting criticism that it was intervening in a civil war to prop up a brutal military regime.

The regional setting also shaped Cuban and American choices during the insurrection. The United States towered as the predominant nation in the Western Hemisphere, but its traditional role under the Monroe Doctrine continued to arouse resentments, and inter-American relations became stormy in the 1950s. As a National Security Council paper lamented early in the decade: "There is a trend in Latin America toward nationalistic regimes." The United States had to "arrest the drift in the area toward radical and nationalistic regimes."[26] Near the end of the decade, Secretary Dulles disapproved of the "tremendous surge in the direction of popular government by peoples who have practically no capacity for self-government and indeed are like children in facing this problem."[27] Fidel Castro hoped to tap the growing anti-Americanism; he fully recognized that he had the potential of becoming a popular symbol of resistance to the United States in the hemisphere. The unwillingness of the once compliant members of the Organization of American States to help the United States temper the Cuban crisis, and the swelling of popular opinion in the hemisphere against military dictatorships like Batista's, also conditioned U.S. decisions.

The Cold War dominated international politics in the 1950s, when American exaggerations of a Communist threat became legion. U.S. diplomats, especially the ambassadors in Havana, dutifully tried to find Communists in the Cuban insurrection. Fixed on the question of a Communist menace, some U.S. officials understudied—and hence underestimated—the profound reasons why Cubans were willing to die for the cause of toppling Batista. At the same time, preferring the order that the Havana regime promised and the Cold War alignment that Batista cemented, they overestimated the ability of Batista and his armed forces to collar the insurrectionists. Given the tenacity of Cold War attitudes and Castro's later conversion to Marxism-Leninism and alliance with the Soviets, *Contesting Castro* investigates in the insurrection period the very tenuous relationship between Cuban rebels and Communists.

The international system of the 1950s was also marked by worldwide currents that were steadily eroding the authority of powerful nations: anti-imperialism, decolonization, revolutionary nationalism, and social revolution. The Cuban Revolution against deeply embedded interests identified with the United States joined this phenomenon of breaking up and breaking away, as did the Vietnamese Revolution against France, the Egyptian Revolution against Great Britain, and the Hungarian Revolution against the Soviet Union. In Latin America, the trend manifested itself in the defiance that nationalists directed against the United States—conspicuous in Jacobo Arbenz's radical government in Guatemala until 1954, the assassination of U.S. ally Anastasio Somoza in Nicaragua in 1956, protests against the Nixon entourage two years later, and the Panama riots of 1959. The challenge to U.S. authority in the hemisphere was but one component of a relative decline of U.S. power precipitated by the rise of the Third World, revival of Japan and Germany, growing competition from the Soviets, difficult allies like France's Charles de Gaulle, and the budget-busting expenditures required to wage the Cold War.[28] In short, power was being redistributed in the international system, and Cuba, at the United States' expense, became a beneficiary.

What increasingly alarmed the managers of U.S. hegemony in the 1950s was that rebel Cubans under Castro's leadership, unlike other Cuban nationalists U.S. officials deemed moderate and friendly, vowed to jettison altogether their dependent status. Defending their superintendency of Latin America, U.S. officials judged that the Castro-led movement against Batista was taking Cuba beyond the historic bounds of acceptable protest. Long before the Communists seized power in Russia in 1917, the United States had adopted a coolness toward most revolutions. From the nineteenth century forward, U.S. leaders opposed revolutions that took a violent course, failed to safeguard private property rights, disrupted the social order by arousing the masses, jilted constitutional procedures, or threatened U.S. security.[29]

Castro's rebels appeared dangerous because they challenged core American values and sought to topple the strategic and economic pillars of the Cuban-American relationship: first, alignment of Cuban foreign policy with U.S. policies and, second, protection for U.S. trade and investment.[30] Castro consistently denounced the Plattist mentality and North American hegemony in Latin America and he vowed to roll back U.S. interests. The United States feared that his uncompromising nationalism would become contagious in the hemisphere. During the late 1950s, moreover, most U.S. leaders did not fear him as a Communist but rather as a neutralist who might abandon Cuban alignment with the United States in the Cold War.

To the very end of Batista's rule in early 1959, U.S. policymakers struggled with a central question: When did the insurrection endanger the hegemonic structure enough to warrant direct U.S. intervention? When did the Cubans exceed the tolerable limits of change? In the end, and too late to preserve North American interests, U.S. leaders learned that the journalist Robert Taber was right: "The Cuban Revolution is, above all else, a declaration of independence from the United States."[31]

What ultimately pulled back U.S. leaders from intervening directly or forcefully to save Batista—and to block Castro—was their presumption that Cuba was so dependent upon the United States that no Cuban would dare challenge Washington or push U.S. interests off the island. Fidel Castro and other Cuban insurrectionists, on the other hand, pledged to purge just such a hegemonic presumption from the U.S.-Cuba relationship. As they lifted their banners and rifles against the Batista dictatorship in the 1950s, Fidel Castro and other insurrectionists recoiled from Cuba's history of U.S. hegemony and dependency and promised liberation from the stultifying past.

I

Binding the Cuban-American Relationship at Mid-Century

1

Dependencies: Batista, Castro, and the United States

THE FBI OPENED a "new case" soon after Fidel Castro boarded a plane in Miami to return to his small revolutionary army training in Mexico. The Miami and New York offices of J. Edgar Hoover's famed investigatory agency then began infiltrating, monitoring, and filing reports on the 26th of July Movement clubs that Castro had helped organize during his seven-week tour of the United States in the fall of 1955.[1] Clubs in Tampa and Key West also drew the FBI's attention, and national headquarters alerted other agency branches to watch for similar groups in Bridgeport, Connecticut, and Union City, New Jersey. Orders went out as well to field offices in Philadelphia, Chicago, and Washington: Consult "established sources" about the 26th of July clubs in those metropolitan centers.[2] Eventually the FBI was spying on clubs in Los Angeles; Newark, New Jersey; and Puerto Rico too.

Bear in mind, the ever suspicious Hoover instructed, "that certain members of these clubs may pose a threat to the internal security of the United States."[3] Perhaps only a routine precaution, Hoover's instructions made no mention of any 26th of July Movement (M-26-7) links with Communism, the dreaded, if exaggerated, Cold War threat of the time.

Castro had no ties with Cuba's Communist party—at least not at this time. His dissident organization actually distrusted the Communists because of their onetime sordid alliance with Fulgencio Batista, the very dictator the rebels intended to topple. The Communists, moreover, initially had opposed Castro's challenge to Batista, dismissing it as mere adventurism.[4] Castro's most conspicuous model in the mid-1950s was the hero of the 1890s Cuban

Revolution, José Martí, not Karl Marx, V. I. Lenin, Mao Zedong, or Josef Stalin.

Still, the FBI, Central Intelligence Agency, and other U.S. agencies searched for traces of Communist contamination throughout the years of insurrection. Mostly, though, officials focused their investigations of Cuban groups in the United States on possible law violations. The Munitions Control Act restricted exports of weapons. The Voorhis Act prohibited foreigners from launching attacks against foreign governments from the United States. And the Foreign Agents Registration Act required "foreign agents" to register with the U.S. government. An anxious U.S. speculation about 26th of July entanglements with Communists nonetheless thrived during the rebellion against Batista.

As for the man whom Cuban dissidents so hated, Fulgencio Batista, U.S. officials knew that his anti-Communist credentials were quite in order. Born in 1901 in the United Fruit Company town of Banes in Oriente Province, Batista was the mulatto son of farm folk. After shifting from one job to another, he joined the army at age 20. In 1933, after the collapse of the Gerardo Machado dictatorship, Batista led the Sergeants' Revolt that helped bring the reformist Ramón Grau San Martín government to power. U.S. officials recoiled from Grau's ardent economic nationalism and from his effrontery in unilaterally abrogating the Platt amendment. Boldly asserting that no Cuban government could long survive without U.S. recognition, Ambassador Sumner Welles began to plot with Army Chief of Staff Batista to overthrow the Grau regime, which, Welles absurdly charged, had become infested with Communists. Welles and his successor Jefferson Caffery flattered Batista that he was the "only individual in Cuba who represented authority."[5] When a solicitous Batista asked Caffery what the United States wanted in exchange for recognition, the ambassador replied: a new government. In January 1934, Batista toppled Grau, who fled to Mexico.[6]

Of medium height, but with a powerful physique, the manicured, well-groomed Batista came to enjoy the good—and the high—life. "Like a panther," he stalked the enemies of his dictatorial regime.[7] A wide, warm smile could not hide the military man's cold, calculating demeanor and ruthless quest for power. Ruling Cuba from the shadows in the 1930s, he amassed considerable wealth through sweetheart deals and fraud. In 1940 he actually put himself up for the presidency and won what was probably an honest election. But after serving four years, Batista stepped down and settled into his Daytona Beach, Florida, estate. He returned to the island in 1948 after election to the Cuban Senate from Santa Clara province. Plotting with army officers and their loyal troops and exploiting a public revulsion against blatant governmental corruption, Batista seized power on March 10, 1952, in a well-planned coup that succeeded in just a few hours.

Carlos Prío Socarrás, the corrupt president whom Batista and his co-conspirators ousted, complained that U.S. intelligence "failed" to warn him about Batista's plans for a coup.[8] But the American Ambassador in Havana, Willard Beaulac, claimed that Batista's seizure of power had come as "a complete surprise to everyone" in the Cuban capital city.[9] Beaulac had learned about the takeover at 6:00 a.m. that March 10, and he quickly told Washington that some might think that the military agreement the United States had signed with Cuba only three days earlier had something to do with the coup d'état. The two events were not linked at all, he insisted, and Washington had to "combat the [mistaken] idea" that they were.

An hour later, Colonel Fred G. Hook, Jr., chief of the U.S. Air Mission in Cuba, conferred with General Batista. "Tell the Ambassador I am 100% in accordance with his wishes," the Cuban leader remarked. "All agreements are in effect."[10] The United States need not worry about its substantial interests on the Caribbean island. Although Beaulac regretted that democracy had been dealt a blow, he told the British Ambassador that "if this had to happen Batista was the best material for the job."[11]

When, on March 22, the new government's Minister of State, Miguel Angel de la Campa, pressed Beaulac for official U.S. recognition, the cautious ambassador lectured that U.S. decisions counted a great deal in the Western Hemisphere. After all, "there were countries where revolutions and upsets might be encouraged by a precipitate act of recognition by the United States." Remembering that the Communist party of Cuba had collaborated with Batista in the 1930s and had endorsed the dictator's unsuccessful presidential candidate in 1944, Beaulac asked if Batista would re-establish his ties with the Communists. Their "freedom" would be "eliminated," Campa assured him. Would conditions for private investment improve? Batista welcomed private capital, Campa said, and "business men were among the most enthusiastic supporters of the new regime" because it would "bring order out of chaos."[12] Five days later the United States extended diplomatic recognition to the Batista government.

Fidel Castro was hardly alone in judging the new regime illegimate and unworthy of international recognition, but he was bolder than most in acting to destroy Batista's handiwork. On July 26, 1953, the 26-year-old Castro and dozens of other young *compañeros* daringly attacked the Moncada Army Barracks in Santiago de Cuba, at the southeastern end of the island. The assault ended in bloody defeat for the rebels. *Batistianos* soon jailed, tortured, or killed the young rebels of July 26. The brutalities suffered by the brash insurrectionists and the publicity generated by the attack soon elevated Fidel Castro and his movement to near folk-hero status among Cubans. U.S. Embassy officials in Havana were less impressed, branding Castro "a ruthless opportunist."[13] They seemed satisfied that the Cuban government sentenced him and his co-conspirators to prison on the Isle of Pines.

In May 1955 the Moncada *fidelistas* gained their freedom under a general amnesty. On June 12 they organized Movimiento 26 de Julio (26th of July Movement).[14] Thinking his life endangered and finding freedom of expression stifled, the now-famous Castro fled to Mexico to plan his next move against Batista. "As a follower of Martí," Castro wrote just before departing Cuba, "I believe the hour has come to take rights and not to beg for them, to fight instead of pleading for them."[15] Fidel Castro vowed to return to Cuba, just as the martyr José Martí had done in 1895.

In Mexico, Castro sought to exploit the talents of Alberto Bayo Giroud, a Cuban-born guerrilla-war expert who had fought in the Spanish Civil War in the 1930s against Francisco Franco and in Nicaragua in the 1940s against Anastasio Somoza. When Castro approached the 65-year-old Bayo, the author of the training manual *One Hundred Fifty Questions to a Guerrilla* said the obvious: Get some soldiers and cash.[16] "I am going to the United States to gather men and money," Castro answered. "At the end of this year, I'll come back to see you and we shall plan what we have to do for our military training. . . ."[17] The trip northward loomed large in the young revolutionary's plot. "We are penniless," Castro noted in late August. "Each of us lives on less money than the [Cuban] army spends on any of its horses."[18] He planned to call on exiles in the United States to help finance a revolution.

After obtaining a tourist visa from the American Embassy in Mexico City, which apparently gave the application no special review, and after borrowing train fare, Fidel Castro in early October 1955 crossed into the United States at the Texas border.[19] The 29-year-old insurrectionist took his revolutionary message to Philadelphia; Bridgeport, Connecticut; and Union City, New Jersey, where Cuban communities existed. In Union City, FBI agents detained and interrogated the young exile at a meeting of sympathizers.[20]

Castro reached New York City on October 23. A black and white photograph of the young man walking across an expansive grassy field in Central Park reveals not a bewhiskered, gun-touting rebel in green guerrilla garb, but rather a well-groomed gentleman in tie, sweater vest, and double-breasted suit. Except for a moustache, he is clean-shaven. With a quizzical, almost imploring look, the tall, broad-shouldered Cuban insurrectionist strides on U.S. soil as Martí had before him.[21] Martí spent fifteen years in the United States; Castro, impatient to get on with the rebellion against Batista, intended a shorter visit.

At the Palm Garden Hall at Fifty-second Street and Eighth Avenue, 800 Cuban exiles and a few observers under cover came to see and hear Cuba's most famous ex-prisoner and unabashed rebel. Dressed up in striped tie, white-collar shirt, and suit—an uncommon uniform for a man whose friends thought him hopelessly disheveled—Castro gave an emotional speech, displaying his oft-noted oratorical skills. Lauding "our Apostle Martí," Castro pledged that "in 1956 we will be free or we will be martyrs." Asserting that his revolutionary movement had "totally disoriented" the Batista regime, a grossly false claim, Castro appealed for "a radical change in every aspect of its [Cuba's] political and social life."

"Radical" might have aroused some murmurs in the crowd, but it is not at all clear what he meant by the term at that time. He also denounced terrorism and assassination as political tactics, and he called for general elections. These words sat more comfortably with most of the Palm Garden audience, many of whom identified with Cuba's reformist Ortodoxo party. Castro himself had been running for Congress on the Ortodoxo ticket when Batista's coup torpedoed the 1952 elections. Castro's New York audience on that November 1 night even heard him praise the elderly Cosme de la Torriente, a moderate politician and architect of an ill-fated attempt to bargain with the Batista government.[22]

Did Castro mute his radicalism so as not to alienate the exiles whose money he had come to solicit?[23] Did he make "all the right noises" so as not to alarm U.S. officials whose support he also sought to woo?[24] The standard charge against Castro is that throughout his life he has said what sells at the moment. The harshest, most polemical accusation is that he has been nothing more than a cynical, self-serving, power-hungry, deceitful manipulator. A less critical and slightly more positive view has it that he has been a supremely pragmatic politician open to almost any avenue that would advance the revolution. The most charitable depiction of Fidel Castro is that he was in the 1950s a leader in training who was reading a great deal, observing other dependent nations' experiences (especially Egypt's), and feeling his way toward a world-view and workable formula for change in his country.

Castro's famous oration at the close of his trial in 1953, "History Will Absolve Me," revealed his inclinations, of course, and some in his Palm Garden audience had no doubt heard about the lengthy, extemporaneous speech that Castro later put to paper when he was in jail and then passed to friends for publication. In this document, Castro repeatedly invoked Martí and called for a just legal system, restoration of political rights, distribution of land to landless farmers, workers' sharing of profits, confiscation of the ill-begotten wealth of corrupt public officials, and nationalization of the monopolistic electric and telephone industries. These goals probably seemed more reformist than radical at the time.[25]

In any case, alert to factionalism in the Cuban exile community in the United States itself, Castro offered patriotism, not a systematic ideology or fixed agenda. At Palm Garden that night, "in a minute," money poured in from enthusiastic Cubans."[26] The New York FBI office sent its summary of the meeting to Washington. No doubt Batista's spies filed a report with Havana. After Castro's Palm Garden speech, three exile groups—Comité Obrero Democrático de Exilados y Emigrados Cubanos, Comité Ortodoxo de New York, and Acción Cívica Cubana—merged to become the Club 26 de Julio de New York.[27] The new organization soon printed and distributed pamphlets of "History Will Absolve Me."

After New York, Castro headed for Florida, long a haven for Cubans abroad. Traveling with him was Juan Manuel Márquez, an adviser and Ortodoxo leader Castro thought "brilliant."[28] Márquez had also gone into exile in Mexico, in his case after a severe beating in Havana. On November 20, Márquez accompanied the M-26-7 insurrectionist to the Flagler Theater in Miami. One thousand Cubans, including Castro's sister Lidia and his son Fidelito, heard the now familiar themes and appeals for money. Castro vowed once again to return to Cuba in 1956 to throw out the "embezzlers."[29]

When, later, the guerrilla-war master Bayo upbraided Castro for harping on 1956 as the year of invasion, because the enemy should never be tipped off, Castro smartly answered with a dictum that he had probably already heard from Bayo himself: The struggle against Batista was more than military—"it is psychological warfare."[30] Not only did Castro hope to promote anxiety in Havana through declarations on timing; he also calculated that dramatic announcements of a 1956 invasion would serve as advertisements for a movement in desperate need of public attention.

Castro knew that the duel with Batista at that moment was in good measure a contest over raising money from sources in the United States. The Batista government received official U.S. military and economic assistance. The 26th of July Movement intended to lobby against this U.S. aid to an illegitimate regime while collecting funds from sympathetic exiles and Americans. That night at the Flagler, dollars filled the collection hats—"alms for the motherland."[31]

To keep the money coming, Márquez organized a club in Miami. Revolutionary clubs also sprang up in Key West and Tampa, where Castro spoke on November 27 and December 7. When he left Miami for Mexico, Castro had collected some $9000.[32] Friends and sympathizers in Puerto Rico, Costa Rica, Venezuela, and elsewhere in Latin America also contributed essential funds.[33] Money later flowed to the 26th of July Movement from Cuba itself—

including contributions from professional and business people, sugar-mill owners, and cattle ranchers, some of them Americans or Cuban managers of U.S.-owned enterprises. While some business people feared rebel reprisals unless they paid "revolutionary taxes," others eventually sought Batista's ouster in order to end governmental corruption and mismanagement. Eventually one million dollars came from José M. Pepín Bosch y Lamarque, president of the Bacardi Rum Company. All of these monies in the late 1950s bought weapons and uniforms, financed airdrops of supplies, bribed army officers, and paid mountain peasants for the food they provided Castro's meager forces after they invaded Cuba from Mexico in December 1956.[34]

The July 26 clubs in the United States never met Castro's expectations. "The whole thing got off to a bad start with Fidel himself," complained the head of the M-26-7 Committee in Exile in the United States. "Fidel tried to please everybody and so left a series of personal contacts and rudiments of organization. These in time became personalistic clans, jealous of each other," Mario Llerena explained in disgust.[35] Most of the clubs did not become active until 1957, and their members constantly feuded. In Chicago and Miami, as FBI snoops discovered, the clubs broke up into factions.[36] The New York City exiles also splintered, compelling an angry Castro in 1958 to order anti-Batista Cubans there to unite under one organization, the Committee in Exile of the 26th of July Movement.[37]

Just before he departed the United States in December 1955, Castro summarized his grievances against Batista. The M-26-7 leader compared the empty pockets of his fellow rebels with the "ill-gotten gains" of the "embezzlers" who had used Batista politics to enrich themselves. As if leading a consumer revolt, Castro protested high electric bills and inadequate telephone service. He claimed that "foreign trusts" were stealing millions of dollars from Cuba, and he condemned the "large sums wasted in gambling, vice, and the black market."[38]

Castro did not mention the United States by name as a culprit, but there could be no mistaking his references. The electric and telephone companies were American-owned, as were many sugar plantations and mills. Substantial North American investments and exports of island sugar to the United States determined Cuba's economic condition. The Batista regime enjoyed official U.S. recognition and public approval, and his military and secret police received U.S. aid. Batista himself owned property in Daytona Beach, Florida. North American mobsters ran the gambling casinos and some of the houses of prostitution that thrived in Havana.

Castro remarked privately in mid-1955 that "the Yankee penetration of Cuba was complete."[39] From this observation, Castro drew a conclusion that strengthened throughout the decade: The revolution against Batista would necessarily become a revolution against U.S. island interests—against the Cuban dependency upon the United States that the Batista regime sustained. If one had to go, both had to go. Some Cuban scholars, no doubt influenced by their desire to improve U.S.-Cuban relations, have preferred to argue that Castro and the rebels were not anti-American, but rather anti-imperialist.[40] Surely this argument begs the question, because the primary target for anti-imperialists throughout Latin America was the United States.

Given the intertwined economic, cultural, military, and political histories of the United States and Cuba, an insurrection against a corrupt dictator, whose close ties with U.S. interests gave the appearance of subservience, could not be anything else but a challenge to the superpower of North America. But Castro could not declare revolution against the United States at the start, before the insurrectionists' triumph over Batista, because he and his rebel movement also depended upon the United States.

Castro and M-26-7 needed help from members of the exile community in North America—their money, their gunrunning, their sons and daughters, their public endorsement, and their lobbying efforts with U.S. officials. Castro also needed the help of U.S. journalists to publicize his movement. He needed U.S. diplomats to press Batista to curb the brutalities of the police and military. Castro needed, too, the assistance of U.S. Embassy personnel to protect jailed 26th of July leaders from death or torture. He needed dollars and communications equipment from any American source possible. Anti-Batista exiles also used the United States as a sanctuary to escape Batista's henchmen, to plot their uprising, and to launch attacks and run supplies. M-26-7 needed as much tolerance as U.S. officials could allow under the law.

Castro also worked to cultivate a tolerant, friendly public opinion in the United States in order to satisfy a major rebel objective: Washington's jettisoning of its support for the Batista regime. As many Latin Americans believed and as experience demonstrated, a U.S. denial of recognition and aid could bring down a government. Neutrality toward a government under attack from its own people might also undermine it. On the other hand, U.S. assistance could help sustain a besieged, unpopular regime like Batista's. As the insurrection gained momentum, moreover, Castro feared that Batista would contrive some incident to provoke a direct U.S. intervention that would save his faltering regime from revolutionary forces.

José Martí had advised caution about the imperial "conquering policy" of the United States in Latin America, and Castro had long embraced Martí's anti-imperialist warnings.[41] Castro knew that Martí had described the United States as the "monster" that could engulf the island.[42] Castro had also read histories of U.S. military interventions in Latin America intended "to defend the most bastardly interests" and to direct "the fate of our peoples as it [the

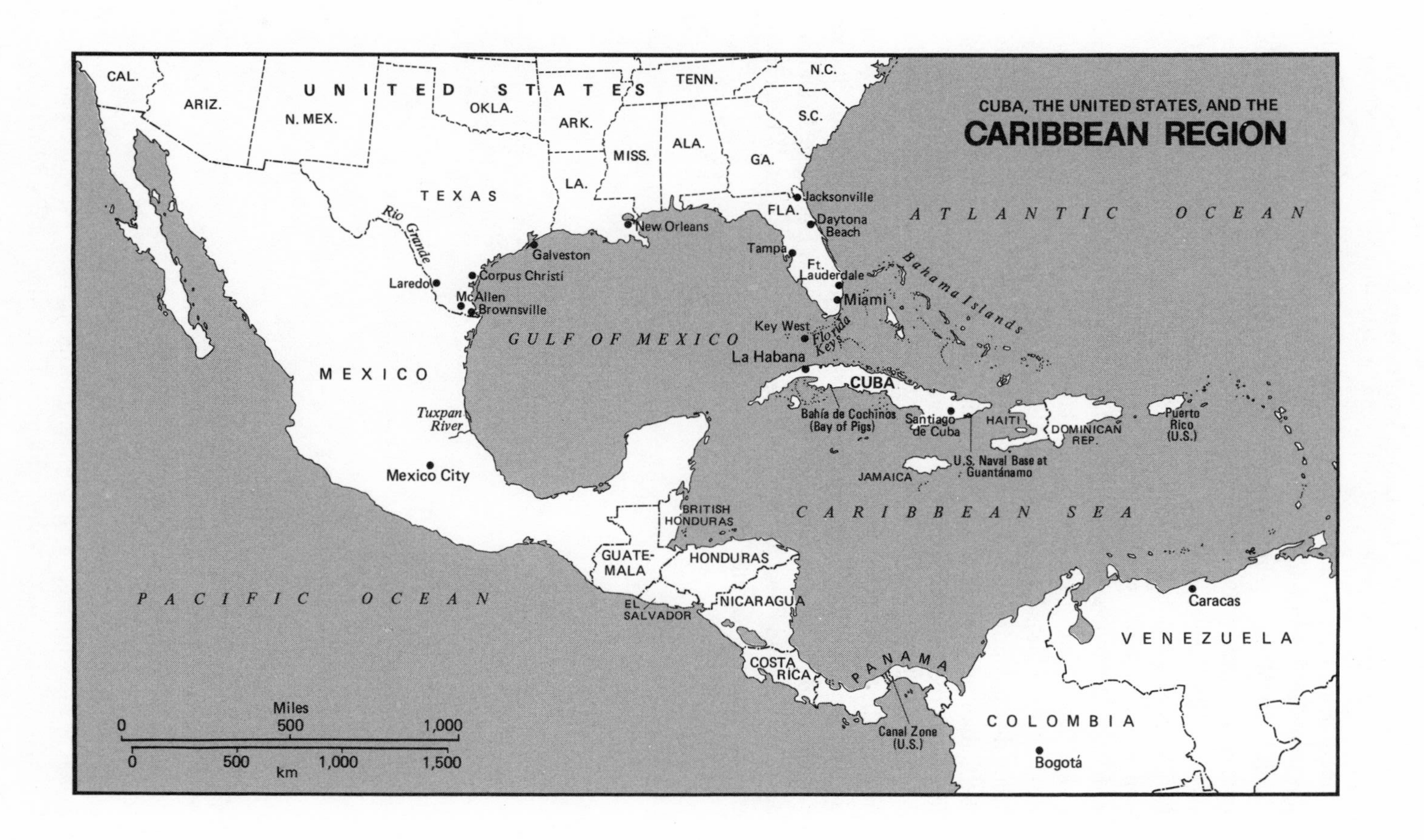
CUBA, THE UNITED STATES, AND THE
CARIBBEAN REGION
UNITED STATES
CAL.
ARIZ.
N. MEX.
OKLA.
ARK.
TENN.
N.C.
S.C.
MISS.
ALA.
GA.
LA.
TEXAS
FLA.
Jacksonville
Daytona Beach
Tampa
Ft. Lauderdale
Miami
Key West
Florida Keys
New Orleans
Galveston
Corpus Christi
Laredo
McAllen
Brownsville
Rio Grande
ATLANTIC OCEAN
Bahama Islands
GULF OF MEXICO
La Habana
CUBA
Bahía de Cochinos (Bay of Pigs)
Santiago de Cuba
HAITI
DOMINICAN REP.
Puerto Rico (U.S.)
U.S. Naval Base at Guantánamo
JAMAICA
MEXICO
Tuxpan River
Mexico City
BRITISH HONDURAS
CARIBBEAN SEA
GUATE-MALA
HONDURAS
PACIFIC OCEAN
EL SALVADOR
NICARAGUA
COSTA RICA
PANAMA
Canal Zone (U.S.)
Caracas
VENEZUELA
COLOMBIA
Bogotá
Miles
0
500
1,000
km
0
500
1,000
1,500

United States] pleased."[43] The anti-American strain in Cuban nationalism flourished in Fidel Castro.

But the M-26-7 rebel, sensitive to how the United States could either thwart or advance the revolutionary movement, initially tamed harsh criticisms of the *Yanquis.* The North Americans had to be used to help dethrone Batista. As Castro said, "to succeed in stopping [pro-Batista U.S.] intervention is in itself to overthrow the tyranny."[44] A prominent Cuban astutely summarized the dependency on the United States that existed for all factions in the island war. Former General García Tuñon told a U.S. State Department officer that each side in the Cuban conflict "is endeavoring to demonstrate that it continues to deserve United States support. The entire history of the Cuban people has conditioned them, whether in or out of Government, to look to the United States for approval, support and/or emulation."[45]

While Fidel Castro rallied Cubans and raised dollars in the United States in 1955, the Cuban government protested his use of U.S. territory for revolutionary activities. U.S. immigration authorities thereupon cut short his stay and canceled his visa.[46] Castro returned to Mexico on December 10 and once again approached Bayo. "I now have the money and the men. When can you start to train my recruits?"[47]

2

Confusionist Cubanism: The Political Mess Before Granma

Fidel Castro and other dissidents vowed to oust Batista, but it was not altogether evident in the mid-1950s that Cuba was ripe for rebellion. The outward signs of order and economic growth reassured many, including U.S. officials, that the fundamental political and economic structure of the island remained intact. U.S. interests appeared safe, and Fulgencio Batista embraced the Cold War crusade against Communism. He broke diplomatic relations with the Soviet Union in 1952, and the following year he outlawed the Communist party. Washington did not complain when Batista called for an election to legitimize his seizure of power.[1] CIA analysts predicted that "President Batista will make sure of winning the November [1954] election."[2] He did. Like everybody else who smelled fraud, opposition parties boycotted the election.

Attending Batista's inauguration ceremonies in February 1955 was a smiling Vice President Richard M. Nixon. Before he traveled to Havana, Nixon received a briefing from U.S. diplomatic officers. Although most Cubans opposed Batista, the army—"the key to the situation"—backed Batista. Cuba harbored "25,000 hard core commies," but the "master politician" Batista had them under control. He "is friendly to the U.S., admires the American way of life, and believes in private enterprise." Still, Cuba suffered "a sense of inferiority which promotes exaggerated nationalism." Cubans, moreover, "tend to feel that the U.S. will bail them out before a collapse and this encourages irresponsibility."[3]

Back in Washington after the inaugural festivities, Nixon reported to

President Eisenhower and the cabinet that the "remarkable" Batista was "strong, vigorous," and "desirous of doing a job more for Cuba than doing a job [for] Batista."[4] Nixon and other U.S. officials observed a reliable client state led by an anti-Communist who had strong support from his military establishment and who stood fast with the United States. "No doubt about it," U.S. Ambassador Arthur Gardner exclaimed, "this country has had a rebirth, and a genuine resurgence" due to the leadership of Fulgencio Batista, "an expert administrator."[5]

While most U.S. officials sang Batista's praises, volatile island conditions were steadily undermining the dictatorial regime. The government's opponents gradually and steadily exploited the flaws in Cuba's political, social, cultural, and economic life. As one U.S. official correctly concluded soon after Batista seized power, "the Cubans seem to be headed for a terrific mess both politically and economically."[6]

In the 1940s and 1950s, cynicism, corruption, scandal, and violence troubled Cuba's political system.[7] The Auténtico party (Partido Revolucionario Cubano), led by Ramón Grau San Martín, president of Cuba for 1944–48, and Carlos Prío Socarrás, president for 1948–52, corrupted politics in the aftermath of Batista's first presidency of 1940–44. Grau and Prío swelled the civil service with political appointees eager to make money for themselves; public officials raided pension funds and the treasuries of the national, provincial, and municipal governments. The British Ambassador in Havana estimated that Prío himself stole $90 million in public funds.[8]

After the Batista *golpe de estado,* Prío fled into exile in the United States, where he used his ill-gotten riches to finance opposition to Batista. In 1954, U.S. authorities discovered a large weapons stockpile in Mamaroneck, New York. The evidence pointed to Prío, and he was fined $9000 for attempting to smuggle arms from the United States into Cuba. To avoid expulsion, the former Cuban president signed an agreement to abide by U.S. laws—in essence to suspend his anti-Batista campaign in the United States. He returned to Cuba in 1955, but Batista once again, in May 1956, forced him from the island. Prío reinvigorated his illegal insurrectionist efforts on U.S. soil by financing oppositionists and supplying them with guns. The U.S. Immigration and Naturalization Service then prepared to expel him, citing the 1954 agreement.[9] Some Cubans became outraged that the United States was actually contemplating prosecuting Prío while supporting Batista. Still, many Cubans thought Prío a scoundrel. Although his wealth bought influence, few of Batista's opponents could be optimistic about a future in which their next leader might be the likes of the Auténticos' Carlos Prío Socarrás.

Under Batista in the 1950s, the political system that Grau and Prío had sullied not only remained tarnished but also suffered illegitimacy and authoritarianism. The Congress and Supreme Court bowed to Batista's wishes. Elections were rigged, and political parties refused to participate in a process oiled by bribes and sinecures and closely managed by the Presidential Palace. Politics still engorged the personal fortunes of unscrupulous leaders and their families. "Not to be rich was a humiliation. It was not a scandal to sell oneself," a Batista minister once remarked.[10] Like Cuban leaders before them, Batista and his cronies showered themselves with governmental funds, apparently including millions of dollars from the national lottery. Cuba's Ambassador to the United States Nicolás Arroyo y Marquéz had once supervised government building projects as Minister of Public Works. Like many other corruptionists, reported a British official, the Cuban diplomat made money "on the side."[11] Then, too, *hacendados* (sugar-mill owners) had to pay off tax inspectors, and merchants had to pay "tax" money to the police in an elaborate system of extortion. Even U.S. officials became entangled in this political culture. Ambassador George S. Messersmith observed some years earlier that "everything which [the American ambassador] says or does is used or misused to serve personal interests when it can be done."[12]

Cosme de la Torriente, an elderly veteran of the war for independence against Spain and respected former public servant, sought to tame the crisis that Batista, Prío, and others were accelerating. Torriente's "Civic Dialogue" movement failed in part because Batista snubbed this moderate effort, further polarizing politics. In late 1956, the political divide became wider with Torriente's death. The already slim hopes for an electoral solution to Batista's rule faded fast.

One of the parties least interested in compromising with Batista was the Ortodoxo party, which had splintered from the Auténticos in 1947 over the issue of corruption. Led by the dynamic Senator Eduardo Chibás until his suicide in 1951, the Ortodoxos attracted young, middle-class Cubans to their ranks. But they squabbled among themselves over whether or not to use violence to unseat Batista, and after Chibás's death no one of stature emerged to direct the party. "The Ortodoxos were leaderless and the Auténticos could not lead," the historian Louis A. Pérez, Jr., has written.[13]

The Partido Socialista Popular (PSP), or Communist party, fared little better.[14] The Communists had cooperated with Batista in the days of Popular Front politics, receiving as a reward a few cabinet seats during the Second World War. But his departure in 1944 undercut their base, especially in the Cuban trade unions. The Auténticos, led by then Minister of Labor Carlos Prío Socarrás, expelled the Communists from the Confederation of Cuban Workers in a furious struggle in 1947. Five years later the PSP opposed Batista's coup, but the Communists nevertheless "atrophied on the sidelines" of the revolt against the regime.[15] The PSP had well-organized cadres and influence in local sugar-mill unions, but, tainted by its prior alliance with Batista and thrown on the defensive in the environment of McCarthyism that afflicted both Cuba and the United States in the 1950s, the PSP failed to attract political allies. The party lost members and became timid.

At first the Communists denounced Castro's 26th of July Movement. They dismissed the Moncada assault as nothing more than a "desperate form of adventurism, typical of petty bourgeois circles lacking in principle and implicated in gangsterism."[16] The Communists also opposed terrorism and sabotage, favoring a "peaceful road" toward removing the Batista dictatorship.[17] Anyway, workers not peasants should carry the revolutionary banner, they insisted. The Communists shifted to support the rebellion being orchestrated from the Sierra Maestra Mountains only after they saw that Castro's 26th of July Movement had a chance of winning—in spring 1958. The rebels always eyed them suspiciously. Individual Communists fought in the rebel army and a small Communist resistance force operated in Las Villas Province, eventually hooking up with Che Guevara's advancing troops. But the PSP worked on the fringes of the popular rebellion against Batista. It could not effectively infiltrate the movement or control Fidel Castro, who remained wary of the party while he hoped to exploit it to advance the insurrection.

The isolated Cuban Communists received little help from the Soviet Union. Although the PSP took cues from Moscow, Kremlin leaders seldom interested themselves in Latin America. U.S. power stood so menacingly there that the Soviets adopted a prudent, reactive stance that honored the law of "geographical fatalism."[18] The official Soviet press harangued against "the stooge" Batista and other U.S.-backed military regimes in the hemisphere, of course, but the Soviet government had only tenuous relations with the PSP and barely noticed Fidel Castro's rebellion.[19] When the Soviets did comment, they disapproved of his violent tactics, scorned his management of an unsuccessful general strike, and highlighted his failure to articulate a systematic political program. Moscow also looked skeptically upon those Latin American leftists, such as Guatemala's Jacobo Arbenz, who expressed a passionate anti-imperialism against the United States but who seemed more reformist than radical or Communist. The CIA-engineered overthrow of Arbenz in 1954 simply reaffirmed Soviet thinking that it was unwise to challenge the United States in its immediate sphere of influence.[20] Even after Castro's victory in 1959, for more than a year, Soviet leaders offered little to the new revolutionary government except congratulations for defying "those sweet-toothed Yankees."[21]

More forthright and violent than the PSP in the anti-Batista rebellion was the Directorio Revolucionario (DR), formed in August 1955 by the student leader José Antonio Echeverría and others from the Federación Estudiantil

Universitaria at the University of Havana. While Castro was organizing his "army" in Mexico, DR activists were staging attacks on the police and demonstrating in the streets at every opportunity. Still another dissident group was the Triple A, financed by Prío's money and headed by Aureliano Sánchez Arango, a former secretary of education. Like many other anti-Batista organizations, this one had to work clandestinely.

As Castro well knew, for M-26-7 to succeed it had to overcome three major obstacles: Batista, the United States, and rival anti-regime groups. The struggling 26th of July Movement was but one among many opposition groups, and this pluralism reflected different interpretations of revolution. "For all the popularity of the term," the moderate anti-Batista dissident Mario Llerena recalled, "the concept remained cloudy."[22] Some anti-Batista Cubans took it to mean no more than the overthrow of the dictator in order to legitimize and purify the Cuban government. Others intended a panoply of reforms to improve opportunities for jobs, landholding, and education. Still others insisted on fundamental, radical change, including the expropriation of property and the ouster of influential U.S. interests. Dissidents issued their passionate manifestos, but they seemed to spend more time debating tactics than doctrines or programs.

Fidel Castro himself seemed to embrace most meanings of "revolution." A Cuban businessman once remarked that Castro seemed capable of going in three directions: toward democratic capitalism, toward the extreme left with policies of nationalization and agrarian reform, and toward "confusionist Cubanism"—by which he apparently meant either a blending of the first two inclinations or a peculiar Cuban propensity for muddled politics.[23] A "lack of clarity," CIA analysts correctly noted, characterized Castro's objectives.[24] Castro's views, like those of Echeverría and other insurrectionists, took shape gradually as he built a movement and as he reacted month by month to Batista's machinations and U.S. acts and decisions. "A revolutionary is formed through a process," the rebel chief himself explained.[25] That process included M-26-7's vigorous rivalry with the many other opponents of the regime.

Batista actually helped Castro defeat rival oppositionists. By continuing to rule through strong-arm tactics, police and military intimidation, and electoral fraud, Batista elevated organized violence—Castro's preference—as the only viable alternative to achieve political change. The government also eliminated Castro's rivals by killing them, imprisoning them, or forcing them into exile. By polarizing politics, Batista also shoved moderates and liberals closer toward the extreme of armed revolt, toward tolerance of violence, toward the 26th of July Movement.

In late 1956, early 1957, moderate upper- and middle-class professionals and business people such as Raúl Chibás (an Ortodoxo leader), Felipe Pazos (former governor of the National Bank), Angel Santos Buch (a physician), and Ignacio Mendoza (a wealthy broker) organized the nonfighting, urban-based Movimiento de Resistencia Cívica in Santiago de Cuba. The Civic Resistance Movement created cells of ten members each and soon expanded across the nation, becoming "a broad front" for the 26th of July Movement—collecting money, sending medicine, food, and other necessities to rebels in the Sierra Maestra Mountains, preparing anti-Batista propaganda, signing up recruits, hiding insurrectionists from regime authorities in safe houses.[26]

For growing numbers of Cubans, violent revolt seemed necessary to oust Batista, cleanse politics, and drive the chiselers from government. Violence, of course, had long plagued Cuban politics. When Castro enrolled at the University of Havana as a law student in 1945, for example, "action groups" dominated the political life of campus and city. These gangs alternately murdered their opponents and debated revolutionary strategies and tactics.

Fidel Castro himself associated with a gang called Unión Insurreccional Revolucionaria, participated in gun battles, and escaped assassination attempts. He may even have killed rivals. Castro also joined a plot in 1947 to overthrow the dictatorial regime of Rafael Trujillo in the Dominican Republic. Setting aside his law studies, Castro took military training on a small island off the coast of Camagüey Province. Probably because of U.S. pressure, the Cuban government moved against the expeditionaries before they could launch their invasion of the Dominican Republic.

In April 1948, in Bogotá, Colombia, where foreign ministers, including Secretary of State George C. Marshall, had gathered to inaugurate the Organization of American States (OAS), Castro and other Cuban students attended an anti-imperialist congress intended to disrupt the meeting. The Colombian police arrested him one night for distributing leaflets. After an assassin's bullet killed a popular leader of Colombia's Liberal party, bloody rioting and looting convulsed the city in what became known as the *bogotazo.* Out of jail then, Castro joined the growling crowds in the streets and was soon leading charges and calling for revolution. Government forces gradually regained control of the battered city, and the 21-year-old Castro escaped from Colombia with the help of Cuban Embassy officers. His exploits gained him wide publicity in the Cuban press.[27]

Back in Cuba, Castro turned to electoral politics and the Ortodoxo party, for, like many other young Cubans, he had grown critical of gangsterism and violence. But then Batista's coup and reign of terror crippled the electoral system, persuading many critics to exercise the option of violence. Although Batista repressed his opponents, he did not crush them. Batista's repression, including censorship of the press and the suspension of constitutional guarantees, was intermittent, imperfect, and inconsistent before early 1958, when he clamped down hard on critics of the regime. Lapses, such as the general amnesty in 1955, gave renewed voice to his detractors. Perhaps craving public approval and legitimacy after his coup, Batista wavered between acting like a constitutional leader and a dictator. "He has thus fallen between two stools," noted one diplomat.[28] The inability of Havana and Washington, moreover, to silence anti-Batista Cubans in the United States and to halt gunrunning from U.S. sites further undermined Batista's efforts to choke off the opposition.

Politically inspired violence rocked mid-1950s Cuba. Bombings, fires, conspiracies, riots, assassinations, arrests, and tortures became commonplace. In August 1955, DR students were jailed after an abortive, Prío-financed plot to kill Batista. During November and December, students and police viciously battled in the streets of Havana. A nationwide strike of sugar workers at the end of the year provoked more violence, more deaths, and more arrests. From Mexico, Castro cheered the strikers and insurrectionists, especially those who painted "M-26-7" on urban walls.

In April 1956, Batista's secret police blunted a military conspiracy by nabbing and court-martialing Colonel Ramón Barquín and other officers. The military attaché for the Cuban Embassy in Washington, Barquín worked with the moderate anti-Batista group called Montecristi, led by Justo Carrillo, an exiled economist and former director of a government agricultural credit bank. That month, too, a DR student was killed after an attempt to seize a Havana television station, and other civilian rebels were mowed down by machine guns when they stormed military barracks in Matanzas. After the Matanzas attack, Batista suspended constitutional guarantees and filled his jails with Prío followers. The government also closed universities and secondary schools.

The Assistant Secretary of State for Inter-American Affairs in Washington gravely observed that Batista had been "weakened" by these events. Still, Batista held the advantage, concluded Henry Holland, because of Cuba's "economic prosperity," public apathy, factionalism within the opposition, Army support for the regime, and labor leaders' preference for negotiations over violence.[29] Just months after this overview, Colonel Antonio Blanco Rico, the head of Cuban military intelligence, was assassinated in the Montmartre Club, one of Havana's gambling casinos run by Flordia mobster Meyer Lansky.

In Mexico, where Bayo was helping to train a rebel invasion force, Castro received funds from the Cuban M-26-7 underground and anxiously urged clubs in the United States to send more money. Castro also issued a host of declarations and lectured Cuban politicians. He dismissed Ortodoxo leaders as "helpless and divided," denouncing them for participating in the doomed Civic Dialogue.[30] Meanwhile, 26th of July Club members in the United States picketed United Nations headquarters in New York and demonstrated in Miami. FBI agents watched their every move and analyzed their speeches. The Department of Justice considered initiating grand jury proceedings against the Florida clubs because they had not registered properly.[31]

As he challenged rivals, Castro nearly saw his own plans quashed. On the evening of June 22, the Mexican Security Police jailed Castro and fellow trainees on charges that they had violated their tourist permits by plotting the assassination of Batista. In custody, Castro denied Batista's claim that he was a Communist and suggested that the tag had been fabricated to persuade U.S. officials to press Mexico to shut down the Castro conspiracy. Available evidence does not reveal if U.S. authorities engineered the arrests. The U.S. Embassy in Mexico City observed shortly after the event that, although some of the arrestees might have "Communist backgrounds," Castro himself did not. Embassy officials reported, however, that the police found in Castro's wallet the business card of Nicolai L. Leonov, secretary to the Soviet Ambassador in Mexico City.[32]

When he heard about the Castro group's plight, Juan Manuel Márquez hurried to Mexico from the United States, where he had gone to raise more money and buy guns. He hired lawyers who successfully worked to prevent Castro's deportation to Cuba.[33] Most of the Cubans were freed within a month. Soon new recruits arrived from Cuba and the United States to join the rebel army. Revolutionary leaders from Cuba journeyed to Mexico, too, including DR's dynamic Echeverría, who met with Castro in the fall of 1956. They signed a vague pact to overthrow the Cuban tyranny. But the two strong personalities could not agree on a merger of their organizations or on the most effective means to overpower Batista.[34] DR and M-26-7 remained intensely independent-minded rivals.

In the fall of 1956 Castro's personal and public lives suffered setbacks. His planned marriage to a beautiful young Cuban exile, Isabelle Custodio, collapsed under the weight of his consuming devotion to the revolutionary mission, while his ex-wife, Mirta Díaz Balart, for whom he still had deep affection, was remarrying. Then, on October 21, Fidel Castro's father Angel died. The relationship of the strong-willed son and the bullying, alcoholic father had engendered more conflict than familial warmth. Perhaps Fidel resented not only his father's authoritarianism but his own illegitimate birth. "Apparently there was some hidden wound relating to his childhood which had never healed," remarked Teresa Casuso, who succored the rebels in Mexico.[35] Angel had also amassed a sizable farm, and the young Fidel, who advocated land reform, was at political odds with his father. What his death meant for Fidel is unknown, for he seldom if ever spoke of the event. Coming as it did with so many other misfortunes, however, his father's death may have unsettled him.[36]

At the same time, Castro's revolutionary plans seemed jeopardized. He had a formidable rival in Echeverría. Mexican authorities lurked about, ready to strike again, and some policemen had to be bribed to safeguard rebel arms caches. M-27-6 was running out of money. Castro may even have asked the Soviet Embassy in Mexico City for arms, but if he did he came away with nothing.[37] Castro had pledged repeatedly that his revolutionary band would invade Cuba before the end of the year, but the prospects appeared bleak.

Then Castro took an extraordinary step: He called upon a man he despised as a corrupt, failed politico of the past, Carlos Prío Socarrás. In September a reluctant Prío traveled to the Rio Grande valley town of McAllen, Texas, where he checked into the Hotel Casa de Palmas. Fidel Castro headed north by car from Mexico City. At Reynoso, Castro splashed into the chocolate-colored Rio Grande. On the U.S. side of the river, friends gave him dry clothes. His illegal entry into the United States successful, Castro found his way to the MacAllen hotel.

The wealthy, middle-aged former president of Cuba and the penniless, young insurrectionist had never met before. Little is known about the encounter, although Prío remembered the meeting for researchers years later. For hours, of course, the exiles talked about rebellion and money. Fidel paced the hotel suite, jabbing the air with now-famous gestures to make his case for the Auténtico leader's money. Prío liked enough of what he heard to pledge $50,000. (Prío later claimed that he gave a total of $250,000 to the 26th of July Movement chief before 1959).[38] "We were willing to do anything for the revolution," Castro later explained.[39] "Pinching $50,000 from an s.o.b. is not theft," he said. "It's a good deed."[40]

With Prío's large contribution, Castro's team searched for a vessel that could carry the revolutionary army to Cuba. A patrol torpedo (PT) boat was located in Dover, Delaware. Although M-26-7 paid $20,000 for the craft, U.S. officials refused to grant a license for its export to Mexico. Prío helped again, sending another $20,000. A Mexican arms dealer then showed Castro a deteriorating pleasure yacht owned by an American living in Mexico City. In October, eager to meet his much publicized deadline, Castro paid Robert B. Erickson $20,000 for the yacht and another $20,000 for a house where the rebels could prepare for their expedition. The *fidelistas* rushed to make the frail *Granma* seaworthy.

Finally, in the early morning hours of November 25, 1956, Castro and 81 compatriots boarded the *Granma* (some 50 rebels had to be left behind) and chugged down the Tuxpan River into a stormy Gulf of Mexico.[41]

3

Sugar, North American Business, and Other Bittersweets

FIDEL CASTRO AND his followers hoped that a successful landing would incite a popular rebellion against the Batista government. The 26th of July warriors calculated that Cubans were so fed up with corruption and violence and so worried about economic uncertainties that they would hail the heroic expeditionary force as a new Martí come to redeem the nation. The *fidelistas* also believed that they could persuade Cubans that the island economy suffered because U.S. interests controlled and manipulated it in their self-interest, trampling on Cuban sovereignty and carrying off the national wealth.

Batista, North American investors, and U.S. officials thought quite the opposite: The strengths of the Cuban economy inclined people to support the regime, especially in 1957 when sugar prices rose. When the *Granma* sailed for Cuba, the island economy was in fact on the upswing and Cubans enjoyed one of the highest standards of living in Latin America. By per capita measurements, Cubans owned more television sets, telephones, and motor vehicles than any other people in Latin America. Cubans ranked just behind Venezuelans in per capita income and near the top in medical care, food consumption, and literacy.[1]

But Cubans were anything but satisfied with their status. "If all the surveys made during this decade in Cuba were laid end to end," remarked a Havana economist, "they would bridge the Florida Straits."[2] The stack of studies revealed that a chronic structural crisis dogged the economy. Everyday life also reminded Cubans that the rhetoric of progress obfuscated reality. A persistent public uneasiness about the future prompted recurrent accusations that foreign interests exploited the island. Cubans also grew embittered because, although incorporated into the U.S. consumer culture, they could not attain the U.S. standard of living to which they aspired. A moral crisis accompanied the economic crisis, for corruption always seemed to accompany the economic development of the island.

Sugar and the United States—Cubans needed them, feared them, loved them, hated them. Sugar and the United States dominated Cuba's economic life. By mid-century extreme Cuban dependency upon sugar and reliance upon U.S. trade and investment had become the predominant features of the island's economy. Cuba ranked as the world's largest producer of the natural sweetener. In the 1950s sugar represented about 80 percent of all exports, and these exports in turn accounted for about one-third or more of total national income. The "spell of sugar" transfixed Cuba even as "diversification" became the popular byword.[3]

The United States took about half of Cuba's sugar exports (reaching 58 percent in 1958) and about two-thirds of all island exports in 1958. Cuba supplied about one-third of the U.S. sugar market. Cuba's shipments of sugar to the United States meant so much, Secretary of State John Foster Dulles once told President Dwight D. Eisenhower, that a reduction in the flow "might easily tip the scales to cause revolution in Cuba and would certainly increase instability and promote anti-American feeling and communist activity in the Carribbean area."[4]

Cuba became intricately tied to the United States in another way: About three-fourths of the island's imports originated in the United States, including large quantities of rice, wheat, and flour. Cuba took about one-third of all U.S. rice exports. Under Cuban tariff policy U.S. products enjoyed an advantage in the island marketplace: U.S. goods were exempted from the 20 percent surcharge on duties that all other countries had to pay.

The interlocking of the Cuban and U.S. economies also derived from U.S. direct investments, which stood at nearly one billion dollars in 1958, up from $686 million in 1953.[5] Although the sugar industry had become increasingly Cubanized, North American-owned mills still produced about 40 percent of the island's sugar, and the largest sugar company, Atlántica del Golfo, was North American-owned. U.S. companies controlled 90 percent of telephone and electrical services and dominated the railway and petroleum industries. Cuban branches of U.S. banks held one-quarter of all island deposits; Chase Manhattan, First National City Bank, and Manufacturers Hanover Trust, among others, operated in Cuba. One-fifth of the national budget was supplied by the taxes North American businesses paid.

Names familiar to U.S. consumers—Texaco, Swift, Goodyear, Remington, Procter & Gamble, Colgate-Palmolive, Borden, International Harvester—also had high visibility in Cuba. Sears, Roebuck and Company ran seven retail stores and Ford had 35 dealerships. The famous King Ranch of Texas expanded operations into Cuba in the 1950s, eventually investing $5.7 million and owning 40,000 acres for cattle raising. One of the largest law firms in Cuba, Lazo y Cubas of Havana, represented many of these enterprises, as well as several U.S. government agencies, the General Services Administration and Department of Commerce, among them.

The growth of U.S. direct investments under Batista impressed everybody—even those people who feared that he was turning Cuba over to North American capital. In 1957, despite the insurrection and rebel sabotage of private property, North American business expanded across the Cuban economy, especially in consumer goods, petroleum, mining, and public service industries, creating what *Business Week* called "boom times."[6] Standard Oil (ESSO) announced that it would enlarge its refinery near Havana. W. R. Grace bought a Cuban paper-container firm; Republic Steel invested in a new mill for scrap; and U.S. Rubber began to build a $5 million plant for the production of 125,000 tires a year. Standard Oil of Indiana began to explore for uranium, and Cuban and North American capitalists started work on four plants, costing $28 million, for processing bagasse (the fibrous sugar cane waste) into paper and hardboard. A new firm of Texas interests, Pan American Land and Oil Royalty Company, gained petroleum concession rights to 1.3 million acres.[7]

The Freeport Sulphur Company invested in a nickel and cobalt plant at Moa Bay, and the U.S. government contracted to buy its production to stockpile strategic metals. Later in the year, the president of the company, accompanied by U.S. Ambassador Earl E. T. Smith, met with Batista. After the interview, Langbourne M. Williams announced that the Cuban government had granted the company tax exemptions so that construction on the new $75 million plant could begin. Several U.S. banks backed the enterprise.[8]

The Batista government, nourishing a welcoming "ambiance" for North American business, invited foreign investors into the country to diversify the economy and cushion it against sugar slumps. The Havana regime offered both investment guarantees against nationalization and lenient regulations governing the convertibility of profits into U.S. dollars. Government loans to foreign developers, state investments in the infrastructure for docks and roads, and tax breaks also attracted foreign capital.[9] The Cuban government even relaxed restrictive labor laws that North American business had long considered obstructionist. Eager to open the new Havana Hilton, for example, Cuban officials permitted the hotel to pass over Cuban chefs in favor of Swiss chefs. As one analyst concluded, "no new investment project has failed to materialize because of labor difficulties."[10]

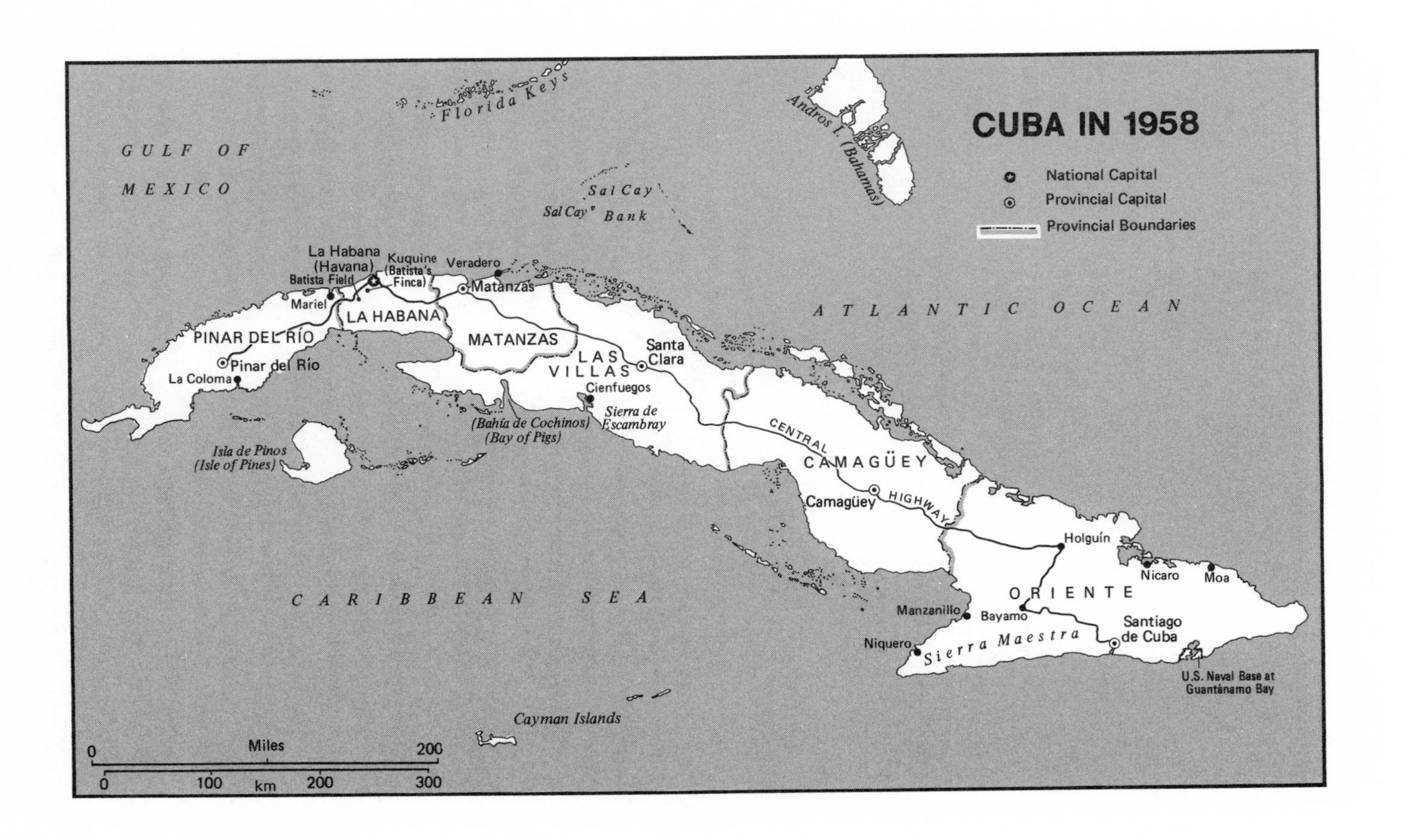
CUBA IN 1958
National Capital
Provincial Capital
Provincial Boundaries
GULF OF
MEXICO
Florida Keys
Sal Cay
Sal Cay
Bank
Andros I. (Bahamas)
ATLANTIC OCEAN
La Habana
(Havana)
Batista Field
Kuquine
(Batista's
Finca)
Veradero
Matanzas
Mariel
LA HABANA
PINAR DEL RÍO
Pinar del Río
La Coloma
MATANZAS
LAS
VILLAS
Santa
Clara
Cienfuegos
Sierra de
Escambray
(Bahía de Cochinos)
(Bay of Pigs)
Isla de Pinos
(Isle of Pines)
CENTRAL
CAMAGÜEY
Camagüey
HIGHWAY
Holguín
Nicaro
Moa
ORIENTE
Manzanillo
Bayamo
Santiago
de Cuba
Niquero
Sierra Maestra
U.S. Naval Base at
Guantánamo Bay
CARIBBEAN SEA
Cayman Islands
0
Miles
200
0
100
km
200
300

Most North American business people warmed to Batista and backed him during much of the insurrection, abandoning him only after his regime had self-destructed. "At least Batista is a person we can deal with," remarked an Otis Astoria executive.[11] By appointing Cubans to boards of directors and handing out bribes and protection money, U.S. investors gained access to authority. One CIA official stationed in Cuba explained that U.S. business cultivated ties with the government because a telephone call to Batista "could fix" a problem.[12] A member of the North American business colony who preferred the dictatorial regime to the chaos that would surely follow any change of government admitted that "we have made our peace with Batista. We know what our taxes and hidden taxes are going to be."[13]

Gulf Oil and Cuban Electric executives held memberships in the Havana Country Club, where they could hobnob with prominent Cubans (including Batista himself) and influential North Americans (including Ambassador Earl E. T. Smith). Batista and Texaco held memberships in the Havana Biltmore Yacht and Country Club, the most North American of the prestigious clubs in Havana, for which Batista spent $800,000 of the national treasury to dredge the club's harbor.[14]

Batista and his friends enriched themselves in part through links with foreign business and the growth in foreign investment. Probably 20 to 25 percent of government expenditures represented graft and payoffs. Batista's personal wealth stood somewhere between $60 and $300 million. In 1959 revolutionary government officials opened his safe deposit boxes and found $20 million. Among the president's "investments" were Cuban Electric Company bonds and 131,500 shares of Atlántica del Golfo. When Batista and his close corruptionists fled the country as 1958 turned into 1959, they took with them—nobody knows how much for sure—some 350 million pesos of the national treasury (one peso equaled one dollar).[15]

The U.S. government strengthened Cuba's economic dependency upon North Americans through vigorous diplomatic representation, Export-Import Bank loans, ownership of the Nicaro nickel plant, and technical assistance projects in agriculture, aviation, mining, and vocational education.[16] Export-Import Bank loans, amounting to $38.9 million in 1951–58, helped Cubans purchase U.S. electrical machinery, shovels, cranes, tractors, generators, and semi-trailers, and develop waterworks and telephone services. Ten of the eleven loans for the period were authorized in 1956–58, at the height of the anti-Batista insurrection. The largest loan of $20 million went to the Cuban Electric Company, a subsidiary of the American and Foreign Power Company. In 1957 the Bank also loaned $16.3 million to the Cuban Telephone Company, another enterprise owned by North Americans.[17]

The largest plant operation in Cuba, employing 3100 persons, was owned by the U.S. government and run by the General Services Administration (GSA). The nickel plant at Nicaro was built during the Second World War to ensure strategic supplies. During 1954–57 the GSA spent more than $40 million to expand the plant's productive capacity by 75 percent—to 50 million pounds of nickel a year. In 1957, U.S. officials, finding it difficult to negotiate a new management contract, facing a large stockpile of nickel, and hearing complaints that the government should not be "in business" when the Eisenhower Administration's clarion call was "free enterprise," decided to sell Nicaro.

U.S. officials had also become alert to the connection between Nicaro and "the political situation in Cuba"—alert, that is, to charges hurled by Cuban nationalists that Nicaro was but another example of Cuba's loss of control over its own economy, and, in this case, made all the more noisome by the fact that the nickel company enjoyed exemption from Cuban taxes.

Located in Oriente Province, in the region where anti-Batista activities surged, Nicaro could not escape the whirlwinds of the Cuban civil war. In late 1956 a nervous guard at the plant shot to death someone he took to be a revolutionary. In November 1957, M-26-7 rebels struck Nicaro, dynamiting electrical cables. The following year rebels overpowered Nicaro guards and seized a road grader and 1000 gallons of gasoline. The plant manager and his assistant were taken captive. By mid-1958 the Nicaro mines could only be worked during daylight hours, forcing a major reduction in production. In the last half of the year, Nicaro's property losses climbed to hundreds of thousands of dollars, including jeeps, bulldozers, and typewriters the rebels used to improve their transportation and communications networks. The destruction of offices and warehouses by Cuban Air Force planes further interrupted Nicaro operations. The Castro government later nationalized Nicaro, then valued at more than $132 million.[18]

Despite appearances of progress and opportunity, the Cuban economy of the 1950s, structure-bound by sugar and dependent upon the United States, showed signs of stagnation. The Cuban gross national product grew by only 1 percent in 1950–58. Boom-bust sugar cycles, inflation, and aging, deteriorating sugar mills that needed frequent repairs and produced at 70 percent of capacity plagued the island. Lagging industrial development revealed the lack of adequate diversity in the economy. The labor force resisted mechanization, while high unemployment, underemployment, and declining purchasing power for many citizens beset the island. The public debt increased from government spending for projects intended to stimulate economic activity and jobs (such as the construction of the tunnel under Havana Bay). Cubans who made money from this shaky economic system—and who surely read some of the warning signs—invested millions of dollars in south Florida real estate or bought U.S. stocks.

"It is hard for a person who comes to Cuba and sees so many signs of building and prosperity to realize," Batista-booster Ambassador Gardner observed, "that only a few miles back from the city hundreds of thousands of people have only the bare necessities of life. Until this sore has been healed by the opportunity to work, Cuba will remain in a restless stage."[19] In 1956–57, more than 16 percent of the labor force was unemployed; another 7 percent was underemployed. The sugar industry hired a quarter of Cuba's workforce, yet these workers had steady employment for only four months of the year (January–April), during cane-cutting. The rest of the time—the "dead season"—they searched for jobs or joined the ranks of the unemployed. An increasing population growth rate also meant that more workers entered the workforce than could be accommodated by new jobs. Fears of unemployment lay behind labor unions' resistance to labor-saving mechanization and their support for labor laws that protected featherbedding and discouraged the dismissal of workers. Although workers were not in the vanguard of the insurrection against Batista and were not Castro's primary source of revolutionary fervor, laborers nonetheless suffered a steady "state of anxiety" that made many of them receptive to rebellion.[20]

Extremes of wealth and poverty and a racial divide also fed discontent. Urban folk outranked rural people (44 percent of the population) in most standard-of-living measurements, from higher literacy rates to better medical care. In the Sierra Maestra Mountains where the invading Castro army took refuge, peasants lived on the margin of subsistence. Many were racked by intestinal parasites. More than one-quarter of Cubans were people of color, descendants of Africans or West Indians pushed to the bottom of society by racial prejudice. Having little stake in a system that discriminated against them, many Afro-Cubans welcomed change.[21]

Wealth was concentrated in the cities, especially Havana, but even there, university-educated, professional middle-class people felt stymied by insufficient employment opportunities. A few big law firms dominated, making entry into the law profession difficult; some lawyers could find work only as typists. Doctors struggled in a poor health-care system or left Cuba for work abroad. "The university [of Havana] was thus not only a factory for the middle class," one historian has noted, "but a factory of discontent." Cuba suffered a "*crisis de professionales.*"[22]

Many Cubans, especially members of the middle class, complained that because of the stagnant economy and inadequate employment opportunities, they were falling behind their North American counterparts.[23] By Latin

American accounting, Cubans were prospering. But when Cubans, bourgeoisie and working class alike, made comparisons about their standard of living, they looked to the United States, not to the rest of the Latin America. Proximity to the United States, easy travel between the two nations, and access to U.S. radio broadcasts, magazines, and newspapers meant that Cubans became familiar with North American products and attracted to the U.S. consumer culture.

After Castro took power, Ambassador Philip W. Bonsal speculated about why revolution had swept Cuba. He emphasized that the "average income in Cuba was about one-third that of Mississippi, the poorest state in the union."[24] Indeed, Mississippi's 1957 per capita income was $1000 compared with Cuba's $374. The overall U.S. per capita income figure of $2000 made Cuba's status look even more dismal.[25] Cubans did not have to know these numbers to sense the great disparity between U.S. and Cuban standards of living. They craved but could not attain the U.S. standard, and this frustration provided yet another reason for their alienation from the Batista regime and its patron the United States.

The sugar question accentuated Cuban uneasiness about the future. Because Cuba had to export sugar, Cuban diplomats spent much of their time seeking to influence North Americans whose decisions on quantity and price could spell either prosperity or distress for the island economy. The United States managed the sugar relationship through a quota system laid out in the Sugar Act of 1948—a "complex, crazy quilt maze of regulations," thought Senator J. William Fulbright of Arkansas, who favored letting Cuba sell without restriction in the U.S. marketplace (earning dollars that the Cubans could spend to import his state's rice, of course).[26] Under the act, the Secretary of Agriculture each year set a tonnage figure for U.S. sugar requirements, and a large portion of that figure was granted to domestic producers to satisfy. The difference between the total U.S. requirements and the amount designated for U.S. producers was allotted to foreign suppliers. With a 98.64 percent share of the foreign allotment, Cuba was significantly favored over the Dominican Republic, Peru, and Mexico (Puerto Rico, the Virgin Islands, and the Philippines were considered part of the "domestic" category). The Sugar Act of 1951 increased the allotment for domestic producers and also reduced the Cuban quota to 96 percent of the foreign allotment.

Cuba became fixed on the United States for its primary market because U.S. sugar prices were higher than world prices, the U.S. tariff on Cuban sugar was lower than that for other nations, and Cuba was often invited to cover recurrent deficits in the U.S. domestic quota. Still, nervous Cubans could never be certain of the size of the U.S. sugar market or price per ton. Nor could they control U.S. government decisions. In 1954–55, for example, at a time when Cuban sugar production and sales were declining after the Korean War, U.S. sugar-beet and sugar-cane interests began to lobby Congress to amend the Sugar Act so as to give these domestic interests an even larger share of the U.S. market.

Batista officials appealed to the Department of State and then to the White House to block such an amendment. Miguel Angel de la Campa, then ambassador to the United States, explained the dire economic and social consequences of another decrease for Cuba. The U.S. Embassy in Havana and the Department of State echoed Cuba's pleas: "The already considerable economic discontent in Cuba" would increase if Cuba's participation in the U.S. sugar market dropped further. "By fomenting unrest, a cut in Cuba's share . . . might strengthen revolutionary elements" and "enable them to overthrow the government." Batista might fall, and "it would be hard to find any other Cuban political figure whose stewardship would be as likely as Batista's to serve the best interests of the United States."[27] A change in the Sugar Act, a Cuban foreign ministry officer pleaded, would be "impossible" to explain to a disgruntled Cuban people.[28] "If the cut is too deep," warned Ambassador Arthur Gardner, Cuba could become "a breeding place for real starvation and misery, giving the Communists a foothold they never had before."[29]

To no avail. Sugar-state members of Congress represented "a very powerful political bloc," witnessed an Assistant Secretary of Agriculture, and their allies in the Agriculture Department carried more weight than did the State Department on what the White House deemed a domestic issue.[30] State Department officials, although sympathetic to the Cuban position, calculated the political odds and advised the Cubans to restrain their protests. Some members of Congress "have been a little annoyed by the Cuban persistance." The Cuban case had become as appetizing as "warmed over potatoes."[31] Ambassador Gardner went so far as to dissuade one Cuban official from visiting Washington in December 1955 by telling the apparently ailing man that the "weather would be so bad" that he might be "risking serious trouble."[32]

After much wrangling, Congress passed the Sugar Act of 1956 in May. Under a highly technical formula to determine quota shares for domestic and foreign producers, Cuba's share once again decreased.[33] Despite the dire warnings, this revision did not stagger Cuba's economy. Cuban sugar exports to the United States in 1956–58 actually increased in volume each year, and Cuba clung to its usual share of about one-third of the American market, because U.S. consumption went up and domestic producers could not meet the demand. World sugar prices also climbed in 1956–57.[34]

However comforting these fortuitous conditions, Cubans remained uneasy about the central reality: The will and whim of the United States—its sugar-eating habits, pricing, producers, Congress, executive branch departments, and presidents determined Cuba's fate. The U.S. sugar-beet industry lobbied effectively in Washington, always taking the position, complained a representative of the United States Cuban Sugar Council, "that whatever is best for sugar producers under the American flag, and more especially within the continental United States, is in the national interest."[35] As well, Cuba's quota faced constant challenge from Mexico, the Dominican Republic, and Peru, whose governments continued to argue for a larger slice of the U.S. sugar market. In Mexico's case, organized Mexican labor worked with U.S. labor unions to gain backing for a bigger quota. One U.S. official observed that Cubans understandably had an "obsession with particular problems (such as a possible loss of advantage to the Dominican Republic in sugar)."[36]

"One lesson that Cuba may safely derive from the entire discussion of the Sugar Act," concluded a prominent Cuban attorney, "is that whatever she obtained was obtained solely because of her own efforts . . . and without any help from any other areas or departments of the United States."[37] Cubans protested that the United States should cooperate more generously with a close ally and reliable economic partner.[38]

To make matters worse, Cuba had to sell to other nations—including the Soviet Union, Japan, and West Germany—the sugar it did not ship to the United States, and that unpredictable global market suffered great variations in volume and price. Global conditions always joined U.S. decisions to determine Cuba's economic health. The Great Depression of the 1930s, for example, sent the island's sugar industry into a slump, whereas the outbreak of World War II brought recovery. Although Cuba participated in the International Sugar Agreements of 1953 and 1958 that attempted to manage sugar sales through price regulations, the International Sugar Council largely failed to discipline the world market. "Extreme looseness, if not . . . futility," distinguished the international regulatory regime.[39]

Besides the sugar problem, Cubans grumbled about another substantial North American element in their economy. Consumers in most nations through history have scorned utilities companies for high prices and poor service. Cubans certainly did in the 1950s. Because U.S. firms owned the major electric, gas, and telephone companies on the island, the complaints became tinged with nationalism. Anti-Batista Cubans particularly targeted these foreign enterprises. Castro's "History Will Absolve Me" speech/pamphlet and the First Manifesto of the 26th of July Movement of July 10, 1955, for example, demanded the nationalization of the companies.[40] During his 1955 trip to the United States, Castro included the utilities in his summary indictment of Batista's Cuba: "the townspeople . . . pay more for electricity than anywhere else in the world and wait ten years for a phone and never get it."[41]

The Cuban Electric Company, owned by the American and Foreign Power Company, which in turn was part of New York's Electric Bond and Share Company, represented the largest single investment on the island, valued by the firm at $256,942,000.[42] Cuban Electric produced 90 percent of the nation's electricity and distributed gas through underground lines. Company advertisements promoted U.S.-made washing machines, water heaters, electric ranges, air conditioners, and other appliances. Compañía Cubana de Electricidad's customers grew from 647,500 in 1955 to 732,400 in 1958. The company expanded its facilities and services throughout the decade, drawing upon loans from the Cuban government and the U.S. Export-Import Bank.

Although most observers ranked Cuban Electric's service good by Latin American standards (blackouts being notorious in other nations), Cubans pilloried the company for charging rates higher than those in the United States and Canada. Ortodoxo party founder Eduardo Chibás had made Cuban Electric rates and profits a prominent political issue. At times the company simply could not expand fast enough to meet the spiraling demand, and old equipment dating back to first installations in the 1920s broke down. The major problem, argued company executives, was high labor costs, taxes, and insurance premiums which could not be offset by an increase in income because the government would not grant a rate hike. But more than questionable service or rates alarmed critics: Cuban Electric Company officers apparently bribed Cuban government officials. In other words, critics held this U.S.-owned enterprise responsible for some of the corruption flourishing under the Batista regime. Castro also had a personal grievance: Cuban Electric once shut off his service because he did not pay his bills.[43]

During the insurrection, rebels sabotaged Cuban Electric property, causing numerous interruptions of service and a loss of at least half a million dollars in equipment. Most of the sabotage took place in Oriente Province, where guerrilla forces held sway, but a bomb rocked a Havana electric station in May 1957.[44] During the next year, explosions damaged transmission lines and transformers. In early 1959, the new revolutionary government slashed electric rates and fired top Cuban officers from the company. Eventually Cuban Electric was nationalized.[45]

The Cuban Telephone Company became even more unpopular than the electric company. This, despite the fact that Cuba had the highest per capita telephone ownership in Latin America—one telephone for every 38 persons, although most of the devices were installed in Havana and the provincial capitals.[46] A subsidiary of U.S.-based International Telephone and Telegraph Corporation (ITT), Cuban Telephone operated under a concession granted early in the twentieth century. Four of seven officers were Cubans, and Antonio Rosado, a graduate of the U.S. Naval Academy and the Massachusetts Institute of Technology, became company president in 1956.

Cubans wanted more, many more, telephones, but Compañía Cubana de Teléfonos failed to keep up with demand. At the end of 1955, more than 52,000 subscribers were waiting for installation of service; in 1956, the company received 60,000 applications, but installed only 1,371 telephones. Cuban Telephone claimed that it could not expand services until the company received a repeatedly requested and long overdue rate increase. Critics charged that the company was badly managed and corrupt, and that dividends were paid to stockholders at the expense of maintenance.

U.S. State Department officials urged the Batista Administration to grant the rate increase. ITT cared not only for profits but for "the interests of Cuba," they argued. Ambassador Campa agreed that a rate hike was due, but he reported that President Batista feared a "strong public reaction."[47] Then, on March 14, 1957, apparently after Cuban Telephone paid bribes, Batista decreed a rate hike of 20 percent. Ambassador Gardner himself attended the decree-signing ceremony, and ITT rewarded Batista with a gold telephone. The timing could not have been worse. Batista issued the unpopular decree the very day after Directorio Revolucionario youths died in an attack against the Presidential Palace. Later, with help from the Export-Import Bank, the company launched an $85 million expansion program to expedite the backlog of requests for service.

During the insurrectionary period, rebels regularly blew up telephone poles and lines. In early 1959 the new revolutionary government quickly canceled the rate increase. In March, Cuban Telephone became the first North American-owned firm to be "intervened"—taken over by the Cuban government. Five months later the unpopular company, symbol of corruption and outside control, was formally nationalized.

Because property owned by U.S. companies stood so prominently on the island and anti-Batista groups sought to oust Batista by sabotaging the interests of those who supported him, other North American enterprises took frequent blows. In the first nine months of 1958 alone, U.S. firms in Cuba sustained property losses of more than $2.25 million. The United Fruit Company, for example, whose reputation throughout Latin America was dark, failed to fend off the rebels. In 1958, M-26-7 raiders sacked company sugar warehouses, kidnapped United Fruit employees, and attempted to levy a tax on every bag of United Fruit sugar.[48] Conspicuous throughout Cuban life, public about its support for a despised regime, and responsible for the U.S. "strangle-hold on the Cuban economy," North American business could not escape the wrath of the insurrection.[49]

4

Curve Balls, Casinos, and Cuban-American Culture

CUBANS CRAVED North American-made telephones, but not the U.S. influence that came with the devices. Cuba "lies almost entirely within the United States zone of influence," British Foreign Office diplomats reported. But "along with a genuine feeling of gratitude and respect for their northern neighbours, many Cubans resent any suggestion of United States patronage, let alone interference in her internal policies."[1] Cubans had long held conflicting opinions of the United States—torn "between trust and suspicion, between esteem and scorn, between a desire to emulate and a need to repudiate," a scholar has written.[2] Such mixed attitudes became conspicuous in the process by which the interlocking Cuban and U.S. economies melded the two cultures. Through their interlaced societies, Cubans and North Americans came to know each other quite well, and because of the large North American presence on the island, Cubans struggled with competing cultural identities.

The Cuban writer Pablo Medina, for one, spotlighted the question. He recalled that in the 1950s "my heroes, my women, my Helens, my Troy was the United States. What did we Cubans have that could possibly rival the glitter of Marilyn Monroe's lips . . . , the purposeful swagger of John Wayne? Or all those battleships and planes crowding the screens?" Medina "more than anything . . . wanted to be an American and live in a suburb." At the same time, another, contrary feeling tugged at him, pulling him toward Cuban culture—Martí's poetry, island music and carnivals, and Havana's impressive El Malecón promenade.[3] The Cuban historian Oscar Zanetti also remembered his fascination with American folklore, including the exploits of Davy Crockett. "Alongside him, we hunted in the woods of the American frontier, and we defended the Alamo until our last breath," Zanetti has written. "Years later, I would understand—with disillusion—that in some of these childhood battles, I had fought on the wrong side."[4]

A Canadian ambassador to Cuba once observed that on the island "the most popular game is baseball" and the Cubans "read news supplied from the United States, see American films in the theatres and go to Florida and New York on their holidays. If they have enough money, they send their children to school in the United States and invest their savings in Florida real estate or hoard them in New York banks as insurance against a financial collapse in Cuba." He added that "Cuban interest in United States domestic affairs runs high; the developments of the United States electoral campaign were followed closely here."[5]

The U.S. Information Service (USIS) became a major source of information about the United States in Cuba. The USIS worked to counter negative Cuban opinion about the seamy side of North American culture and to build Cuban respect for free enterprise, Cold War anti-Communism, political stability, and U.S. national security objectives. USIS officers, working from the U.S. Embassy, "placed" and "generated" stories in Cuban newspapers. They ran picture stories of U.S. achievements in atomic energy and medicine. The film *The Nautilus Crosses Under the Pole* screened on Cuban television. By preparing and placing stories about a U.S. satellite tracking station in Cuba, the USIS heightened interest in U.S. exploration of outer space. In a secret report to Washington, agency officers claimed that on coverage of a meeting of Latin American foreign ministers, "the USIS version was accepted and printed by *Excelsior, El Mundo,* and *Diario de la Marina.*" The newspaper *Alerta* also cooperated, printing a USIS story "which labeled Russian education as a tool of the state."

The USIS donated a substantial book collection to Camagüey's Lyceum, a prominent cultural center. The U.S. agency also sent the secretary of the institution to the United States on a "leader grant." Nearly half of the activities of the Lyceum, the USIS reported, "will be reflecting American cultural developments" in a city that "exercises great influence at approximately the center of the island." In Havana, meanwhile, the USIS-supported Cuban-American Institute taught classes in English to some 5000 enrollees. The USIS also maintained a reading room in Santiago de Cuba and planned to build a student center near the University of Oriente in that city. Sample U.S. textbooks in labor economics and industrial relations were donated to Cuban universities. The USIS also opened exhibits of North American art at Havana's Lyceum and Lawn Tennis Society. Racks at airports offered USIS pamphlets. And every month colorful USIS posters—"*Mundo Gráfico*"—went up across the island.

To promote labor-management cooperation and buttress the anti-Communism of the Confederation of Cuban Workers (CTC), the large labor organization that backed Batista, the USIS and the U.S. Embassy's labor attaché teamed up to distribute to key labor figures and employers copies of articles and reports on U.S. labor practices. The CTC worked with the AFL-CIO to promote the Inter-American Regional Organization (ORIT), North American labor's instrument for countering Communist influence in Latin American unions.[6] Although the AFL-CIO questioned CTC's "misalliance" with Batista, the huge U.S. organization, like ORIT, deferred to its Cuban "friends."[7] In cooperation with the AFL-CIO, the U.S. government also funded the training of Cuban trade unionists in the United States in such subjects as labor administration and collective bargaining. The Embassy crowed in early 1958 that these U.S.-trained labor leaders were beginning to emerge in key positions, such as in the Havana Bus Workers' Union.

Through a number of other programs, Cubans enrolled to study in U.S. institutions, including universities, and upon their return became part of a network the American Embassy regularly cultivated through letters, shipments of books and magazines, and invitations to special events. U.S. companies also provided scholarships to Cuban students for study in the United States to publicize the benefits of the "free enterprise system." Batista closed Cuban universities during the insurrection, making it difficult for U.S. Embassy personnel "to work through student organizations." USIS officers nonetheless judged "student scholarship placement . . . an excellent instrument in our hands."[8]

Still, in the end, the USIS failed to sustain a hearty Cuban endorsement of U.S. institutions, ideas, and preferences for order, which after all really meant support for the Batista regime. "It's not that USIS didn't do enough," explained the chief USIS officer in Cuba, Richard G. Cushing. "The people wanted a change."[9]

U.S. culture, ideology, and products had long penetrated Cuba through other avenues than the USIS.[10] Commercial U.S. radio and television broadcasts into the island cultivated Cuban receptivity to U.S. styles and goods, many of which were advertised by Hollywood movie stars. For the 150,000 Cubans employed by U.S. businesses, North American tastes and customs became especially familiar. U.S.-made refrigerators and Coca-Cola became popular, and ownership of a *Cadilá* (Cadillac) signified high social standing. Before Christmas, large quantities of North American pine trees arrived on the island. Miami became a magnet for Cuban shoppers. In 1948 it was to Miami and New York that Fidel Castro took his new bride for their honeymoon.

Some 8000 Americans lived in Cuba, forming the predominant part of the "ABC community" of American, British, and Canadian citizens.[11] These three groups lived primarily in Havana, where yacht clubs, with both Cuban and North American members, imitated the life-style of the U.S. social elite. The Rotary, Lions, and Knights of Columbus sponsored clubs in Cuba. The successful proselytizing work of U.S. missionaries begun half a century earlier was measured in the great number of Protestant parishioners, churches, and schools throughout the island, including Methodists and Baptists at the eastern end. "Jazz-tinged" Cuban music and "Cuban-tinged" jazz sounded through dance halls in both the United States and Cuba as band leaders like Tito Puente and Noro Morales blended rhythms to produce the Mambo craze of the 1940s and 1950s. Columbia Records released the LP *Mambo with Morales,* and Victor followed with *Mucho Puente.*[12]

As Cubans and Americans played, prayed, worked, and socialized together, their languages became mixed as well. Many Cubans became bilingual or spoke a blend of Spanish and American English. Cubans adopted "sandwich" and other English languge words into their vocabulary. "*Jonrón*" meant home run, and "*doble plei*" double play in baseball games. The development of "Spanglish" since 1898, some observed, represented the supreme blending of the two cultures.

Baseball especially bonded Cuba and the United States. *Beisbal* became a symbol of both cultural intersection and outside influence. The U.S. Ambassador himself presented "The Ambassador's Cup" each year to the outstanding Cuban baseball player selected by Cuban sports writers. "Cuban fans can hold their own with Milwaukee fans any day," remarked Ambassador Earl E. T. Smith.[13]

Introduced into Cuba in the 1860s by U.S. sailors and by Cubans returning home after living in the United States, baseball soon went professional. The first national championship was held in 1878.[14] In the early twentieth century, African-Americans who could not play in the United States because of the color line joined Cubans on teams that competed against touring U.S. major leaguers, including Ty Cobb. Cuba fielded strong teams in the Central American Games and Pan American Games.

In the 1940s, Brooklyn Dodgers president Branch Rickey determined to break the segregation barrier. He first considered the black Cuban shortstop Silvio García. When Rickey asked García what he would do if a prejudiced white American slapped him, the Cuban snarled: "I kill him."[15] The Dodgers chief picked Jackie Robinson instead. In 1947 Rickey shifted his team from Florida to Cuba for spring practice because he wanted to ease Robinson into the line-up where baseball was integrated—where, in short, Robinson might be shielded somewhat from the expected U.S. racist uproar over the young black's appearance in a Dodgers uniform. Still, Rickey segregated his players in Havana. White players took rooms at the elegant Hotel Nacional; Robinson and a few other African-American players had to sleep at a seedy hotel. With the fiery Leo Durocher as his manager, Robinson played a few games in El Gran Stadium, where Cubans of color especially recognized him as a "first."[16]

Cuban baseball players such as Saturnino Orestes Arrieta Armas "Minnie" Minoso gained fame in both nations. Beginning as an outfielder for a Matanzas team, Minoso went on to star in the 1950s as both hitter and fielder for the Chicago White Sox. Another Cuban, Edmundo "Sandy" Amorós, enjoyed a moment of celebrity when he played left field for the Brooklyn Dodgers. The Dodgers had signed the lean ballplayer in 1952 after the club spotted him during their barnstorming tour of Cuba, and he played for them for seven years. During the seventh and deciding game of the 1955 World Series, Amorós raced to grab a Yogi Berra opposite-field line drive, turning the catch into a double play that preserved a 2-0 victory over the New York Yankees.

Many American baseball players who later became stars in the U.S. major leagues earned experience and kudos in the Cuban Baseball League. Wally Moon, Dick Sisler, and Don Blassingame played for the Havana team; Brooks Robinson took the field for Cienfuegos, and in the 1957–58 season he shared honors for most home runs hit; Willie Mays played for Almendares; and Roy Campanella and Don Newcombe, later Brooklyn Dodger greats, wore the uniform of the Marianao club.[17]

At the 35,000-seat El Gran Stadium, the Havana Sugar Kings baseball team competed in the International League, a Triple A competition. The Sugar Kings entered the U.S.-based league in 1954. Although a majority of the players on the Havana team were Cuban, players from the United States also took the field for the Sugar Kings. Even in the throes of rebellion and with their team finishing last in 1958, Cubans flocked to games. (In July 1960, troubled U.S.-Cuban relations persuaded the Sugar Kings' owner to transfer the Havana franchise to Jersey City.[18])

Fidel Castro himself had earned some baseball notoriety by the late 1940s. North American major-league scouts noticed his considerable pitching talents for the University of Havana baseball team. Known for "a wicked *bleeping* curve ball," Castro seemed a good prospect for professional baseball in the United States. "He could set 'em up with the curve, blow 'em down with the heater," recalled a scout for the Pittsburgh Pirates.[19] But because his fastball was not overpowering, Castro became known as a "smart" player who kept batters guessing. In a November 1948 game against a team of touring major leaguers, Castro struck out the all-star Hank Greenberg and gave up only three hits and no runs.

In 1949 the New York Giants offered Fidel Castro a contract with a $5000 signing bonus. He did not jump at the opportunity. His law studies came first, he explained. "We couldn't believe he turned us down," recalled a Giants negotiator. "Nobody from Latin America had said 'no' before."[20]

One night in the early 1950s, when Don Hoak, later a National League third baseman, played for the Cienfuegos team, Castro and a few hundred demonstrators carrying anti-Batista banners dashed on to the diamond. Castro captured the mound, demanded the pitcher's glove and ball, and began to hurl. Hoak was the unamused batter who remembered the right-hander's "great windmill flourish." From the "hipper-dipper windup" came a curve ball, a fastball, and another fastball. One ball, two strikes. Hoak had had enough and jawed the umpire to oust the renegade pitcher from the park. Riot police soon dispersed the protesters; shoved from the mound, Castro gave up the ball. "With a little work on his control," thought Hoak, "Fidel Castro would have made a better pitcher than a prime minister."[21]

While many Cubans embraced baseball and their cultural interactions with North Americans, others complained. One Cuban poet remarked that "even our bad taste was imported. We were not even allowed to discover our own bad taste; it was brought to us from Miami and New York." Pablo Armando Fernández also recalled that on the sugar plantation where he was born and raised, U.S. history was taught and the U.S. national anthem was sung at school. Only much later did he learn that the words were not about Cuba.[22] Another Cuban once observed that even the island's rum and Coke drink—the *Cuba Libre* (Free Cuba)—depended upon Coca-Cola, an imported U.S. product.[23] These comments may not have been typical, but many Cubans worried that their nation, especially under Batista, was becoming dependent upon the United States in almost every category—economic, political, and cultural. Perhaps this is what one observer meant when he wrote that the Cuban rebellion "is rising from the heart and the head, not from the stomach."[24]

Cuban uneasiness with the tourist industry illustrates the point. As in so many other examples, an anti-Yankee message became interlaced with grievances against Batista. Cubans who excoriated the Batista regime as politically illegitimate, morally bankrupt, and subservient to foreign economic interests also made the case that U.S. crime bosses, in concert with the Cuban president, soiled the nation's reputation by turning Havana into the Las Vegas of Latin America, the Monte Carlo of the Caribbean, the brothel of the New World.

In the 1950s the Batista government vigorously promoted tourism as a hedge against the always precarious sugar trade.[25] Fancy hotels with gambling casinos went up, and U.S. mobsters shared profits with government officials. Everyone seemed to know that Batista brother-in-law and Army general Roberto Fernández y Miranda skimmed money from the slot machines—and from Havana's parking meters. Organized crime not only fueled corruption in high offices; it also streamlined a vile commerce in prostitution and narcotics and set up abortion mills. "With money," remarked a Havana resident, "you could buy anything, anything, *anything!*"[26]

In search of rum, rhumba, and much, much more, some 300,000 tourists each year flocked to the island to partake in Cuba's bustling life of outdoor cafes, bars, nightclubs, brothels, and casinos. Sugar prices might fall, but Havana's 270 brothels always seemed to thrive. U.S. Navy ships regularly disembarked sailors for weekend liberty. In 1956 alone, 375 U.S. naval vessels carrying 84,000 personnel visited Cuban ports. That year alone these American bluejackets spent at least $2.5 million in Cuba.[27]

Like many other prominent Americans, the young Democratic U.S. Senators John F. Kennedy of Massachusetts and George Smathers of Florida partied on the island. In December 1957, Kennedy and Smathers journeyed to Havana. They visited Ambassador Smith, a Kennedy friend from Palm Beach, and his wife and former model, Florence Pritchett Smith, with whom the Massachusetts politician had had a love affair some years before. The pleasure-seeking senators apparently never discussed the rebellion, although Smathers, as he himself put it, "had made a career of Cuban problems."[28] Instead, golf, sailing, nightclubs, and women occupied their time. Crime boss Meyer Lansky's widow claimed later that during that trip her husband helped locate women to satisfy Kennedy's now-famed sexual athleticism. "Kennedy wasn't a great casino man," remembered Smathers, "but the Tropicana nightclub had a floor show you wouldn't believe." The Florida senator who became staunchly anti-Castro recalled that once Cubans "started looking after you . . . , why it was just elegant."[29]

North Americans of all classes rode the ferry from Key West or took one of Pan American's many inexpensive flights from Miami to Havana, a huge city of more than one million people. Package deals, advertised through teasing postcards of sandy beaches and beautiful women, attracted sex-and-sun seekers. "Gamblers' specials" flew to the capital city from the United States in just 55 minutes; passengers could fly in at night to party and gamble and then depart on a morning plane. Cockfights, jai-alai *frontones,* and horse races joined the casinos in attracting wagerers. Nightclubs featured U.S. performers such as Frank Sinatra, Abbott and Costello, and Elvis Presley.

The gaudy Tropicana staged lavish floor shows of leggy, scantily clad Cuban dancers gyrating with precision to Caribbean beats.

The popular television comedian Jack Paar took his program to Havana in July 1958 to capture the city's magic. Rumors floated that revolutionaries intended to kidnap him, so the Batista government provided Paar with a car, driver, and bodyguard. Everything seemed to go wrong. Ernest Hemingway refused to go on the show, and the movie star Errol Flynn could not be found. "The show itself was the worst disaster in Havana since the sinking of the *Maine*," Paar recalled. Broadcast from the Tropicana nightclub and the Havana Hilton, the program stumbled through missed signals. When Paar shouted "El stoppo del musico," the rhumba band kept playing. Paar tried to make light of the moment. He implored his audience not to think of the performance as a show. "Just think of it as international bad relations. It's a lend-louse deal." Later, in 1959, Paar interviewed Fidel Castro, "the only man I know who talks longer on television than I do."[30]

In Havana, away from the neon lights, tourists admired the elegant Spanish colonial architecture, El Morro Castle at the harbor's entrance, and the handsome mansions of Fifth Avenue. Visitors could stroll on the Malecón, a graceful promenade protected by a seawall that winds along the waterfront. There sat the tall U.S. Embassy building and a monument to the USS *Maine,* both a short walk from the Hilton Hotel and the University of Havana. La Habana Vieja (Old Havana) was by every account beautiful, but the newer parts of the city were "unrepentant Manhattan."[31] Vacationers to Cuba especially prized the legendary sands and clear warm waters of Varadero Beach, just two hours from Havana.

The Cuban writer Guillermo Cabrera Infante remembered Havana's "yellow radiance enveloping us with the luminous halo of night life, the fatal phosphorescence that was so promising: street life and free days all." Havana's light "came from the city itself, created by Havana to bathe in and purify itself of the dark that remained on the other side of the wall."[32] Imposing and simple, graceful and tacky, elegant and tawdry, seductive and repellent, light and dark, Havana both charmed and disappointed.

While sipping *mojitos,* Cuba's rum mint drink, tourists might get a glimpse of Ernest Hemingway at La Floridita Bar, where he made the daiquiri famous and a reserved table always awaited him. "Havana is more like Miami Beach all the time," observed the author of *The Old Man and the Sea* (1952), a story inspired by his observations of Cuban life.[33] Others spotted Errol Flynn, the U.S. movie industry's heartthrob Casanova, at his favorite Old Havana restaurant and bar, La Bodeguita del Medio. "Best place to get drunk," he scrawled on a menu.[34] The aging Flynn, then in his fifties, had earned a reputation for drunkenness and womanizing in the Cuban capital, where in 1957 he even made *The Big Boodle,* a forgettable melodrama with a cops-and-robbers plot. On New Year's Eve, 1958, the night that Batista fled the Presidential Palace, Flynn suffered a slight leg wound, perhaps from strafing by a government plane. The actor was then traveling with rebel forces, who had attracted his fancy. In his last film, *Cuban Rebel Girls* (1959), Flynn narrated the story of Cuban women who rebelled against Batista.[35]

North Americans did not invent the corruption, vice, and excesses of life in Havana. Cubans themselves avidly played the numbers games in the legal National Lottery and illegal *bolitas.* But when U.S. mobsters began to collaborate with Batista to develop tourism and revive the gambling industry, the public perception of U.S. responsibility for the debasement of Cuba strengthened. Most of the major hotels in Havana were U.S.-owned and most housed casinos leased by unsavory U.S. gambling interests.[36] "Who did you expect to be running the games down here?" snickered one gambler. "[Secretary of State] John Foster Dulles?"[37]

In 1954 the Cuban government began collecting from each gambling casino an initial payment of 25,000 pesos, quarterly payments equal to 20 percent of profits, and a 2000-peso monthly fee. Four years later the government doubled all of these contributions. Hotel Law 2074 (1955) granted tax exemptions to the tourist industry and encouraged applications for casino licenses. Both U.S. gangsters and Cuban politicians reaped the benefits.

In the years 1952–58, some 28 new hotels and motels opened on the island. In Havana's Vedado district, the Capri (1957), Riviera (1957), and Hilton (1958) hotels towered as the most expensive investments. Even the grand Hotel Nacional was refurbished after a subsidiary of Pan American Airlines assumed management and installed a new casino. The 374-room Riviera, twenty-one stories high, encased in turquoise mosaic, and sporting a windowless, egg-shaped casino, had the best of everything, including central air conditioning and a fine restaurant. The Batista government provided $6 million of the $14 to $18 million invested in the complex.

The Riviera made money for Meyer Lansky, the Florida underworld boss who, tailed by FBI agents and monitored by the Internal Revenue Service (IRS) in the United States, found Cuba more hospitable. High rollers from the United States enjoyed his expenses-paid trips. The Riviera became his dream establishment. The palace opened on December 10, 1957, with a floor show featuring Ginger Rogers. She could wiggle but not sing, complained Lansky, always the perfectionist. Lansky lived in an elegant penthouse at the top of the hotel. There he cut deals with Cuban government officials. For the purposes of filing taxes with the IRS, Lansky listed himself as the hotel kitchen manager.

For years Lansky had run the gambling tables at Havana's Montmartre Club. Big-time gamblers frequented the well-managed place, for unlike many Cuban houses in the early 1950s, Lansky's games were honest. After tourists complained loudly against blatant cheating (blackjack was dealt from a hand-held deck rather than from a box, for example), Batista worried about gambling's bad press and declining tourism. In 1953–54 the Cuban president hired Lansky as a consultant to reform the industry. With the help of Batista's henchmen, Lansky forced "clean" games on casino operators and drove out the razzle-dazzle crooks.

Organized crime figures from Las Vegas, Miami, Cleveland, and elsewhere in the United States moved into Havana. Jake Lansky served as pit boss at the Hotel Nacional, where the manager was another Lansky associate, Sam Tucker. "Black Jack" Thomas Jefferson McGinty, Jimmy Blue Eyes, Joseph Silesi (aka Joe Rivers), Joe Bischoff (alias Lefty Clark), Fat the Butch, and other *pistoleros* associated with gambling in the United States also showed up in Havana. Why were such gangsters permitted to run free in Havana? an American once asked Ambassador Earl E. T. Smith. "It's strange," the diplomat answered, "but it seems to be the only way to get honest casinos."[38]

When the racketeer Albert Anastasia was gunned down in the United States in 1957, rumors flourished that Lansky had ordered the kill because Anastasia had been attempting to horn in on Lansky's lucrative Havana interests. In November of that year, to bring peace to the underworld, and apparently to bring order to Havana gambling, the chief mobsters (but not Lansky) gathered at a mansion in the town of Apalachin, in upstate New York. Local police moved in and arrested some fifty. Meyer Lansky was later picked up in New York City, but a judge quickly released him after the police tried to stick Lansky with a charge of vagrancy. To this day it is not known what was discussed or decided at Apalachin, but Cuban casinos had to have been on the agenda.

Opened in March 1958 and managed by Hilton Hotels International of Chicago, the 32-story, 630-room Havana Hilton was built with $24 million from the Retirement Fund of the Cuban Cooks and Waiters' Union (AFL-CIO) and from the Cuban government. Batista picked a Cuban, Roberto Mendoza, to operate the Havana Hilton's casino, but English, not Spanish, was the language of business. Signs on doors read "PUSH" and "PULL," service personnel wore uniforms with "Bellboy" over their chest pockets, and no North American could get lost with signs pointing to "Bookshop," "Barber," and "Swimming Pool." Tourists in the Sugar Bar at the top of the hotel enjoyed the spectacular panorama of Havana below.

At the $5.5 million, 226-room Capri, the Hollywood star George Raft fronted for a syndicate headed by Santos Trafficante, Jr., who sometimes went by the name of Louis Santos in Cuba. Raft's career as movie gangster was virtually over at age 62 in 1957, so he welcomed the job at the Capri.[39] His boss Trafficante, son of a Florida mobster, also managed gambling at the Sans Souci nightclub. Charles "The Blade" Tourine oversaw the Capri's slot machines and other games of chance. The Nevada state government became so alarmed with the success of Trafficante and others in attracting gamblers away from Las Vegas that in 1958 it prohibited holders of Nevada gambling licenses from operating casinos in Cuba. A few hoods (but not Lansky and Trafficante) then sold their Havana interests.

Ambassador Arthur Gardner found much of the tourist business disgusting, although he seemed more critical of the ill-behaved partygoers than of the gangsters and political corruptionists. The tourists, he reported in 1956, "are bent only on pleasure and never think of Cuba except in terms of fun, rum, and nightclubs. . . ."[40] One group of vacationeers particularly rankled the ambassador. Soon after arriving on a cruise ship, they arrogantly insisted upon an interview and cocktails with Batista himself at the Presidential Palace. The Cuban president obliged the North American visitors. An aghast Gardner described the scene: The "unkempt, untidy" U.S. citizens wore informal sports shirts and "swarmed into the Palace as though they were goint to a movie or a peepshow."[41]

Proud Cubans resented this rude aspect of their interaction with North Americans. Former Prime Minister Manuel Antonio Varona, for one, appealed to Senator Estes Kefauver of Tennessee, whose investigating committee had exposed racketeering in the United States. Denouncing "big gambling" interests as purveyors of vice and allies of a tyranny, Varona argued that "these strange elements are extremely dangerous to the Cuban people."[42] Castro's representative in the United States, Ernesto Betancourt, pledged that once in power, the rebels would boot the North American vicemongers out of Havana and return respectability to Cuba. Even the managers of the new Havana Hilton, he remarked in early 1958, might have to face a special tribunal.[43] The gambling lords in Havana may have dismissed Fidel Castro as "Fiddle," but they could not have ignored the rebel's condemnation of their vice trade as a cancer inserted from outside that had to be cut out of the nation.[44]

Still, Meyer Lansky's imperious view of Cuban politics prevailed among gambling interests: Any Cuban politician could be bought off. Should Batista fall, his replacement would have his price too. North Americans, as always, would prevail. Some Cubans disagreed. "You just go ahead and send your toughest gangster down here," said one. "I guarantee you that even a second-rate Cuban politician will run rings around him."[45]

Anti-Batista Cubans conspired to destroy the casinos. Just before opening ceremonies for the Riviera Hotel in 1957, for example, police quashed a plot to blow up the building. Then, to prevent rebel sabotage at the hotel, the police posted undercover agents disguised as waiters.[46] M-26-7 tried to scare North American tourists away. In March 1958 in Los Angeles, 26th of July Movement members distributed mimeographed leaflets in both English and Spanish. "The tourist business is a large and important one in the Cuban economy," read the handouts. A tourist boycott would help "oust the dictator."[47]

In early 1958, Cuban hotel business began to slump because of sunless days, U.S. recession, and the insurrection. The Havana Hilton opened on March 23, 1958, but just a few days later only 44 guests occupied the huge building. Soon the Hilton laid off 400 employees, the Riviera 270.[48] With a symbolism surely not lost on the observant, the hotel industry and Batista were collapsing together.

5

Supplying Repression: Military, CIA, and FBI Links

OFFICIAL U.S. MILITARY and intelligence links with Batista's regime joined political, economic, and cultural ties to bind the worlds of Cubans and North Americans. The military links particularly drew fire from nationalistic Cubans, for Batista's menacing police and armed forces sustained the dictatorship, and U.S. aid helped sustain the dictator's military, which became more repressive and brutal with each turn of the rebellion. Fearing that Latin American nations might "drift" toward "neutralism" in the Cold War, U.S. policymakers had worked after World War II to make hemispheric military establishments dependent upon U.S. equipment, weapons, and training. If such dependency proved successful in Cuba and elsewhere, the United States could forestall neutralist "blow[s] to U.S. prestige" and "defections" that "would seriously impair the ability of the United States to exercise effective leadership of the Free World."[1]

Before Batista's 1952 coup, the United States, as part of its Cold War effort to gird the globe with military partners and bases, signed agreements with Cuba to install U.S. Army, Navy, and Air Force missions on the island and to provide military equipment under the Mutual Defense Assistance Act. Because the accord for aid was signed on March 7, 1952, just three days before Batista's seizure of power, some Cubans sniffed collusion.

Soon the Batista government filed requests for a host of weapons and military goods—submachine guns, recoilless rifles, hand grenades, incendiary bombs, rocket launchers, armored cars, T-33 jet-trainer aircraft, radio equipment, trucks, and more. For some of these items Cuba paid cash; others on the long list came under the U.S. Military Assistance Program (MAP). In fiscal year 1953 Cuba received $400,000 in military assistance, in 1954 $1.1 million, in 1955 $1.6 million, and in 1956 $1.7 million. During the height of the insurrection, 1957 and 1958, military aid expanded to $2 million and $3 million respectively.[2] By Cold War standards these were not large amounts, but Cuba was a small country, and such U.S. assistance buttressed the regime's repression.

The personnel of the U.S. Military Assistance Advisory Group (MAAG) formally began work in Havana in 1955. Some 40 Army, Air Force, and Navy advisers, officers and enlisted personnel, established channels with the Cuban armed forces.[3] Cuban officers went to the United States to study and train, to be integrated into the U.S. military system. In January 1955, for example, the Cuban Navy Chief of Staff and other naval personnel went on a two-and-a-half week inspection of U.S. Navy training facilities. Cuban officers graduated from the Naval Command College at the Naval War College, Newport, Rhode Island. For 1950–58, 38 Cuban Army personnel, 171 Navy officers and enlisted sailors, and 211 Air Force personnel received training in the United States under the military aid program.[4]

From the start, Ambassador Willard Beaulac recognized the "delicate" political problems in Cuba arising from U.S. military ties with the dictatorship. He urged a "quiet and discreet" linkage.[5] It seldom was. In early December 1956, for example, shortly after Castro's army landed in Oriente, the USS *Canberra* visited Havana from the American naval base at Guantánamo. The ship fired a twenty-one-gun salute, and after cocktails at the Presidential Palace, Batista, Ambassador Gardner, and Cuban and U.S. brass attended a luncheon on the 16,000-ton vessel. "President Batista," the U.S. Embassy's report noted, "was rendered full presidential honors on arrival and departure," and he received a special demonstration of the ship's guided missile launching gear.[6]

Batista exploited these highly visible displays of military alliance to demonstrate U.S. backing of his regime. He also counted on U.S. officials' winking at the restrictive terms of the mutual security agreement and at blatant Cuban violations. Aid was supposed to be used for "hemispheric defense," but U.S. diplomats and military officers knew that their assistance strengthened the arsenal Batista used not to contain foreign enemies but to repress his critics. Although U.S. officials publicly explained that the Cuban-American military relationship provided defense of coastal sea communications, maritime routes, and the Panama Canal, ensured that Cuba's military would be at U.S. disposal during a major war, and guaranteed U.S. access to Cuban bases, privately they acknowledged that the regime marshaled the aid for internal security.[7]

In August 1953, for example, a Cuban patrol boat arrived in Norfolk, Virginia. It took on equipment to increase its "effectiveness" in "anti-revolution patrols."[8] Cuban naval vessels "tend to be retained in ports for domestic political reasons," reported the U.S. Embassy the next year.[9] Embassy staff understood that Cuban armed forces financed under the U.S. military assistance program "will not leave Cuban soil but will be devoted to maintaining internal security."[10] Any "realistic appraisal," reported the MAAG chief in 1958, "boils down" to the fact that the Cuban military's primary mission was internal security.[11]

Everybody seemed to acknowledge that the argument that military aid flowed to Cuba for "hemispheric defense" was spurious. As a former State Department official told a meeting at the Council on Foreign Relations, military aid served purposes more political than military. Such assistance, including the military missions, helped squeeze out European arms merchants, orient Cuban military officers toward the United States, and promote internal political stability.[12] In short, to shore up Batista and his armed forces was to perpetuate a preponderance of U.S. power in Cuba. Whatever the language of the law, then, the case for supporting Batista's armed forces became compelling: a stable, orderly Cuba that quelled insurrectionists and Communists would better serve the U.S. goals of hegemony and Cold War victory.

Numbering almost 30,000 in 1958, backed by 18,000 reservists and by the Rural Guards and National Police, Batista's army became the president's primary instrument of power.[13] It also became a major symbol of abuse because of its undisguised practices of corruption, torture, and murder. The protection of human and political rights was not one of its missions.

Although Batista's armed forces received U.S. training and weapons and enjoyed celebratory toasts at numerous public ceremonies, MAAG officers had orders not to accompany Cuban units into combat. Nor were U.S. officers supposed to advise the Cuban military on how to subdue the rebels. Yet the line was often crossed in training programs, because U.S. officers taught the methods needed to defeat an enemy—any enemy. Commander F. W. Zigler, chief of the U.S. Naval Mission, 1954–57, developed close relations with Batista, giving at the very least the appearance of collusion. MAAG officers also correctly read their orders to mean the maintenance of good relations with their Cuban counterparts.[14] Ambassador Earl E. T. Smith frankly acknowledged the "close daily contact between Mission personnel and Cuban officers. Mission offices are located in the Cuban military staff headquarters."[15]

One U.S. naval officer recalled that the U.S. Naval Mission "did not actively assist the Cuban Navy in fighting against Castro and/or rebels.

However, in helping to keep planes flying and ships operating, they indirectly assisted."[16] The Cuban Navy had 35 ships, mostly patrol craft.[17] The U.S. Army Mission trained and supported a special Military Assistance Program infantry battalion which battled M-26-7 forces in the Sierra Maestra Mountains.[18] In the Sierra, too, U.S.-equipped Cuban government airplanes strafed and bombed, sometimes indiscriminately. Some 70 percent of the Cuban Air Force's officers had received MAP training.[19]

Cuban dissidents frequently denounced the close Batista-U.S. military relationship, holding the United States responsible for much of the blood spilled during the insurrection.[20] They angrily demanded termination of U.S. military assistance and sales and withdrawal of the missions, and they provoked incidents to spotlight U.S. military backing of the regime. At the same time, rebel leaders gradually came to understand that U.S. military advisers never gave Batista's legions adequate counterinsurgency training. The advisers failed to impart the lessons of counterguerrilla warfare learned in Greece, the Philippines, and Malaya. Nor were U.S. personnel able to instill the morale or supply the leadership the Cuban military needed to quell the insurrection. Batista appointed his political cronies to high ranks (a professional officer corps never developed because Batista feared that it might turn against him), corruption prospered, and individual soldiers never received adequate training in mountainous terrain.[21]

Continued U.S. possession of the Guantánamo Naval Base, through which flowed U.S. military equipment to Batista, also drew political fire and entangled the United States in Cuba's internal disputes. Guantánamo always had the potential of becoming a "whipping boy" in Cuban politics.[22] Under the Platt amendment, which U.S. officials forced Cubans to insert in their 1901 constitution and then incorporated into the Permanent Treaty of 1903, the United States leased territory around Guantánamo Bay in virtual perpetuity. Cuba granted the United States use of the area for an annual fee of $2000 in gold (in 1959 recalculated at $3,403), but many of the other terms were vague. The agreement would last, for example, "for the time required for the purposes of coaling and naval stations." The United States acknowledged Cuba's "ultimate sovereignty" over the reservation, but gained "complete jurisdiction and control."[23] Although Washington and Havana abrogated the Platt amendment in 1934, the United States clung to Guantánamo. The new agreement read that the United States could remain at the naval station as long as it wanted to or until both nations agreed to modifications in the accord.[24]

The United States valued Guantánamo because of its location near major shipping routes. The base helped protect the avenues of approach to the Panama Canal, which lay some 800 miles away. Used as a staging area for interventions in Cuba and Latin America in the twentieth century, the base also permitted the United States to wield military muscle in Cuba. During the Second World War, base operations expanded to stalk Axis submarines in the Caribbean. Good weather, excellent harbors for large warships, and an ocean floor that drops off quickly to permit deep-sea operations made Guantánamo a prime training site for the U.S. Atlantic Fleet in the 1950s. The Guantánamo installation became part of a Caribbean base complex that included Panama, Trinidad, and Roosevelt Roads in Puerto Rico.

The oldest U.S. base on foreign soil, Guantánamo was home for some 2500 sailors and marines (and another 2500 family members). They called it "Gitmo" after the official abbreviation GTMO. Almost every day some 4000 to 5000 Cubans walked through guarded gates to work at the base, which became a central force in the regional economy. The picturesque, 45-square-mile base of hills and water included a naval station, naval air station, supply depot, hospital, and communications facilities. Baseball and softball fields, tennis courts, a roller-skating rink, a 27-hole golf course, bowling alleys, outdoor movie theaters, basketball and volleyball courts, swimming pools, and sandy beachers washed by tropical waters made the base seem more like a vacation spot than a military tour of duty. And Guantánamo City, sixteen miles away, provided brothels and bars.

Because it sat at the southeastern tip of the island surrounded by rebel strongholds, the base also became ensnared in the revolution. Anti-Batista workers on the base spied for the rebels, and M-26-7 followers stole equipment. A few sons of U.S. service personnel escaped to the mountains to join Castro's army. Marines traveling outside the base were harassed and in some cases captured.

GTMO had a vulnerability—its dependency upon Cuba for fresh water. The base's drinking water had to be piped from the Yateras River, which flowed five miles away, inside Cuban territory, and the base had a storage capacity for only one day's supply. During the insurrection, U.S. officials often feared that the rebels would sever the base's water supply. At one point, to howls of protest from anti-Batista Cubans, U.S. marines actually took up guard duty at the Yateras waterworks. As the insurrection neared success in late 1958, Castro rebels periodically interrupted the water flow to the U.S. base they so reviled.[25]

Cuba also joined the Cold War space race when it participated in Project Vanguard, the U.S. Navy's project to launch a man-made satellite into outer space, where it would orbit the earth at 18,000 miles an hour. By grabbing headlines with their *Sputnik* satellite in the fall of 1957, the Russians made this project urgent for the United States. An elaborate, worldwide tracking network of camera and radio stations had to be created to monitor the speed, direction, and height of the U.S. satellite. Batista Air Field, built by the United States during the Second World War and then turned over to the Cubans, became one of five observation stations in Latin America. The Army Corps of Engineers constructed the station in 1957 on land donated by the Cuban government. A U.S. Army Signal Corps officer commanded 26 U.S. personnel at the installation. Because of numerous misfires and mechanical failures, the first successful Vanguard rocket launch of a satellite had to wait until March 1958. From Cape Canaveral, Florida, a slender missile lifted into orbit a "man-made moon" loaded with instruments to measure temperatures and test sun-powered batteries.[26]

Far more controversial than Project Vanguard's presence in Cuba was the shadowy activity of the Central Intelligence Ageny. Although the extent to which the CIA was active in Cuba remains security-classified, enough evidence exists to sketch the outlines of its operations. The agency helped Batista initiate the Buró Represión de las Actividades Comunistas (BRAC).[27] "I was the father of BRAC," CIA Director Allen Dulles admitted.[28] The Bureau never came up to U.S. standards for effectiveness, but for a besieged regime it was one more weapon to use against critics. The organization investigated alleged Communist participation in sugar strikes, and the Minister of Labor used a BRAC report to block the employment of Communists in the Cuban Telephone Company. BRAC agents also raided the headquarters of the Communist newspaper *Hoy*.[29] At times BRAC may have served as a means to co-opt foes. The veteran Cuba-watcher Ruby Phillips of the *New York Times* speculated, from sources unnamed, that one of the reasons why BRAC made few arrests of Communists in early 1958 was that the Communists had struck a deal with Batista: If they refrained from supporting Castro, Batista would stop harassing them.[30]

When CIA headquarters official Lyman B. Kirkpatrick, Jr., visited Batista in 1957 on a mission to strengthen BRAC, the Cuban president, to Kirkpatrick's shock, brought in a photographer to snap pictures that showed up in the next day's Havana newspapers. "Thus I found myself being used to bolster a shaky and increasingly unpopular regime," Kirkpatrick lamented. As he learned the next year, BRAC was almost entirely targeting M-26-7 and using violence in interrogations, "despite our constant protests."[31]

The CIA itself placed several agents in Cuba, the major ones under cover in the U.S. Embassy in Havana and in the Santiago de Cuba Consulate in Oriente Province. The CIA officers maintained contacts with both Cuban government officials and dissidents and developed a network of informers. As in other parts of the world, the CIA probably recruited journalists as agents; some of the journalists who hiked the Sierra Maestra to interview the mountain insurgents may have been secret CIA informers.[32] The CIA station chief in Havana from 1954 to the middle of 1958 was William B. Caldwell, listed officially as an ambassy "political officer"; James A. Noel, also under cover as a "political officer," succeeded him in August 1958. Another CIA officer was Robert D. Wiecha, who served as a vice consul in Santiago de Cuba from September 1957 to early 1959.[33] The U.S. naval attaché in the Havana embassy seemed surprised that their cover was so "shallow," for "everybody in town who had any interest in it knew who they were."[34]

When Consul Park F. Wollam arrived at the Santiago post in early 1958, he noticed a room full of communications gear, which he took to be U.S. Army field telephones. "What's that?" he asked. A few days later the equipment had disappeared. "Well, there was only one place it could have gone," recalled the State Department official Robert A. Stevenson—as "a wire" into the Castro camp in the Sierra Maestra Mountains to maintain contact between the CIA and rebel headquarters.[35] "The CIA was trying to get access (or penetration into) the Castro crowd for intelligence purposes. That was their job."[36] Wollam said later that "I took the pious attitude that if I was supposed to know, Havana [the embassy]—where I had been getting briefed for a week—would have so informed me, and I heard no more."[37] U.S. equipment also helped the Sierra insurrectionists to set up Radio Rebelde to challenge Batista's censorship in the propaganda war. Apparently several boxes of radio parts were sent by a Cuban rebel leader from New York to the Santiago consulate, which then passed them on to M-26-7.[38]

A longtime student of Cuban affairs, the journalist and Castro biographer Tad Szulc, basing his argument on slender evidence and unattributed, confidential sources, plausibly suggests that in 1957–58 the CIA funneled $50,000 through Vice Counsul Wiecha to the Castro organization in Santiago. Szulc finds the story "mysterious." Another writer hints that the U.S.-based company Interarms, which was sometimes used by the CIA to channel weapons abroad, sold arms to pro-Castro groups in the United States.[39]

A host of unanswered questions clouds this issue: Did high-level Eisenhower Administration officials (not part of the CIA) or the U.S. Ambassador approve the aid, or did the CIA launch an aid project on its own to hedge its bets for the future by maintaining relations with both sides in the conflict? Many U.S. Ambassadors have complained for years that they could not control the free-wheeling CIA.[40] Ambassador Smith himself became alienated from the political and CIA officers of the Havana embassy, judging that they had become anti-Batista and pro-rebel against his wishes. In any case, as will be developed throughout this book, U.S. officials established numerous contacts with insurrectionists in the cities and the mountains. Castro grew wary of these interactions, but he apparently maintained them. He once wrote from his hideout: "And if they make demands? We'll reject them."[41]

The FBI had its Cuban connections too, although the U.S. government has kept the story buried through secrecy. In the United States, agents spied on rebel sympathizers and bugged their telephones. In Miami, for example, U.S. authorities monitored the phone calls of the headquarters of the 26th of July Movement.[42] The agency had an office of five personnel in the U.S. Embassy in Havana, and the legal attaché funneled information to Director J. Edgar Hoover. The FBI assisted the Servicio de Intelligencia Militar (SIM), Batista's vicious and feared special police.[43] As the American consul in Santiago observed, SIM preferred "not to take prisoners."[44] One of SIM's men was Bernard Barker, who served as a liaison with the FBI. An American born in Havana, Barker later, in 1959, moved to the United States after Castro came to power. Then he began to work for the CIA, ultimately helping to plan the Bay of Pigs invasion. He gained notoriety in 1972 when he was caught during an attempt to break into the Democratic National Committee's office at the Watergate complex in Washington.[45]

The image of the FBI in Cuba resembled that of other U.S. institutions and interests on the island: outsiders wielding influence and propping up a violent, illegitimate regime. Pull out the U.S. props, Cuban nationalists argued, and an elaborate structure of power built by North Americans and their sordid island allies would crash and crumble. Only then could the long-awaited *Cuba Libre* be realized.

II

Confronting the Insurrection

6

Thunderstorms: Castro's Granma *Rebels and the Matthews Interview*

FEW OF THE anxious men crouched on the *Granma* on November 25, 1956, could have been contemplating the myriad characteristics of mid-century Cuban-American relations.[1] They were preoccupied with an overcrowded, unreliable boat trying to glide out of a Mexican harbor without being detected by Batista's spies. Before long, stormy seas and heavy rains tormented the M-26-7 men. They got seasick; equipment failed; food and potable water ran short. Seven days of hell. On December 2, three days later than planned, the yacht ran aground in a muddy, tangled mangrove swamp at the southeastern end of the island near Niquero.

Now wearing the olive-green uniforms of the 26th of July Movement, the bedraggled warriors had to wade for hours with weapons held overhead to reach Playa de Las Coloradas. The beach was distant from the jeeps, trucks, food, and guns the movement underground had assembled for the M-26-7 expeditionaries. Worse still, Castro heard on his radio that government forces had crushed the uprising staged by comrades in Santiago de Cuba on November 30 to coincide with the scheduled landing of the *Granma*. What Castro and his compatriots did not know was that Batista agents had learned of the *Granma*'s departure from Mexico and that the Cuban military had airlifted troops to Oriente Province to await the rebel expedition.

Barely on the beach, the insurgents soon had to scramble for cover. An airplane, no doubt directed by a warship in nearby waters, had spotted them. Much of the invading army's equipment remained on the grounded yacht. The rebels ran desperately for three days, reaching Alegría del Pío. One afternoon, as they lay exhausted in a field, hungrily devouring sugar cane, government troops closed in. Like an "inferno," a volley of gunfire quickly killed a number of the insurrectionists.[2] Others scattered, only to be captured and murdered. Batista's men caught and executed Juan Manuel Márquez, Castro's valued adviser.

Ernesto "Che" Guevara, the 25-year-old Argentine doctor who had joined Castro's army in Mexico, took a bullet in the shoulder. "Everything seemed lost." Then he thought about "an old Jack London story in which the hero, knowing he will freeze to death in the cold wilderness of Alaska, leans against a tree trunk and gets ready to die with dignity."[3]

Only 20, 16, or 12 insurrectionists—the number is disputed—survived the ordeal, including Guevara, Fidel Castro, and his brother Raúl.[4] With help from local peasants, the remnants of the M-26-7 army moved deep into the rugged Sierra Maestra Mountains to regroup and plan "the Revolution." The rebel force signed on some peasant volunteers, but still it numbered only 29 members by the end of December. Who—except perhaps Fidel Castro—would have predicted at that moment that this tiny, dependent, poorly outfitted, and sickly band of insurgents would end up in charge of the government in Havana only 25 months later? "One may ask," reported a foreign diplomat who believed that Batista was "impregnable," "what this revolutionary hoped to accomplish by landing an ill-equipped assortment of seasick men in the middle of nowhere at a time when the high demand for sugar . . . forecast the same prosperity for 1957 and 1958 as has already been obtained in 1956 and when the President [Batista] had just announced a distribution of $70 million to the sugar workers and a Christmas bonus of $30 to the lower-paid civil servants."[5]

It is not surprising that neither President Dwight D. Eisenhower nor Secretary of State John Foster Dulles took much notice of an incident along Cuba's coast. Dulles was in the hospital struggling with the first signs of the cancer that would kill him, and Eisenhower was coping with more dramatic events in Eastern Europe and the Middle East. A French journalist remembered that in Europe, too, Castro's expedition stirred minimal interest—"we were preoccupied with the Polish 'October,' with the Hungarian tragedy, with the Algerian war, and with the Suez adventure."[6] Still, the story of the Cuban rebels' landing and near annihilation made the news. M-26-7 sympathizers, watched by FBI agents, picketed the United Nations in New York, while inside the building some speakers noted similarities between the Cuban and Hungarian uprisings.[7] From Havana, Canada's Ambassador informed his superiors in Ottawa that the Castro expedition had failed to ignite a national rebellion because the opposition was splintered, Cubans still had confidence in Batista, economic conditions were "too good," and the regime, backed by the army, was simply "too tough a nut to crack."[8]

U.S. officials in Havana and Washington received news about the expedition with skepticism—skepticism that Castro and his men had survived their encounter with Batista's troops; skepticism that any survivors could withstand the pounding the Cuban military was giving them from the air; skepticism that, without supplies, they could continue to elude their pursuers; skepticism that M-26-7 was alone responsible for the island-wide disturbances their landing spawned (Were not Communists stirring up the trouble? Or, perhaps, was not Rafael Trujillo's Dominican Republic responsible?); skepticism that the Cuban people, apathetic, abhorring violence, and embracing order, would support the M-26-7 rebellion.[9] The U.S. naval attaché in Havana calculated the numbers and reported his doubts: A 62-foot yacht could not possibly have carried 83 men![10]

Such suspicions derived in large part from U.S. officials' reliance upon Cuban government reports, which, in language known to be pleasing to Americans, emphasized that Communists engineered the *Granma* enterprise and civil unrest. Embassy personnel also got mixed reports from other sources; one day Castro was alive, the next day dead. Batista's imposition of press censorship in Oriente made discovering the truth all the more difficult.

U.S. diplomats seemed certain about only a few conclusions. They expressed confidence that Batista was calmly and effectively handling the crisis. They expected him to prevail. They judged that the sporadic bombings throughout Cuba, including those against Cuban Electric Company stations and lines, would subside, especially after the press, feeding on rumors, stopped exaggerating the importance of the terrorist attacks. Absent from the initial U.S. analyses was any probing of the reasons why some Cubans were taking great risks to dislodge the Batista regime and why some Cubans thought radical or reformist change necessary.

The U.S. Embassy, in the Cold War terms of the times, at first took the Batista line that Communists were behind the sabotage and terrorism disturbing the island's peace. But the embassy staff soon learned, as the British Ambassador in Havana observed, that this view was a "facile over-simplification."[11] Embassy officers initiated a study project "to ascertain Castro's Marxist-Leninist proclivities—or lack of them." They discovered "no credible evidence to indicate that Castro had links to the Communist party or even had sympathy for it."[12] Still, during the period of insurrection, American officials searched more for signs of any Castroite flirtation with Communist dogmas than for the profound sources of Cuban discontents.

In the United States the FBI expedited its surveillance of the clubs that

Castro had founded during his 1955 trip. The anti-Batista organizations had become more active through demonstrations, including the picketing of the White House.[13] In January, utilizing FBI-gathered information with an eye to prosecution, the Department of Justice began to take grand jury testimony on the activities of Cuban exiles in the United States. One Cuban called to testify in Miami was the former President of Cuba, Carlos Prío Socarrás. State Department observers at the proceedings thought Prío "very shaky" and advised the Justice Department that they had no objection to an indictment of him and his associates.[14]

Gradually, in the weeks after Castro's landing in Oriente Province, Havana embassy officials began to worry that the island might be entering a serious crisis, that Batista was vulnerable, and that the *fidelistas* were not only surviving but succeeding. One of the rebels even managed to leave Castro's Sierra Maestra headquarters and evade Batista's army to address a New York rally of anti-Batista Cubans at the Whitehall Hotel.[15]

Bombs exploded at nightclubs, schools, businesses, and government buildings in and around Havana. Because of press and radio censorship, rumors and panic flourished as exaggerated, undocumented stories of sugar-cane field burnings and explosions circulated among uneasy islanders. If the regime was not in trouble, went the popular question, then why was it so intent upon suppressing information, including that offered by such U.S. publications as *Time* magazine, the *New York Times*, and the *Miami Herald?*[16] Even censorship could not block the growing evidence that mountain peasants were giving succor to the rebels, who, on January 17, assaulted a small military garrison at La Plata and captured valuable weapons and ammunition. "We had our first successful battle," recalled Castro, "when no one believed we were still alive."[17]

Batista's repression of government critics increasingly alarmed many Americans. Reports of murders at the hands of the police and military mounted day by day. If the government failed "to pin down, neutralize, and eventually dispose" of the rebels in the mountains, concluded the U.S. Embassy staff, it would appear weak and thus encourage further anti-regime ferment.[18] By mid-February, Ambassador Arthur Gardner, noting the "recurrent killings" by police and army and the growing Cuban revulsion against the regime's "extreme methods and brutality," advised Washington that the government faced "very grave pressure" that was building up like "an enormous snowball."[19] Gardner was not at all pleased that that "rabble rouser" Fidel Castro had unleashed so much trouble.[20]

Gardner's and the embassy's desire for a rapid return to normalcy became obvious. U.S. officials seemed determined to demonstrate that the storm lashing Cuba was not disturbing Washington's fundamental support for the

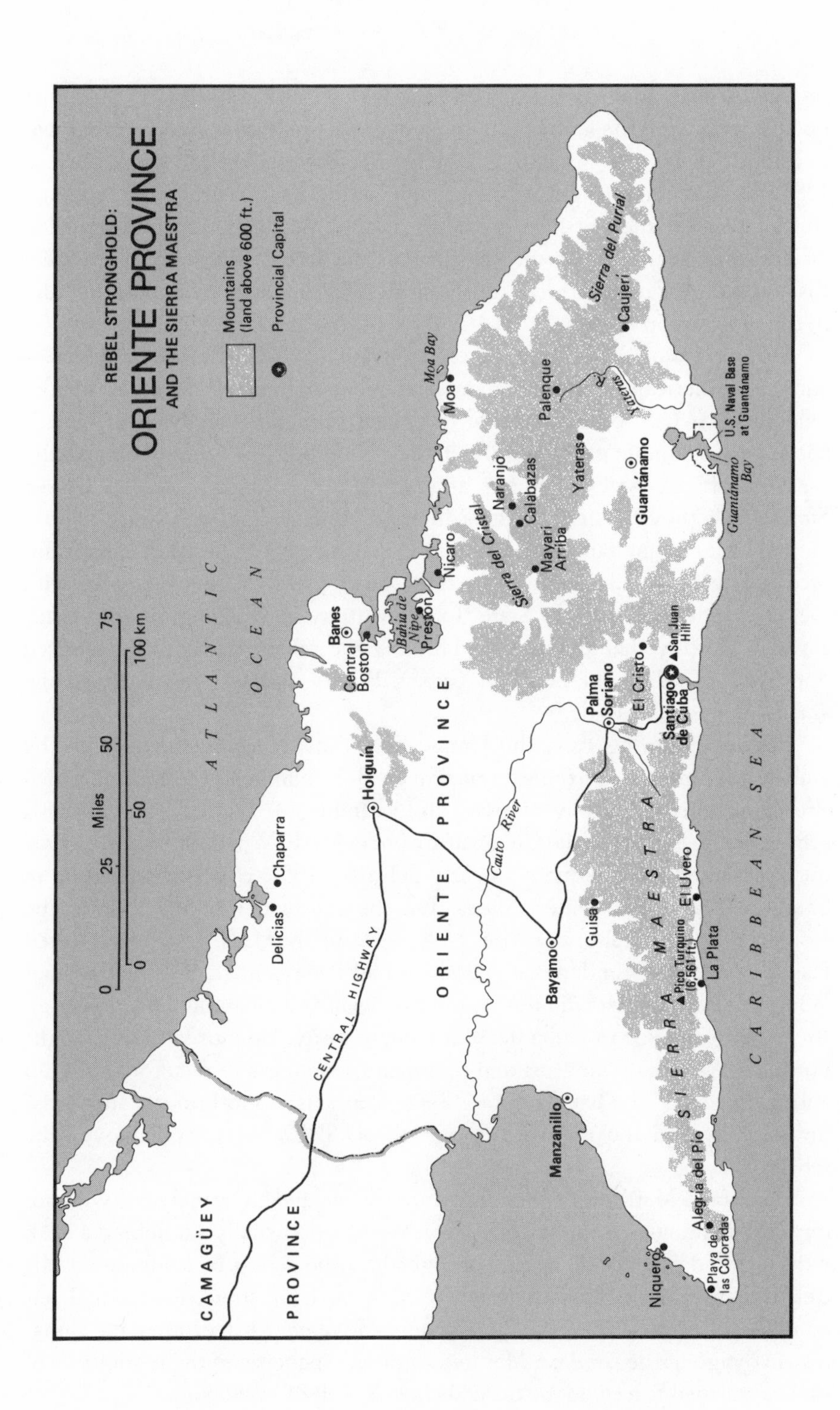
REBEL STRONGHOLD:
ORIENTE PROVINCE
AND THE SIERRA MAESTRA
Mountains (land above 600 ft.)
Provincial Capital
Miles
0
25
50
75
0
50
100 km
ATLANTIC
OCEAN
CARIBBEAN SEA
CAMAGÜEY
PROVINCE
ORIENTE PROVINCE
SIERRA MAESTRA
CENTRAL HIGHWAY
Cauto River
Delicias
Chaparra
Holguín
Banes
Central Boston
Bahía de Nipe
Preston
Nicaro
Moa
Moa Bay
Sierra del Cristal
Mayarí Arriba
Calabazas
Naranjo
Yateras
Palenque
Yateras
Sierra del Purial
Caujerí
Guantánamo
Guantánamo Bay
U.S. Naval Base at Guantánamo
Palma Soriano
El Cristo
Santiago de Cuba
San Juan Hill
Bayamo
Guisa
El Uvero
La Plata
Pico Turquino (6,561 ft.)
Manzanillo
Alegría del Pío
Niquero
Playa de las Coloradas

regime. Ambassador Gardner, who took his post in October 1953, had long openly applauded the Cuban government and its president, and Gardner was not about to temper his enthusiasm for the Batista "order" now. Born in 1889, the manufacturing executive fought in the First World War as a tank corps captain, earned a bachelor's degree from Yale University, and served as an assistant secretary of the treasury in the late 1940s before President Eisenhower rewarded his loyalty to the Republican party by appointing him to the embassy in Havana.

Seldom from Gardner or other high-ranking U.S. officials would Cubans or Americans hear indignant statements against Batista repression. Seldom from U.S. officials would they hear thoughtful analyses of Cuba's social and economic problems and its moribund politics. Quite the opposite, for U.S. officials saw their jobs as shoring up the regime. In the United States itself they toasted Batista's Interior Minister Santiago Regy, whose national tour in January 1957 under a Department of State leadership grant received uncensored coverage in the Cuban press. When Rey met with Deputy Under Secretary of State Robert Murphy in Washington, the visitor assured Murphy that Batista was blunting the rebellion. Rey also pledged improved coordination between Cuba's BRAC and U.S. agencies in the anti-Communist crusade.[21]

Besides honoring Rey, the United States made arrangements with the Cuban government for the sale of sixteen B-26 bombers, the first of which President Batista officially received on December 14. The U.S. military also sent seven M-4 tanks to the Cuban army under MAP. When Batista ceremoniously accepted the Shermans in early February 1957, criticism welled up in both Cuba and the United States. On the fourteenth of the month, the Korean War-tested aircraft carrier USS *Leyte* and four U.S. destroyers visited Havana. Cubans wondered if the ships had been sent to bolster Batista.[22] While embassy staffers did not report any negative Cuban public opinion, an American journalist in Cuba heard it straight away. He mused that throughout Latin America "the arms and the open friendliness to dictators build up anti-Yankeeism."[23] That *New York Times* journalist soon broke a story that further unsettled the regime, angered the U.S. Embassy, and buoyed the rebels.

Herbert L. Matthews's seemingly impossible and supremely risky feat of physical stamina—certainly for a 57-year-old who usually sat behind a desk writing for the editorial page—brought him the scoop of a lifetime.[24] His trek into the Sierra Maestra established what conflicting reports had left muddled: the survival of the 26th of July Movement leader after the disastrous voyage of the *Granma*. Matthews not only penetrated the roadblocks of Batista's armed forces but also the dictator's wall of censorship.

Always alert to the importance of public relations in advancing his movement, Castro planned the event carefully. He asked that a *foreign* member of the press be brought into the mountains. A Cuban reporter would not do; Batista censors would kill the story. A North American woman would not do; she would be too conspicuous. To Havana, Castro sent one of the *Granma* expeditionaries, René Rodríguez, who met with the covert Civic Resistance leader Felipe Pazos. Pazos walked into the office of R. Hart Phillips, the *New York Times* correspondent on the island. In a whisper he asked for Ruby Phillips's help. "You have contact with Fidel Castro!" she burst out. "I cannot believe it!"[25] She then met with Pazos and Faustino Pérez, a Havana rebel, at the Bacardi Company, not far from the Ministry of Defense. Phillips said she would find an appropriate journalist. Her office mate Edward ("Ted") Scott of the *Havana Post* and the National Broadcasting Company had just received a letter from Matthews on another matter, and he recommended the seasoned journalist for the expedition. Phillips cabled New York and instructed Matthews to hurry to Havana. She gave no reason.[26]

The choice made eminent sense. Matthews was the Latin American affairs specialist on the *Times* editorial board. He spoke Spanish, and he already knew danger, having covered the Spanish Civil War in the 1930s and Italy during the Second World War. Yet the lanky, thin-faced man did not inspire confidence that he could endure the travail of hiking into the rugged Sierra Maestra, the island's highest mountain chain. Pazos looked at the slightly balding, gray-haired man and wondered if he was "apt at mountain climbing."[27] Matthews would not be deterred; he knew a rare opportunity when he saw it, and he thrived on opportunities.

Born in New York City in 1900, Matthews served with the U.S. Army in World War I. He graduated with Phi Beta Kappa honors from Columbia University and immediately, in 1922, went to work for the nation's premier newspaper. As he moved up the *Times* ladder, he wrote six books (he wrote six more in the years after his interview with Castro). Always forthright about just causes and the shame of U.S. backing of dictators in Latin America, Matthews did not flinch from criticism that he was undermining U.S. foreign policy by giving attention to anti-Yankee nationalists. He must have looked upon the Castro interview as a chance to test his ideas, if not relive his past—Batista as Franco and Mussolini, another tyranny challenged.

Seeking to make his mission look "more innocent and natural," Matthews invited his wife Nancie to join him.[28] On the night of February 15, the Matthews drank daiquiries and ate Moro crab at the Floridita while waiting for the unpunctual M-26-7 escorts to arrive. At the last minute Herbert borrowed a camera. Accompanied by three Cubans, the Matthewses headed east by car from Havana. The journalist disguised himself as an American tourist by dressing in garb fit for a fishing trip. They stopped several times for Cuban coffee before entering Oriente Province. Once, a soldier stopped them at a patrol station, eyed the passengers, and motioned them on. "The absolute dope," Nancie remarked to her co-conspirators.[29]

On the afternoon of February 16, after 16 hours on the road, the weary travelers arrived in Manzanillo, a city but 25 miles from the Sierra Maestra. They rested in the home of Pedro and Ena Samuel, pro-Castro teachers. The Cubans responsible for sheparding Matthews to Castro soon argued among themselves about the journey because army patrols had been seen in the vicinity. Finally at seven in the evening they left by jeep for the Sierra. They coordinated their cover story: If stopped, they would identify Matthews as an American planter who could not speak Spanish. An army guard did halt the vehicle. Skeptical, he nonetheless bought the tale and waved them on. The jeep wove its way through sugar and rice fields and across rivers. Around midnight they abandoned the jeep and began to walk.

Matthews found the going tough. Crossing cold stream water, he became chilled. Finally reaching the rendezvous point, the soaking wet and muddy journalist, now set upon by ravenous mosquitoes, had to wait for a new team of rebel scouts. After two hours the rebel signal—a soft double whistle—sounded and was answered. Matthews moved again through dense vegetation that reminded him of a Brazilian tropical forest.

With the morning light of February 17 came Fidel Castro to the grove where Matthews and his guides rested. Matthews had already been taking notes. He thought the *barbudos* (bearded ones) very young. He noticed they carried a mix of American-made weapons. A few of the rebels spoke English. Castro squatted on Matthews's blanket. They ate and they smoked Havana cigars. For three hours they spoke quietly in Spanish. Castro was "a powerful six-footer, olive-skinned, full-faced, with a straggly beard"—a "great talker" whose brown eyes "flash."[30]

During the conversation, by Castro's design, a soldier appeared and reported having made contact with the "Second Column." There was no such rebel unit, but Castro succeeded in getting Matthews to think that insurrectionist columns and camps were spread throughout the region. Raúl Castro marched several soldiers—the same soldiers—back and forth in front of the Matthews-Castro huddle to plant the impression that the rebel forces were larger than they really were.[31] Before Matthews departed, a rebel used the borrowed camera to snap a picture of the journalist and Castro puffing on cigars.

In broad daylight on February 17, Matthews made his way through thick undergrowth, mud, and fields to a farmer's house, where he hid while his guides fetched a jeep. They successfully crossed an army roadblock. In the afternoon, a dirty, exhausted but triumphant Matthews reached his nervous wife in Manzanillo. They went by car to Santiago de Cuba. Halted at a roadblock, the passengers faced a guard full of suspicion. But he let them go on. "Had that soldier searched us he could have been promoted to general," thought Nancie.[32] From Santiago the Matthewses flew back to Havana.

"I knew I had a sensational scoop," remembered the *Times* writer, and he was eager to return to New York before a leak endangered him.[33] But he had scheduled an interview with the leader of the Directorio Revolucionario, José Antonio Echeverría, who hinted that he and his co-conspirators would soon launch a major project to eliminate Batista. Matthews also took time to visit Ernest Hemingway on the writer's *finca*.

Matthews's notes—"a keg of dynamite"—would surely draw the attention of customs inspectors at the Havana airport.[34] After all, Fidel Castro had signed and dated them to verify the meeting for doubters in the United States. Nancie solved the problem by stuffing the notes in her girdle. Ruby Phillips sighed with relief when the plane took off on February 19. Once in flight, Nancie retrieved the crumpled notes, and her husband began immediately to write his story.

Castro, meanwhile, met with underground M-26-7 leaders to prepare a new manifesto calling for economic sabotage. They planned to release the document to coincide with publication of the Matthews articles. With a good deal of pre-publication advertising, the Matthews articles appeared in the *New York Times* on February 24, 25, and 26. The 26th of July Movement had dispatched a sympathizer to New York with instructions to duplicate copies of the Matthews columns and then to mail them to prominent Cubans on the island. This circumvention of government censors put Matthews's account in the hands of thousands of Cubans.[35] Then, Batista inexplicably lifted his censorship decree on the twenty-seventh. Soon Cuban radio stations and newspapers carried the stunning news of Castro's existence, perseverance, and intransigence.

Matthews not only reported that Castro was alive but also that the rebel and his youthful army were thriving. The writer deemed Castro a formidable opponent for the regime because he could exploit public discontent with Batista's repression, governmental corruption, and economic uncertainties. Although Castro had become the most famous insurrectionist, wrote Matthews, he was only one of many oppositionists, who included the "best elements" in Cuban society.[36] In the articles, Castro himself appeared as an inspiring, fanatical, and courageous leader with a vague program. But could this army of 90 to 100—a much exaggerated number, Matthews later learned—defeat Batista's forces? Matthews thought not; Batista would finish out his presidency into early 1959.

Matthews also reported that anti-Americanism was growing in Cuba. Although Castro had said that "you can be sure we have no animosity toward the United States and the American people," he complained bitterly about Batista's use of U.S.-supplied weapons "against all the Cuban people."[37] Matthews identified three sources of anti-U.S. sentiment: American arms shipments to Batista; Ambassador Gardner's open admiration for Batista; and U.S. business backing of the regime because it protected foreign interests.

Ted Scott wrote Matthews that the journalist had become "a stench in official nostrils" on the island. "But the Opposition is proposing institutional reforms so that you can be elected President."[38] The Minister of National Defense, Santiago Verdeja, soon announced that the Matthews articles were chapters in a "fantastic novel." The journalist could not possibly have interviewed "the pro-Communist" Castro.[39] The *New York Times* then published the Matthews-Castro photograph. The regime looked foolish.

U.S. officials recognized that the Matthews stories undercut Batista and U.S. support for the regime. Taming his anger, Ambassador Gardner called the articles "unfortunate."[40] The Havana embassy feared "some awkward questions," and in a series of reports to Washington, claimed that the *Times* writer was not presenting a "proper balance." The Batista government "does not deserve the completely negative tone" of the articles. But, more, Batista "has the situation fairly well under control." The economy was "sound and prosperous." Cubans enjoyed a higher standard of living than ever before in their history. And the armed forces remained loyal to Batista. Embassy analysts asserted that Matthews had become too self-impressed with "the romantic nature of his experiences." His "one-sided picture" overestimated the strength of the Castro movement and underestimated that of the regime.[41]

The Acting Assistant Secretary of State for Inter-American Affairs basically agreed. R. Roy Rubottom, Jr., complained that Matthews's articles "would build Castro up into a hero somewhat beyond his real proportions."[42] The U.S. official most responsible for defining U.S. Cuba policy over the next few years, Rubottom was born in Texas in 1912 and went on to earn degrees from Southern Methodist University and the University of Texas. In the 1930s he worked for an oil company and the University of Texas administration. During World War II he served in the U.S. Navy, leaving with the rank of commander. After a short stint as a bank vice president, Rubottom entered the Foreign Service in 1947. His first post was Bogotá, where he witnessed the rioting that disrupted the OAS meeting.

During the early 1950s he held various Latin-American affairs posts and served in Spain. In May 1956 he became Deputy Assistant Secretary for Inter-American Affairs, in September the Acting Assistant Secretary, and in June 1957 the Assistant Secretary.[43] During the Matthews embroglio, Rubottom joined other U.S. officials in standing fast behind Batista. No U.S. official offered a probing analysis of Cuban discontents. The familiar theme echoed from the embassy and the State Department: Batista had the crisis under control and his enemies at bay.

In private one official dissented. "I'm in a strange position here," Richard G. Cushing wrote in a personal, "strictly" unofficial letter to Matthews. A U.S. Information Service officer in the embassy since early 1954, Cushing complimented the journalist for his "dandy" articles and remarked that "your series caused a lot of seething indignation in high quarters."[44] Cushing, who soon became a point of contact in the embassy for anti-Batista dissidents, later revealed his disaffection from U.S. policy and Ambassador Gardner, who was "sold on Batista." "Often I had the unpleasant feeling in the pit of my stomach," the USIS officer mused, "that our foreign policy was faulty, and that I was in error, in the larger scheme of things, to be involved in propagandizing it."[45] The U.S. Embassy staff, in fact, was beginning to divide on the central question of U.S. backing for an authoritarian regime.

For the struggling M-26-7 band, the Matthews interview provided desperately needed publicity to alert followers at home and abroad that the rebellion was surviving. The news of Castro's escape from Batista's clutches served another purpose Castro surely calculated: It accented *his* leadership of the rebellion against Batista; it put Castro's rivals on notice. The publication of the interview reinforced popular suspicions that the Batista government was enamored of lying. It generated speculation that Batista might not be so powerful after all if his legions could not stamp out a small guerrilla army and could not even prevent an American journalist from visiting with its *jefe*.[46]

Matthews's articles of early 1957 and after also helped advance the image of Fidel Castro as a folk hero of remarkable energy, daring personal sacrifice, and passionate commitment to a noble cause. No doubt Matthews's work shaped anti-Batista opinion in the United States and thus satisfied Castro's aim of rolling back U.S. support for the regime. In the months after his adventure and scoop, the journalist grew sensitive that "I am now so involved in Cuban affairs" that people would necessarily read his reporting on the insurrection as biased. He claimed that his articles "have literally changed the course of Cuban history," but he told the *Times* publisher Arthur Sulzberger that "we must be very careful to remain within the bounds of strict journalism."[47] Although Matthews's reports on Cuba came closer than most to understanding the sources of the revolt—certainly closer than the U.S. Embassy's—many thought that he had crossed the line from journalism into partisanship.

Matthews, who visited Cuba again several times later, became "a scapegoat and whipping boy in our national debate on what to do about Castro."[48] In the 1960s his *Times* colleagues themselves would not permit Matthews, an editorial-page writer, to prepare stories on Cuba for the news department. One of them explained that Matthews had become too "subjective" and "emotional" on Cuba. The journalist had "destroyed his usefulness as a reporter on Cuba." The issue was "credibility."[49]

Until Matthews's death 20 years after the Sierra Maestra interview, he endured a standard charge levied against him by *batistianos,* U.S. right-wingers, anti-Castro exiles, and U.S. Ambassadors to Cuba Arthur Gardner and Earl E. T. Smith: He "tremendously influenced" the Department of State to back away from Batista and he "made a hero and a Robin Hood out of Castro."[50] Robert Hill, an ambassador to Mexico, asserted that *New York Times* reporters and State Department officials conspired to "put Castro in power."[51] The conservative pundit William F. Buckley castigated Matthews in a 1961 article on Castro titled "I Got My Job Through the New York Times."[52] One Cuban claimed that for Fidel Castro, Matthews was the "equivalent of an army division."[53] Batista himself said that the Matthews articles proved of "considerable propaganda value to the rebels."[54]

Some reporters heaped responsibility on Matthews for salvaging the Castro movement. Jay Mallin of *Time,* for example, declared that "if no U.S. [news]papers had entered Cuba, it is entirely possible that Fidel Castro, blacked out by Batista censorship, would today [1959] still be in the Sierra Maestra."[55] This superficial comment claims too much for the Matthews stories.[56] The clumsy way in which the Batista government handled them and the grisly Batista human-rights record itself undermined the regime. No American journalist, moreover, manufactured a Fidel Castro that Cubans and some Americans did not already know or at least imagine. Castro was a popular figure well before Matthews stumbled his way to the rebel's mountain hideaway. No American journalist persuaded the embassy in Havana or the State Department to dump Batista. No American journalist defeated Batista's formidable military. No American journalist initiated the Cuban rebellion or fueled the economic, social, cultural, and political conditions that sped it. As the *Hispanic American Report* explained, attributing Castro's success to the *New York Times* and Matthews's reporting "is as absurd as blaming a meteorologist for a thunderstorm."[57]

7

Ambassador Gardner and the Propaganda War

On the very day Herbert L. Matthews interviewed Fidel Castro, February 17, 1957, three young Americans sneaked out of the U.S. naval base at Guantánamo Bay, where a 26th of July Movement cell operated and sometimes pilfered arms. Charles E. Ryan (ago 20), Victor J. Buehlman (17), and Michael L. Garvey (15), sons of U.S. personnel, disappeared leaving no clues. Their parents suspected that they had joined rebel forces in the Sierra Maestra. The three had in fact hooked up with a Santiago contingent of some 50 insurgent recruits. They reached Castro in mid-March.[1]

Cuban government officials lied again in another public relations snafu, looking incompetent and ridiculous. Servicio de Intelligencia Militar announced that the three youths had been seen at a Havana bar; Batista allowed that they were in Miami; and Cuban military officers in charge of catching Castro claimed that the runaways could not possibly have eluded government forces stationed around the mountains.[2] The *New York Times* exploded all official statements by publishing a picture of the young men in 26th of July uniforms. The newspaper also printed letters the new rebels had addressed to President Eisenhower and Ambassador Gardner asking them to halt U.S. arms shipments to Batista.[3]

The U.S. Embassy once again had to deal with an event the rebels counted as good propaganda. Yet Gardner tried to persuade Washington that that incident was "likely to prove detrimental to the rebel cause." He reported that "the boys left their homes in search of adventure and not because of strong political convictions." Gardner remonstrated that the boys' enlistment in the mountain army was atypical of "American feeling toward Cuban politics in general and toward Castro in particular."[4] The American Consulate in Santiago began to pass letters to the three youths through rebel contacts, while U.S. authorities studied whether their U.S. citizenship should be yanked because of their enrollment in a foreign army.[5]

Soon the Directorio Revolucionario's attack upon the Presidential Palace in Havana on March 13 elevated the violence of the anti-Batista movement. The student-led, urban-based DR, probably using Prío money, sought nothing less than the assassination of Batista—"the decapitation of the regime."[6] Attaining this goal would surely deliver another: a shift in leadership of the insurrection from M-26-7 to the urban-based DR, from Castro to Echeverría. DR members and other anti-Batista oppositionists had grown increasingly critical of Castro's "bossism," and they feared that he might become another dictator.[7] Just before the assault, the conspirators also talked about "American meddling in Cuban affairs."[8]

DR commandoes stormed the palace and seized a Havana radio station. After fierce fighting, Batista's forces regained both sites. DR leader Echeverría was killed in a skirmish with police. In all, 40 attackers died. Government authorities rounded up, tortured, or killed other DR members and oppositionists. Among them was Ortodoxo party president Pelayo Cuervo Navarro, who was probably murdered on Batista's orders. One American, a tourist from New Jersey, was killed when his Hotel Regis room was sprayed with gunfire.

From the Sierra Maestra, Castro denounced the DR attack; the real fight, he insisted, was in the mountains, not in the streets of the capital city.[9] Because the Revolutionary Directorate took such punishment during and after the attack, Castro emerged the victor. After the foiled attack, "we had scarcely any arms, homes, money or cars" except an old black Oldsmobile and brown Pontiac, recalled one DR member.[10] Batista had elminiated one of M-26-7's primary rivals, Echeverría. "Had the attack succeeded," the veteran Cuba-watcher Tad Szulc has written, "it would have left Fidel Castro in his mountains as a suddenly irrelevant factor in the revolutionary equation."[11]

Ambassador Gardner and embassy officers assessed these events in typical fashion: Batista's control remained steady, with only "mopping up" ahead; the "majority of Cubans are politically apathetic"; and "Batista still stands a good chance of finishing his term."[12] The day after the tragedy, Gardner went to the bullet-ridden palace to celebrate the granting of a rate increase to the Cuban Telephone Company. An American reporter wrote: "The effort of Gardner to help Batista convey an impression both at home and abroad that things were normal in Cuba at that moment neither ingratiated him and the State Department with the people of the country nor enhanced the popularity of the Cuban Telephone Company."[13]

Other highly visible public displays of U.S.-Cuban cooperation also seemed designed to bolster the regime. In the weeks following, for example, Gardner and high Cuban officials shared the platform to dedicate a plaque to the memory of American pilots. The U.S. Chief of Naval Operations, Admiral Arleigh Burke, visited Havana as a guest of the Cuban Navy and exchanged medals with his Cuban counterparts. The visiting mayor of Kansas City decorated the Cuban president with the city's Gold Medal for Batista's contribution to "democracy and progress in the Americas." Rear Admiral Robert B. Ellis, the commanding officer at the Guantánamo Naval Base, made formal calls on Batista and the Minister of National Defense. Ellis enjoyed a dinner party at the Tropicana and generally helped in "cementing the good will and fine relationships" between the base and the regime.[14]

At the Cuban government's urging, the Havana embassy also pressed Washington to monitor and interrupt the suspected revolutionary activities of former President Prío, who was still prohibited by his agreement with the Immigration and Naturalization Service (INS) from conducting an anti-Batista campaign from U.S. soil. The grand jury investigation of Prío and others went forward, producing on March 27 an indictment of Juan Orta, a 26th of July Movement leader in Miami. The Cuban government kept Washington abreast of Prío's insurrectionary acts—such as the outfitting of a plane to fly from Miami to drop live phosphorous on cane fields and buildings.[15] The INS nonetheless decided against deporting Prío; the agency lacked credible evidence. But, more, Prío's "influence in Cuban revolutionary circles appears definitely on the wane." Any attempt to remove Prío would only make him a martyr and "bring further charges from anti-Batista elements of U.S. favoritism toward the present regime."[16]

Ambassador Arthur Gardner and the embassy sharply stated their fundamental position on Cuban-American relations in a long memorandum dated April 1. Sent to Washington, the report judged that the "authoritarian" Batista regime had used "increasingly repressive measures" to fend off critics. Still, it was expected that Batista would complete his term into early 1959. Gardner made the case that Batista was sensitive to public opinion by arguing that, if he were not, he would have adopted even tougher methods. Most important, no other Cuban leader could lead a stable government. "From the standpoint of the United States, Batista's continuance in office is probably in the best interests of the U.S. in that we would not like to see a return to the gangsterism, license and corruption of the two preceding administrations." To be sure, some graft and policy brutality marred the regime's record, but "conditions have been fairly orderly." Batista was friendly to the United States and he "advocates private enterprise." In case Washington had forgotten, the report emphasized that Cuba ranked high as a partner because of its strategic location in the Caribbean, the Guantánamo base, sugar, shipments of the strategic raw materials nickel and chromite, North American investments, and the government's fervent anti-Communism.[17]

The U.S. intelligence community's forecast a few weeks later differed from the embassy's report. Discounting a Communist threat to the Batista regime, analysts concluded that "we do not believe that the Cuban government can fully restore public order or check the emergence of new civilian opposition elements." Indeed, "there is only an even chance that the Batista regime will survive the period of this estimate."[18] This view, no doubt based at least in part on the Havana CIA chief's advice, revealed the growing division among U.S. officials in Havana over the future of Batista. Although Washington officials were receiving mixed messages, they still sided with Batista as the surest protector of U.S. interests.

North American journalists aggravated this schism in official U.S. ranks and Batista's troubles. Matthews himself continued to write critical reports and editorials for the *New York Times,* and "Matthews copy-cats," as the Havana embassy tagged them, trekked across Cuba in pursuit of stories about the rebellion.[19] U.S. newspapers became more anti-Batista by the week.[20] In early March, SIM arrested two American journalists connected with the *Birmingham News* and National Broadcasting Corporation (NBC). Disguised as geologists, George Prentice and Anthony Falletti had traveled through Oriente Province gathering revolutionary literature and interviewing and filming regime critics in the region where the rebels were strongest. SIM slammed them in jail for a few days and confiscated their films and notes. Pressed by the U.S. Embassy, NBC, and the Inter-American Press Association, SIM released the journalists for a quick exit from the island. The Batista government, which had been touting its ending of press censorship, looked hypocritical at best.[21]

While Lawrence E. Spivak scouted ways to get Castro on his weekly television program, *Meet the Press,* a veteran leftist observer of Latin American affairs also journeyed to Cuba.[22] Carleton Beals wrote about insurrectionist "terrorists" and Batista's "flat-tire politicians." Highlighting the U.S. place in Cuba, he did not fail to mention that he walked down Havana's Theodore Roosevelt Street and that near his hotel the Bank of Boston building was bombed. His article in *The Nation* criticized the rebels for lacking an ideology, for failing to explain "what kind of Cuba they are trying to create." He chastised Castro as a "quick-tempered, impatient, imperious, violence-prone adventurer." Beals did not interview Castro; instead he formed his perspective from talks with urban-based revolutionaries who rivaled the 26th of July Movement leader. Batista could derive little comfort from the Beals critique, however, because Beals soon wrote more favorably of the Castro rebels.[23]

The regime strove to counter such negativism by granting special interviews with Batista to handpicked journalists. In early May, for example, the Cuban government invited David Sentner, Washington Bureau Chief for the Hearst newspapers, and Holmes Alexander of the McNaught Syndicate to talk with Batista.[24] But public relations disasters continued to plague the regime.

In late April, underground M-26-7 activists Celia Sánchez and Haydée Santamaría escorted Robert Taber, a journalist hired by the Columbia Broadcasting System (CBS), and Wendell Hoffman, a CBS photographer, into the Sierra Maestra. Mario Llerena, a university professor and publicist who had taught Spanish at Princeton and Duke and now served M-26-7 as a spokesperson in the United States, accompanied the CBS men to Havana. He arranged a cover story: They were Presbyterian missionaries on assignment to photograph their denomination's schools on the island. After passing customs without difficulty, the North Americans, laden with equipment, started their 600-mile, five-day journey to Castro's camp. They made it through several government checkpoints. Once they had to hide in a coffee grove when soldiers passed near. Haydée Santamaría thought the exhausted Americans "could hardly go on."[25] The 38-year-old Taber, however, had suffered rough conditions before when he covered Vietnam and Guatemala. He and Hoffman pushed on. Taber came at a time "when we were not feared so much," Che Guevara remembered. He hoped that Taber, like Matthews, might embellish the "strength" of the mountain army to make it appear more fearsome than it really was.[26]

One of Taber's actual missions, assigned to him by the U.S. Consulate in Santiago after "discreet conversations" with rebel contacts, was to bring out the three young Americans who had left the Guantánamo base to join the rebellion.[27] Although the three had helped propagandize the movement against Batista in the U.S. press, rebel leaders concluded that they lacked the necessary revolutionary commitment.[28] After spending several days interviewing Castro and other guerrillas, Taber and Hoffmann, on May 9, walked out of the Sierra with Garvey and Buehlman. The two young men looked healthy but much thinner. Charles Ryan, the oldest of the three, stayed with the rebel army. He later fought beside the *barbudos,* and in October returned to the United States to speak at fundraisers for the 26th of July Movement. He repeatedly called for an end to U.S. arms shipments to Batista.

CBS aired *The Rebels of Sierra Maestra—Cuba's Jungle Fighters* on both radio and television on May 19. The Taber-Hoffman show included an interview with Fidel Castro atop Pico Turquino, the highest point in Cuba. Behind the M-26-7 leader stood a monument to José Martí. No *batistiano* could deny this time that Castro was in the mountains and that North American journalists could not have gotten past government barricades. Speaking in English for the U.S. audience he was seeking to woo, Castro leveled familiar charges against the regime. Was there any Cuban hostility toward the United States? Yes, Castro remarked, and it stemmed from U.S. support for Batista through weapons shipments. As usual, Castro did not lay out the details of a political platform.

Scores of Cubans flew to Miami to view the television program. But they soon discovered that the CBS show was blacked out in southern Florida. Miami station WJTV explained that commercial contracts obligated it to screen an episode of *Lassie* instead. For a few Cubans, Prío and his former prime minister Manuel Antonio Varona among them, the station did provide a private, closed-circuit showing. Rumors circulated that the Cuban government had bought the time slot to block the program. To no avail. *Life* magazine (including the Spanish-language version) carried the story of the Taber interviews with pictures, as did *Bohemia* in Havana.[29]

Once again Castro scored a victory in the propaganda war. Batista suffered another public relations embarrassment, not just because Castro got his message out of the mountains, but because two North American journalists and several M-26-7 rebels, lugging heavy recording gear, managed to dodge Batista's military. Just how strong and effective was the dictator's army, anyway? The Matthews story had made the regime look foolish; the Taber story made it look inept. Eager to fire up his troops to "get Castro or else," Batista had the CBS film shown at Camp Columbia to 1000 soldiers, some of whom were being ordered to the Sierra Maestra.[30] One wonders what effect the film had on Batista's young troops.

While Castro staged interviews with the U.S. press, Cubans in the United States continued to publicize the movement and raise money. They gathered in their numerous clubs, held rallies, picketed, and lobbied State Department officials and U.S. journalists. All in all, Cuban exiles created an effective propaganda machine that stimulated anti-Batista opinion in the United States and permitted them to sidestep Batista's controls on information. One anti-Batista Cuban whom the State Department welcomed warmly was Rufo López-Fresquet, a leader of Acción Libertadora, a group of moderate professionals who supported Castro's insurgency as the only immediate instrument for overthrowing the dictatorship. Perhaps State officers looked to him as a potential third force, for the economist-publicist sought not only to overthrow Batista but eventually to block Castro. His organization conspired to free Colonel Ramón Barquín from the Isle of the Pines prison, to oust Batista before Castro "becomes too strong," and then to "force Castro to deal with Barquín." López-Fresquet criticized Ambassador Gardner, disparaged Prío as a corruptionist eager to buy a revolution, and judged Castro an ambitious, "not normal" man who had the makings of a "caudillo."[31]

After one meeting at the Department of State, a delighted López-Fresquet told Herbert Matthews that he was "surprised" that State officials in Washington "were very well informed now and that they were sympathetic to the anti-Batista point of view."[32] Although some officials regretted Batista's repressive behavior, López-Fresquet was reading too much into the conversation. The United States was not ready to abandon Batista in 1957. No planning got under way to consider alternatives to Batista. U.S. officials still considered him a reliable ally, and he still seemed capable of quelling the insurrection.

The M-26-7 clubs in the United States, meanwhile, continued to agitate. A new club organized in Chicago. Anti-Batista Cubans picketed (others fasted) in front of the United Nations in New York with placards reading "Cuba the Hungary of America." The Patriotic Club of July 26 in Miami held rallies throughout the year, including appearances by Castro's sisters Lidia and Emma. M-26-7 demonstrators marched in front of the White House in May to demand the cessation of arms exports to Cuba. The following month in Miami, several Cuban groups temporarily shelved their rivalries in a unity rally at the Flagler Theater. An audience of 1000 heard Prío blast Batista. U.S. officials took note that Prío, once again, was violating his agreement.

Violence flared on occasion. On May 31 in Miami, some 100 Cubans paraded to the José Martí bust in Bayfront Park. According to the FBI, when the "near riotous" procession moved toward the Cuban Consulate in the Pan American Bank Building the police tried to break up the crowd. A mêlée of fistfights and clubbings followed, and 50 Cubans were hauled off to jail. A few weeks later, in New York's upper Manhattan, the brother of an officer of Batista's police was beaten by several Cubans. They too were arrested.[33]

U.S. officials grew alarmed at the acceleration of anti-Batista ferment among Cubans in the United States. Prío became the object of renewed scrutiny on May 24, after the yacht *Corinthia,* a U.S. flag vessel, landed on the north coast of Oriente Province near the Nicaro nickel plant. The craft had departed from Miami. Financed by Prío, the expedition planned to create an army in the mountains to rival Castro's. After several days, the Cuban military captured and executed most of the invaders.[34]

In Washington, State Department officials speculated that the *Corinthia* expedition "constitutes a breach of our neutrality laws."[35] They joined INS officials in thinking about whether Castro's sisters Emma and Lidia, on visitors' visas, and the hunger strikers in front of the United Nations, should be permitted to remain in the United States.[36] From Havana, Ambassador Gardner appealed to Washington to shut down exile Cuban activities in Miami.[37]

Prío was but one source of the violence that mounted in Cuba. Batista had many enemies who resorted to bombings to destabilize the regime. Apparently government agents themselves set off bombs in order to justify arrests and killings of targeted dissidents. During the night of May 4 five bombs rocked Havana; one exploded under an automobile parked near the swank Hotel Nacional. For days after, explosions rattled commercial sites in Havana and across the island. In the early morning hours of May 28, terrorists blew up the main conduits near the Havana plant of the Cuban Electric Company. Although the company restored power to suburban areas, central Havana went without electricity and telephones for 57 hours, forcing businesses to close.[38] Two days later Gardner told Assistant Secretary Rubottom that conditions in Cuba were "becoming more and more grave and the tensions greater and greater." He went on: "From a business man's point of view, [the] U.S. can't withdraw support for the regime." If Batista would only use more force, argued Gardner, he would gain ground on the opposition.[39]

The Batista forces stumbled again and again. On May 28, at the military garrison in El Uvero, M-26-7 warriors won a costly battle (six rebels died) and carried away government weapons.[40] Then the Cuban military airlifted MAP-supplied troops and American bloodhounds to the Sierra Maestra for a showdown with Castro that never occurred. Instead, to separate the rebels from their supporters, government soldiers forced thousands of peasants into retention centers, where they suffered. Some Cubans likened the removal to Spanish General Valeriano Weyler's brutal "reconcentration" policy during the 1890s. Anti-government Cubans protested that Batista's troops used U.S. weapons.[41] After spending two weeks in June on the island, Herbert Matthews wrote another series of critical articles, hammering on Batista's islandwide unpopularity and the growing anti-Yankee sentiment.[42]

Arthur Gardner's forced retirement compounded Batista's problems. President Eisenhower had political debts to pay after his re-election in the fall of 1956. Following custom under the spoils system, the administration decided to replace sitting, noncareer ambassadors with a new set of political appointees. Gardner, like others, had submitted a pro forma letter of resignation. But when he learned that his resignation was actually accepted, he began to campaign for an extension at his Havana post. His behavior and his presumption that U.S. policy was safe only in his hands annoyed his superiors and ensured his departure. He went over the head of Secretary of State John Foster Dulles to the president himself, who advised Gardner that "there is nothing personal in the matter at all."[43] The ambassador next appealed to Secretary Dulles, who patiently explained in a flattering letter that the administration could not deviate from common practice. Gardner also marshaled help from the Batista regime, which dispatched to the Department of State its U.S. public relations consultant, Edward Chester. Cuban Ambassador Campa sought out Rubottom to express Cuba's "regret" over the departure of Gardner, "a good friend of Cuba."[44]

Gardner worried that his exit from Havana in May could be interpreted, especially by opposition groups, as a repudiation of the Batista regime and of himself—as U.S. rejection of an "orderly change of government."[45] The State Department hierarchy and the White House staff, not sharing Gardner's sense of crisis and fed up with his incessant lobbying in violation of custom, activated his letter of resignation. Just before he departed Havana in June, Gardner threw himself a large farewell party at the Country Club. Prominent Cuban military, police, and SIM officials attended. Then the ambassador left Havana in a huff.[46] He got even later.

Did Gardner's unwilling retirement signify a change in U.S. policy toward Cuba? A few years later, American conservatives, including former Ambassador to Mexico Robert C. Hill and Gardner himself, hurled specious charges that Gardner's removal meant that lower-echelon State Department officers had decided in early 1957 to abandon Batista. Gardner claimed that Herbert Matthews had persuaded Rubottom and the director of the Office of Middle American Affairs, William Wieland, to take a pro-Castro stance and that they conspired to depose Batista by unseating Gardner.[47] As was often the case, the ambassador had it wrong. Gardner had to leave his post because the Republican spoils system demanded his departure; other politically appointed ambassadors at other sensitive posts also retired.

Gardner had to leave Havana, moreover, because he had become an embarrassment. One prominent Cuban remembered that Gardner "entertained lavishly and well, but he did not speak Spanish and he moved almost entirely within a small circle of wealthy, English-speaking Cubans."[48] By all accounts, including his own, the ambassador could not bestow enough praise on Batista. For every Batista misdeed Gardner offered an apology. "I'm glad Ambassador Gardner approves of my government," Batista was heard to say, "but I wish he wouldn't talk so much about it."[49] Prizing the apparent order that Batista maintained in the island, Ambassador Gardner developed only a shallow understanding of the tumult sweeping Cuba. He never seemed to understand why Cubans were willing to die to oust Batista and why the United States, as Batista's buttress, might become a target of Cuban ire. He acted "more like a businessman than an ambassador," lamented the Cuban magazine *Bohemia.*[50]

8

Violence Victorious: Ambassador Smith Meets the Rebellion

GIVEN THE VIOLENCE enveloping Cuba and the unsavory image of the Batista regime, Havana seemed to scare off ambassadorial candidates to replace Gardner. The Eisenhower Administration considered Charles "Chip" Bohlen, a career Foreign Service Officer and former Ambassador to the Soviet Union, but he went instead to the Philippines, where his talents were needed to handle sticky questions about U.S. bases and a rural insurgency. H. Freeman Matthews, another seasonal diplomat and the sitting Ambassador to the Netherlands, seemed a possibility, but he turned down the job. The post was then offered to someone with little diplomatic experience, Thomas A. Pappas, a company executive and prominent Massachusetts Republican of Greek descent. He too declined.[1] President Eisenhower then rang up his brother Milton Eisenhower, president of the Johns Hopkins University and a noted specialist on Latin America. Would Earl E. T. Smith do for Cuba? Milton judged him "a gentleman and a good businessman."[2] The President tapped Smith.

Yale graduate, Army veteran, wealthy investment broker, and a high society figure known for his athletic prowess in golf, polo, and boxing, Smith had become a prominent Republican in Florida politics and the friend of Democratic party notables. His only government service was as a dollar-a-year-man on the War Production Board during the Second World War.[3] The six-foot, five-inch, gray-haired and balding 54-year-old Smith had visited Cuba, believed he knew the island well, and remarked that he had long sought the Havana post. He once claimed that the American ambassador was "the second most important man in Cuba; sometimes even more important than the President." The U.S. ambassador was "a symbol of both power and friendship."[4] His successor, Philip W. Bonsal, regretted that Smith "did not perceive that the position of the President had deteriorated to the point that being his runner-up was not what it had been."[5]

The Senate Foreign Relations Committee held a perfunctory hearing on the appointment. Smith displayed little grasp of Cuban issues. Asked by Senator J. William Fulbright if he had any views on relations with Cuba, Smith simply answered: "No, I have no definite feeling about it." The senator mentioned the "rather explosive" conditions on the island and asked for Smith's impressions. "I look forward to this with interest" came the banal reply. Surely Smith's performance did not satisfy the senators, but Smith's was a political appointment and Cuba was a U.S. client state that still seemed manageable. The committee members expressed no sense of urgency, and Senator John F. Kennedy, one of Smith's friends, spoke up for him as an excellent appointment. Later in the year, Kennedy became Smith's guest of honor at the Havana embassy's Christmas party.[6]

Confirmed by the Senate on June 3, Smith prepared himself for Havana by calling upon Herbert Matthews. Middle-American Affairs Director William Wieland recommended that Smith have lunch with the *Times* journalist, who had become about as expert on Cuba as anyone in the United States. A sensible step, it nonetheless later became the stuff of right-wing vituperations that Wieland and other State Department officers had fallen under Matthews's spell and had thus plotted with the *Times* writer to ensure Batista's fall. Actually, according to Wieland, the suggestion for the luncheon meeting came from New York Republican Senator Jacob Javits, passed along by Under Secretary of State Christian Herter and Assistant Secretary Rubottom.[7]

Later, during the "who lost Cuba" debate that was so reminiscent of the "who lost China" fracas earlier in the decade, Smith himself charged that, in mid-1957, State Department conspirators kept President Eisenhower and Secretary of State Dulles in the dark by withholding from them official reports that the 26th of July Movement had been infiltrated by Communists.[8] The documentary record at the time of Gardner's departure and Smith's arrival contains no such reports. Nor does the record reveal a plot to oust Batista in order to install Castro. State Department officers speculated about whether Batista could last into 1959 and they reported on the violence he was causing in his nation, but in mid-1957 they were not abandoning the dictator. They were always on the outlook for a popular, moderate third force that could restore order. But none had yet emerged, and State officials certainly did not want the young mountain rebel Fidel Castro to become Cuba's next president.

At the time of Smith's appointment, in fact, officials in Washington did not seem particularly alarmed by events in Cuba. The Middle East claimed far more attention, for the Eisenhower Doctrine had just been launched to establish a greater U.S. presence in the region after the Suez crisis. About Cuba, President Eisenhower heard no alarm bells from his advisers. He frequently talked with his brother Milton and others about Latin America, so the president probably knew the basics about Cuba. Secretary Dulles and Under Secretary Herter knew that Cuba had descended into political chaos, for they held staff meetings, talked with journalists such as Matthews, and read Rubottom's frank memoranda that Batista courted trouble with his "brutal retaliatory tactics."[9] Yet no extraordinary steps were taken to select as ambassador a diplomat of known mediating skills, an experienced troubleshooter, or a Latin American specialist who spoke Spanish. The new ambassador got the job because he had achieved prominence as a wealthy Republican.

The hegemonic presumption governed. Despite the turmoil wracking Cuba, the economic climate remained attractive to foreign investment. Batista still had the support of the Cuban military; many dissidents rested in cemetaries and jails or had fled abroad; Havana was generally quiet; and Castro's small army was confined to a remote mountain region. From Washington's perspective, Cuba was merely up to its old political tricks again. Cuba's fundamental relationship with the United States did not seem endangered. The basic structure of dependency remained solid. If and when the rebellion showed signs of becoming a successful revolution against U.S. interests—if and when the revolution became "more than just a back-door harassment," as Rubottom once said—U.S. officials would act.[10] But not until that time, and that moment had certainly not arrived by mid-1957.

After interviewing a number of U.S. officers, Matthews predicted that Smith "will get very different instructions than Gardner."[11] And he did. Smith was directed to "alter the prevailing notion in Cuba that the American Ambassador was intervening on behalf of the government of Cuba to perpetuate the Batista dictatorship."[12] The instructions addressed diplomatic style more than policy substance. That is, the ambassador should be less fawning toward the regime in order to counter Cuban critics who were becoming loudly anti-Yankee. Ambassador Smith should burnish el Tío Sam's image on the island.

At the start of Smith's tenure, the United States did not conduct business with Cuba as usual, but almost. The two nations signed a trade agreement to reduce U.S. tariffs on tobacco—this a help to Tampa, Florida, cigarmakers who used Cuban leaves—and to lower Cuban import taxes on U.S. tinplate, motors and other products. Havana and Washington also agreed on the terms for a convention to eliminate double taxation.[13]

At his first press conference on July 24, Ambassador Smith praised Cuba for having created BRAC to stand with the United States against the Communist menace, although he did not claim that the opposition to Batista was Communist-inspired. In a familiar subterfuge, he explained that U.S. military assistance to Cuba supported hemispheric defense against the Communist threat. Then he said the obligatory words: The United States would not intervene in the internal affairs of the island. In the context of a question which noted that Gardner had been very close to the regime, Smith remarked that he was prepared to talk with anyone, even oppositionists, so long as they approached him through "normal channels." He would not participate in "clandestine meetings."[14] These remarks could not have alarmed Batista stalwarts or cheered critics of the regime, for many dissenters—especially known 26th of July followers—would likely have been arrested by SIM if they emerged openly to seek an embassy interview. At that very time, in fact, the police were hunting down Frank País, M-26-7's national organizer and chief urban leader. A popular figure in Oriente Province, País moved desperately from one house to another in Santiago to avoid arrest.

Although the guerrilla army depended upon País to sneak supplies and weapons into the Sierra Maestra, Castro eyed him as a challenger to his political leadership. País once complained that Castro was becoming a *caudillo,* and Castro became piqued when País wrote to him with plans for the reorganization of M-26-7 and the formulation of a platform. Castro quickly took the political offensive with the "Sierra Maestra Manifesto," issued on July 28. The declaration pledged free elections, rejected a military junta as an alternative to Batisa, offered social and economic reforms, and warned against foreign intervention or mediation in Cuban affairs. The United States, insisted the document, must cease its arms shipments to Batista. In reporting this manifesto, the U.S. Embassy correctly spotted "bickering" within the movement. From contacts with oppositionists, embassy officers learned that within the anti-Batista opposition "there is an increasing tendency to question his [Castro's] leadership."[15]

When the police gunned down Frank País on July 30, the Batista regime once again removed a rival to Castro's leadership. Angry residents of Santiago protested in the streets and sparked a general strike. The rebellion heated up while Castro's position as rebel *jefe* strengthened.[16]

Ambassador Earl E. T. Smith had announced soon after his arrival in Cuba that he planned to learn about the island's different regions through visits. He picked Oriente Province—Santiago de Cuba, Nicaro, and Guantánamo—for his first trip. The announcement created some speculation in both official and rebel circles. Oriente had become the hotbed of anti-Batista sentiment, and embassy officials knew that Santiago harbored ardent Castroites. Wouldn't the ambassador's plane have to fly over the rebel-held Sierra Maestra? Was the visit "supposed to convey an appearance of normalcy" on the island, wondered Vilma Espín, a founding member of M-26-7?[17] Was Smith heading into this volatile environment, especially so soon after the death of Frank País, to demonstrate solidarity with the regime? Batista, after all, personally approved the ambassador's visit to Santiago when he met with Smith on July 23. Or, did Smith anticipate trouble that would provide him with the opportunity to offer a very public display of a more distant U.S. posture toward Batista?

Did Smith understand that dissidents would seize the moment to stage, at the least, a protest rally aimed at embarrassing the regime? Did he not know that the Santiago police would not tolerate anti-Batista demonstrations, especially when the American ambassador was in town? Did Smith not sense that he was entering a highly charged setting where business had been shut down because of a general strike? Surely his staff was aware of all of this; if so, they gave him bad advice—unless, of course, his advisers welcomed the chance to distance the United States from the Batista administration.

Ambassador Smith and his entourage, including the three U.S. military attachés, arrived in Santiago the day after País's death and just hours before a funeral march to honor the slain insurrectionist. The North American visitors received the obligatory keys to the city at the municipal building. As the customary speeches of friendship sounded in the hall, women demonstrators, dressed in black, paraded outside shouting "*Libertad!*" The police, military intelligence officers, and members of Batista-backer Senator Rolando Masferrer's private army began beating and arresting the demonstrators. The police chief personally pistol-whipped women. Fire hoses were also turned on the crowd. Shrieks soon interrupted the ceremonies inside. As Smith emerged from the building, some women, drenched from the hosings, rushed forward but were roughly manhandled. More than 30 people were arrested.

Set upon by reporters, Smith first remarked that it was "unfortunate that some of the people of Santiago de Cuba took advantage of my presence here to demonstrate and protest to their government."[18] At a later press conference, after consulting with embassy staff, the ambassador added that "excessive police action" was abhorrent to him and that he regretted that his presence had ignited violence. He urged release of the arrestees. In private, he directed his staff to appeal to local military authorities for their freedom. That night, at a reception attended by Santiago's most prominent citizens—lawyers, doctors, merchants—several wives approached Smith to tell him with pride that they had participated in the morning's protest rally. Such news should have begun to educate Smith about the depth of the anti-Batista rebellion.[19]

When M-26-7 guerrillas in the mountains heard about Smith's condemnation of police brutality, they cheered. Smith's remarks, the police beatings, and the País funeral in Santiago combined to touch off a political firestorm across Cuba. Terrorists planted bombs, tore up rails, burned schools, torched tobacco warehouses, and cut electric lines. U.S.-owned property was not spared: A bomb exploded in the Havana office of Owens-Illinois. Spontaneous strikes erupted, and rebel leaders called a nationwide strike. The Batista government soon suspended constitutional guarantees and reimposed press censorship. Rumors buzzed about a military coup. Bitter toward Smith, Batista's henchmen insulted, attacked, jailed, and deported North Americans.[20]

Government officials pressed their hunt for oppositionists too. On the last day of the month, the police arrested Raúl Chibás, a prominent Ortodoxo leader who had gone over to M-26-7, and Roberto Agramonte, Jr., son of a former Ortodoxo leader then in exile in the United States. The men had been in hiding in anticipation of a trip to the United States to raise money for the revolution. Even Ernest Hemingway became a victim of the island strife. Soldiers entered his *finca* outside Havana in search of an insurrectionist. The next morning the famed author found his small dog dead, its skull smashed by a rifle butt.

In another round of the propaganda war, Batista officials heaped criticism on Ambassador Smith for meddling in Cuban affairs. Ambassador Campa filed a complaint with the State Department, and former Ambassador Gardner took the occasion to telephone Herter to pass on a message he said came indirectly from Batista: Smith may very well be declared persona non grata. In the midst of the Smith affair, a U.S.-registered, twin-engine Beechcraft airplane crash landed near Havana on August 7. Its passengers escaped by automobile. The Batista government rekindled its charges that Prío was violating his U.S. agreement by financing rebel activities.

In the United States, Matthews praised Smith in the pages of the *New York Times.* Pickets in front of the White House applauded the ambassador, while members of the 26th of July Club in New York placed a large Cuban flag on the crown of the Statue of Liberty, where it flew for half an hour before police hauled it down. Secretary Dulles publicly backed his new ambassador, calling Smith's Santiago comments "well-balanced" and "very human."[21]

Smith vigorously defended himself, and he exploited the opportunity to state to Cuban leaders that, one, the United States was neutral in the civil strife, and that, two, the U.S.-Cuban relationship was not disturbed by the Santiago incident. He told Foreign Minister Gonzalo Güell that U.S. policy was "non-intervention in the strict sense of the word." After the meeting, a heartened Smith informed Washington that the Santiago crisis "has cleared the air." Everybody now knew, he concluded, that the U.S. Embassy "is following a middle-of-the-road policy."[22] He was wrong: Cubans knew something quite different—that the United States was not following a neutral, even-handed policy.

British diplomats astutely read the crisis surrounding Smith. The Smith visit to Santiago, the British Ambassador informed London, was "extremely ill-timed." Would Smith be recalled? Not likely, remarked a Foreign Office officer who understood the preponderance of U.S. power in the Washington-Havana relationship, because "Batista cannot afford to antagonize the U.S. Government over this." And although Smith had followed State instructions to demonstrate a certain U.S. coolness toward the dictatorship, "he may be reluctant to repeat the risk in the future."[23]

At his own initiative, Smith talked with Batista for two hours on August 21 in the Presidential Palace. Batista baldly declared that since Smith was accredited to his government, he should support Batista. Smith reminded the Cuban President that the United States had come under criticism for being too friendly to the regime and especially for supplying MAP weapons. When Smith said that he intended to be neutral, Batista allowed that he understood. In the end, Smith reassured Batista that cooperative Cuban-U.S. relations remained "unimpaired." Smith reported that Batista had regained his "generally highly favorable" attitude toward the United States.[24] Herter expressed relief that Smith had "gotten back into the good graces" of Batista.[25]

The very day that Smith was smoothing relations with Batista, the State Department determined to dampen Cuban revolutionary activities in the United States. Washington instructed U.S. embassies around the world to "reduce the number of Cuban exiles coming to this country" by exercising "extreme caution in issuing visas" to them.[26]

Hardly had the Santiago episode and its aftermath quieted down when the United States became entangled in yet another dramatic event. On September 5, young naval officers aligned with M-26-7 and Auténtico militants rebelled at the naval base at Cienfuegos, in Las Villas Province. They seized the facility, distributed weapons to civilian rebels, and the combined forces took the city. But planned, simultaneous revolts at other naval stations failed to get started. Cuban Army units rushed to Cienfuegos, while Air Force planes strafed the city. Casualties climbed into the hundreds and rebel prisoners were tortured and murdered. In the end Batista's forces retook Cienfuegos.[27] Surveying the damage, the U.S. Embassy informed Washington that Batista was "victorious but weaker." The foundation of his power—the military—was beginning to crumble.[28]

To crush the Cienfuegos revolt, Batista forces had used U.S. weaponry, including B-26 aircraft purchased under the Mutual Defense Assistance Agreement of 1952. Article I, Section 2, of this agreement specifically prohibited Cuba from using such military equipment for purposes of internal security, although, as we have seen, U.S. officials had long ignored the provision. Ambassador Smith knew that U.S. weapons had been helping to keep Batista in power, but he warned Washington that any official U.S. public criticism of the "improper use of MAP equipment would seriously weaken [the] Batista regime."[29]

In front of the United Nations in New York anti-Batista placards again appeared, and American newspapers criticized the role of the U.S. military in the Cuban civil war.[30] In February the United States had delivered seven medium tanks to Batista's army. In the spring, Washington had disapproved Cuban requests for M32 incendiary bombs and napalm bombs. But the Eisenhower Administration had approved a request for six light tanks and 20 armored cars, which, the embassy had advised, should "arrive [in] Cuba without US markings," because the sale would "undoubtedly cause further criticism by [the] opposition alleging that [the] US [was] supporting [an] unpopular and repressive regime."[31] In May the Cuban government had ordered eight more M-4 A-3 medium tanks. This request troubled U.S. officials. They were finding it more and more difficult to wink at Cuban violations of the military agreement because Batista was so conspicuously using U.S. weapons.[32]

After the Cienfuegos revolt, "acute" public criticism persuaded Assistant Secretary Rubottom to act, although members of Congress hardly murmured a word in the House or Senate about this issue or any other Cuban question during the entire year of 1957.[33] Still, State officials detected opposition and decided to defer action on the request for the eight medium tanks on the grounds that the Cuban military would surely direct the weapons against regime opponents.

A defiant Batista thereupon canceled the order and hinted that if the United States could not meet his military needs, he would find other sources. How could the United States refuse to supply him with arms yet permit Prío to send weapons to revolutionaries in Cuba? Batista asked.[34] The Cuban president also disbanded the MAP unit armed with U.S. weapons and trained for hemispheric defense, distributing its soldiers to several units at war with rebel mountain forces.[35]

Peeved and uncertain about the reliability of U.S. weapons supplies, Cuban officials began to shop around for arms in Canada and Britain.[36] They asked British contacts to quote prices for automatic rifles, light tanks, hand grenades, and military aircraft. "Treat this tentative enquiry as secret," cautioned a British diplomat, because until this time the United States had a lockhold on armaments trade with Cuba.[37] The Americans, warned an officer in London, will certainly object to British weapons sales to Cuba as an intrusion into the U.S. sphere and a challenge to the U.S. effort to standardize arms in OAS nations.[38]

The United States, in an apparent attempt to balance the negative decision on the tanks with something positive in order to lower Batista's temperature, continued to ship arms to Cuba. Shortly after the tank deferral, for example, Washington filled a Cuban order for 100 Colt .45 pistols.[39]

U.S. authorities also bore down on exile activities—especially those of Carlos Prío Socarrás. The *New York Times* picked up stories from exiles around Miami that they were "having a rough time with the police, the FBI and other authorities. Some of them are even afraid that they will be expelled from the country."[40] Rubottom informed Herter that State officers intended to "crack down on him [Prío] for his abuse of American hospitality by inciting trouble while in Florida."[41] At a meeting with Prío on September 20, Rubottom's subordinate William Wieland reminded the former Cuban President of repeated accusations that he was trying to overthrow the Cuban government from U.S. territory. Prío stoutly denied the charges and complained that U.S. authorities were harassing him. When Wieland asserted that U.S. policy was one of non-intervention in Cuban affairs, Prío countered that he knew better, because as president of Cuba he had witnessed "the exercise of U.S. tutelage whenever it served U.S. interests to do so." The United States could intervene against Batista if it really valued democracy on the island.[42]

A month later, Attorney General Herbert Brownell, Jr., convened top officials from a number of agencies to confer on the Prío case after Secretary of State Dulles, upon Rubottom's urging, had asked for a major investigation. The evidence against Prío was then clear enough. The problem was that witnesses would not testify in court for fear of violent reprisal at the hands of the rebels. U.S. officials expressed frustration at not finding a way "to get at Prío."[43]

From Havana, Ambassador Smith argued that if the United States wanted Batista to lift his suspension of constitutional guarantees, U.S. authorities had to collar Prío first. Rubottom balked at such logic; he did not like the client tail's wagging the hegemonic dog. Batista, he protested, was demanding unacceptable terms from the United States by offering to restore constitutional guarantees only if Washington curbed Prío. Still, to satiate Batista and to prepare for a grand jury investigation, U.S. officials began to approach the Cuban government for help in identifying witnesses and collecting evidence against Prío, especially on the *Corinthia* episode.[44]

9

Expanding Contact with the Rebels

THE *BARBUDOS* REMAINED in the mountains and the Batista forces hesitated to go in after them. All the while, the 26th of July Movement guerrillas gradually enlarged their "Free Territory" in Oriente Province, gathering recruits and weapons. With handsewn triangular shoulder patches identifying them as members of the rebel army, the Castro forces patiently cultivated the support of local *guajiros* (peasants), who supplied food, refuge, and their young sons as lookouts and messengers.[1] Rebel sympathizers in the United States smuggled weapons into Cuba by hiding submachine guns, bullets, and pistols in automobiles shipped by ferry from Key West to a Havana auto agency. Female activists traveled from Miami to the island with small arms concealed under their full skirts.[2]

Bombings, killings, attempted assassinations, and harsh countermeasures continued to shatter island peace in late fall 1957. In Havana, on November 8, no fewer than 20 bombs exploded. The terror left no one safe. In Santiago de Cuba the dynamited tower of a radio station fell on the Rancho Club Motel where Texaco refinery workers lived. The incident "could have produced disastrous results for the American colony in this city."[3] No Texaco employees died, but the near miss alarmed North Americans elsewhere on the island.

Economic sabotage became a rebel priority. Fidel Castro announced in November that his insurgents intended to burn sugar-cane fields no matter who owned them. U.S.-owned property would not be spared. The Cuban-American Sugar Company in eastern Cuba ordered rifles and tear gas bombs for protection and asked the State Department for assistance should some 50

Americans at the company's Chaparra and Delicias facilities have to be evacuated.[4] By mid-December 1957, Summer Pingree, an American plantation owner in Oriente Province, had had to put out 40 fires on his holdings.[5] The electrical cables of the Nicaro nickel plant were bombed. "It is owned by the U.S. government," crowed the rebel fighter René Ramos Latour. For this reason, the explosion "caused a big stir."[6] Cuban dissidents also sought to disrupt the lucrative tourist trade. In Key West, signs erected by M-26-7 activists warned American tourists that Cuba was "unsafe to visit; bombing of theaters and other places of amusement are frequent." The "Cuban Gestapo" would arrest visitors who talked to ordinary Cubans.[7]

Assistant Secretary Rubottom fumed over the threatened destruction of the Cuban economy. Batista's opponents, he insisted, were pursuing "a ridiculous and dangerous" policy.[8] Castro defended the sabotage with historical analogy: "When the North American colonists dumped tea into the sea, didn't they do exactly the same thing?"[9]

North Americans, in a host of ways, became players in the Cuban drama. An enlisted U.S. Navy man, Robert Riggs, was arrested for carrying arms to the rebels. Cuban officials delivered him to the Guantánamo naval base, where he faced court-martial proceedings.[10] In the United States, law officers made repeated seizures of weapons destined for the rebels—in Memphis, Tennessee; Laredo, Texas; and Miami, among other places. In mid-November, in the Florida Keys, U.S. agents halted the yacht *Philomina III,* removed its cargo of arms, medicine, and uniforms, and arrested 31 Cubans.[11]

Such results pleased Colonel John E. Kieffer, president of Universal Research and Consultants, Inc., the Washington, D.C., public relations firm Batista hired to promote his regime. Kieffer published *Report on Cuba,* a periodical brimming with stories about Castro's Communist proclivities, boom times in the island economy, and the inability of the rebels to marshal a popular following. Kieffer heralded Batista's promise to hold free elections—this at a time when Cuban government authorities perpetuated press censorship, kept the universities closed, and harassed, jailed, tortured, and murdered both moderate and radical critics of the regime. Some conservative U.S. magazines and newspapers nonetheless adopted the Batista point of view, William Buckley's *National Review* among them.[12]

Anti-Batista groups continued to use the United States as a safe haven to organize followers, raise money, dispatch supplies, and issue propaganda to evade Batista's censorship. They won some friends in the left-liberal U.S. press. *The Nation* magazine, which had published Beals's articles earlier, translated and reprinted from *Cuba Libre,* a Costa Rican publication of the 26th of July Movement, an article by Fidel Castro himself. In this piece,

Castro chastised the United Fruit Company and other foreign enterprises for holding so much arable land while 200,000 rural families owned not "a square foot." Castro vowed to improve the lives of landless small-scale farmers by disciplining the "possessors of capital" who "keep the people bowed under ox-yokes." He also promised to advance industrialization, to confiscate the fortunes of "grafters and embezzlers," to nationalize the U.S.-owned electric and telephone "trusts," and to give "generous asylum, brotherhood, and bread" to democrats in other hemispheric nations who were struggling to overthrow tyrannical governments.[13] Such a program appealed to liberals and leftists in the United States but obviously not to U.S. officials in Washington. They found in Castro's trenchant message too many radical components that, if successfully implemented, would diminish U.S. interests not only on the island but in the hemisphere.

Castro's M-26-7 representative in Washington, Ernesto Betancourt, also pressed the M-26-7 cause against Batista-ruled Cuba. He regularly visited the State Department to lobby for an end to U.S. arms shipments to Cuba. Other exiles picketed, held rallies at the Palm Garden and Flagler Theater, issued declarations, and planned clandestine expeditions. The FBI routinely watched and reported these comings and goings, especially alert to the incessant feuding among exile groups. The young militants of the Revolutionary Directorate and the 26th of July Movement still distrusted one another, and the old guard led by Prío, who financed much oppositionist activity, sought to block Castro while it clung to dreams of restored power.

An ever vigilant Castro continued to worry that leadership of the opposition would be wrested from him. To monitor his rivals and to improve the M-26-7 message for the American audience, Castro appointed a four-person committee in the United States headed by Mario Llerena. Joining him were Carlos Franqui, a journalist, Lester Rodríguez, a veteran of the Moncada attack, and Raúl Chibás. When they were not squabbling with other exiles in Miami, these agitators spotlighted U.S. arms to Batista to argue that U.S. "neutrality" was a fraud.[14]

On the surface, exiles in Miami displayed unity in October by forming the Junta de Liberación Cubana. Followers of M-26-7, DR, Federación Estudiantil Universitaria, the Ortodoxos, Auténticos, and others signed up. Because he had the money, Prío seemed to dominate the coalition. U.S. officials warned Betancourt that another step—such as the formation of a government-in-exile—would violate U.S. law and likely "result in the expulsion of the participants from this country."[15] They need not have worried, because the Miami Pact did not last long.

Upbraiding the 26th of July leader who endorsed the agreement, Castro torpedoed it. Although Castro always welcomed Prío's financial support, he refused to join the politico in any kind of alliance. Castro especially feared that Prío and his cohorts would successfully place a military junta in power, effectively removing Batista and blocking Castro. The mountain commander also complained that the pact had failed to make a case against U.S. interference in the rebellion. Some M-26-7 leaders, in fact, speculated that U.S. officials had engineered the unity agreement. Che Guevara believed that they were "pulling the wires behind the scenes," and "we unfortunately have to face Uncle Sam before the time is ripe."[16] No evidence has surfaced to suggest that U.S. authorities prodded the Cubans into the unity accord. July 26 leaders nonetheless thought that U.S. officials were intervening—one more piece of a growing M-26-7 indictment that the United States was manipulating Cuban affairs, this time by favoring other anti-Batista insurrectionists over the mountain rebels.

In mid-December Castro blasted the Miami Pact and other anti-Batista groups. His tirade, made public in January, castigated exiles for "fighting an imaginary revolution" while "the leaders of the 26th of July Movement are in Cuba, making a real revolution."[17] Castro's angry letter embittered other rebels who for months had been skirmishing with police in Cuba's urban centers, attacking government buildings, watching friends die at the hands of SIM, sitting in jails awaiting their next torture. Other insurrectionists, moreover, had sent aid to the Sierra Maestra and were no less commited than the arrogant Castro to the overthrow of Batista and the inauguration of a new Cuban order. Faure Chomón, a DR leader who had participated in the assault on the Presidential Palace, countered Castro: The self-important mountain rebels hardly constituted the only Cubans fighting the regime. Thousands were in the cities, only a few hundred in the mountains. He reminded Castro that the DR had been quite active before the *Granma* ever left Mexico. And where had the 26th of July been when DR members stormed Batista's palace? While Castro was in the "lofty mountains of Oriente," Chomón jabbed, the DR was meeting Batista head-on.[18]

Betancourt admitted to State Department officials that Castro's read-them-the-riot-act letter was "abusive in its language." But Castro's bitter "views were expressed under very trying conditions."[19] This was not the first suggestion—from Cubans or North Americans—that Castro, when under stress, could explode. But it would be a mistake to interpret Castro's renunciation of the unity pact as a mere outburst from a volatile, irrational man. Castro's blast was calculated to weaken his rivals and to elevate M-26-7 and himself to opposition leadership.

While the Cuban oppositionists practiced their divisionist politics, Cuban-American relations under Ambassador Smith seemed to be returning to business as usual. After all, Castro was still confined to his remote mountains;

Havana was only intermittently inconvenienced by bombings and sabotage; and the hegemonic hubris assured North Americans that this particular Cuban political mess, like others before it, would be cleaned up in due course. U.S. investors continued to fund Cuban projects, and an agreement to supply Cuba with nuclear fuel for nonmilitary uses, part of Eisenhower's Atoms for Peace program, went into effect in October. The following month, the Export-Import Bank authorized its credit to the Cuban Telephone Company and an agreement providing for guarantees against the inconvertibility of profits entered into force. The two nations also successfully concluded negotiations on a convention to conserve shrimp resources in the Gulf of Mexico.[20] Republican Senator George Aiken of Vermont visited Cuba at the end of the year and, remarkably, found no Cuban resentment against U.S. economic interests. Any Cuban government, he reported, would by necessity be friendly with the United States.[21] "Revolution? What Revolution?" TV variety-show host Ed Sullivan was heard to say after his trip to Cuba about the same time.[22]

Had U.S. observers not been so taken by the Smith-Batista propaganda of strong U.S.-Cuba relations, they could easily have discovered layer after layer of Cuban discontent. The awarding of a U.S. Legion of Merit to Cuban Air Force Colonel Carlos Tabernilla in November ranked as a particularly conspicuous and controversial demonstration of U.S. support for the Cuban regime. No other previous gesture demonstrated such insensitivity to Cuban public opinion and to the impact of U.S. decisions on Cuban sensibilities. The 38-year-old Colonel Tabernilla, son of the Batista crony and Army Chief of Staff General Franciso Tabernilla, had ordered the bombing and strafing of Cienfuegos in September. Now the United States was honoring him, and no press censorship got in the way of this story of U.S. support for the regime. Batista made sure that photographs of the luncheon, attended by U.S. General Truman H. Landon, appeared in Havana newspapers.[23] The U.S. Embassy, obviously not anticipating the furor that followed, first judged General Landon's trip "highly successful." Indeed, "he was warmly received by Cuban military, governmental and civilian figures."[24] Annoyed State Department officials in Washington, apparently not informed beforehand that the decoration was being offered, found the timing "unfortunate."[25]

Still, just a few days later, Marine Corps General Lemuel C. Shepherd, head of the Inter-American Defense Board, met with Batista at the Presidential Palace. In an exchange of toasts, Shepherd hailed Batista as a "great President" and a "great soldier."[26] State officials scrambled to quiet these conspicuous military *abrazos* while they themselves repeated U.S. pledges of neutrality toward the insurrection.

U.S. journalists covered the Tabernilla episode and other examples of U.S. links with the Batista regime. Matthews unrelentingly used the *Times*'s editorial pages to contrast Cuban disaffection from Batista with Castro's tenacity.[27] He was hardly alone. John O'Rourke of the *Washington News,* Don Hogan of the *New York Herald Tribune,* Jules Dubois of the *Chicago Tribune,* and many others plumbed the sources of the Cuban rebellion and came away doubting that Batista could survive.[28] The Inter-American Press Association condemned the regime for imposing censorship. Handicapped by Batista's control of information, Ruby Phillips of the *New York Times* paid tipsters—some on the government payroll—to send her news from several parts of the island. "Every day it becomes more important to have the proper contacts on the government side and the opposition," she told her New York boss. "This is a matter of long established relations and, fortunately, I have them."[29]

U.S. journalists such as Phillips often received calls from rebels who passed on information about dissidents who had been arrested. She would ring up the chief of police or Batista's secretary, make a simple inquiry, and hint that she would write a story. The fear of publicity made some of Batista's henchmen hesitant to torture or murder prisoners. In this way North American reporters probably saved lives, and the rebels were appreciative. Castro once told Matthews that "you and your colleagues have performed a greater service for the United States than all its diplomats and heads of military missions."[30] Near the end of the insurrection, when it had become clear that the rebels were closing in on Batista, Fidel Castro picked a wild orchid in the Sierra Maestra. Within days a rebel sympathizer walked into the *Times* Havana office and presented a box to Ruby Phillips. Inside was the flower, a rebel tribute to an American who had lived in Cuba since 1923 and reported for the *New York Times* since 1937.[31]

U.S. press coverage of human-interest stories often projected images critical of both Batista and his U.S. backing. During the Christmas season in late 1957, for example, a Cuban Navy officer, Luis Miranda León, stowed away aboard the Cuban freighter *Bahía del Nipe.*[32] Before he could disembark in New York, however, Cuban authorities arrested him, charging that he had participated in the Cienfuegos uprising. The attorney Sheldon J. Kravitz asked U.S. State Department and Justice Department officers to grant Miranda political asylum. A firm "no" came back. Miranda then slashed his wrists with a razor blade. All legal avenues to gain the sailor's release having been closed, Cuban exiles—many of them DR and M-26-7 members—stormed the East River pier. They battled police, who arrested 31 of them for disorderly conduct and incitement to riot. "U.S. Refuses to Save Cuban Rebel on Ship" blared the *New York Post* headline.[33]

After a search for bombs, the ship carrying Miranda departed for Havana. The navy officer arrived in Cuba two days after Christmas to face a court-martial. "I have just received a perfect example of why Communists are winning the 'Cold War,' " remonstrated one critic in the United States. "Our Government buried its eyes, mind, and heart while a Cuban rebel was returned to Cuba, to Batista, and death." Indeed, "no shipment of grain, no expenditure of millions of dollars of foreign aid would have won the hearts and friendships of the Cuban people and Latin Americans as would offering this Cuban political asylum."[34]

Although Ambassador Smith staunchly defended the regime and argued that Cuba's problems could be resolved *with* Batista, some members of his Havana embassy staff and officials in the Department of State in Washington were beginning to think otherwise. Contacts with anti-Batista Cubans both in Cuba and in the United States no doubt spurred this shift. State officers in Washington met regularly with Betancourt, Llerena, and others. From them they learned details about rebel intentions, squabbles within the opposition, and Cuban opinions of U.S. policies. U.S. officials also used these meetings to explain policies and decisions and warn exiles to abide U.S. laws.[35]

In Havana and Santiago de Cuba, U.S. officials developed links with oppositionists. Some contacts developed out of crises involving Americans, as in the case of the three American runaways from Guantánamo. Others developed because U.S. officials simply wanted to gather accurate information and to anticipate events that might damage U.S. interests. Some contacts became channels through which the United States might exert influence. Ties between U.S. officials and Cuban insurrectionists meant that the former could also help the latter in emergencies. Haydée Santamaría told a journalist that when the war was over she was going to give a bouquet of flowers to a U.S. official who had served the revolutionaries in this way. In January 1958, after Cuban authorities arrested Armando Hart, a lawyer and M-26-7 underground leader, Vice Consul Robert Wiecha inquired about his well-being. Through this display of concern, the CIA agent probably saved Hart's life—at least Castro thought so. In a visit to the Department of State, Felipe Pazos thanked U.S. officials for helping to protect his son Javier, who had been captured with Hart.[36]

Richard Cushing, the Public Affairs Officer in the embassy, built ties with Rufo López-Fresquet and his American wife, who fed him intelligence.[37] López-Fresquet recalled that he had "frequent contact" with embassy officials. Besides Cushing, the Cuban met with John L. Topping, the embassy's First Secretary and chief political officer.[38] Ambassador Smith later claimed that Topping and CIA Station Chief William Caldwell were not only anti-Batista but pro-Castro.[39] A Havana backer of the Prío-funded Second Front of the Escambray recalled that CIA agent Jack Stewart, undercover in the embassy as a political officer, hung around cafes to talk with members of the underground. Fluent in Spanish, Stewart even met with anti-Batista Cubans in his own apartment. On one occasion the CIA officer apparently used the embassy's diplomatic pouch to smuggle a rebel letter out of the country.[40] A member of the liberal Montecristi Movement, Andrés Suárez, who later revealed CIA contacts with oppositionists, regularly lunched with Basil Beardsley, another CIA agent in the embassy.[41] The assistant naval attaché, Lt. John M. Davis, also cultivated contacts with pro-Castro people in the civilian community.[42] Before his arrest, Armando Hart informed Castro that he too had "contact with people close to the embassy."[43]

The journalist Robert Taber did not think that such U.S.-rebel links were "officially sanctioned," because Ambassador Smith had apparently prohibited his staff from contact with revolutionaries.[44] Smith must have suspected these ties, however, because some of his staff's information could only have come from rebel sources. The ambassador himself, fixated on the Communist issue, urged the CIA to place a spy inside Castro's forces in order to discover any Communist infiltration.[45] Later, in summer 1958, when Raúl Castro's forces kidnapped several U.S. citizens, Smith helped set up direct talks between U.S. officials and M-26-7 revolutionaries.

U.S. efforts to establish direct contact with the rebels in Santiago de Cuba apparently came as early as July 1957, when Frank País informed Castro that an American diplomat wished to see the rebel leader in the Sierra Maestra. Castro waxed enthusiastic: "It is a recognition that a state of belligerence exists, and therefore one more victory against the tyranny." Castro would accept no demands, he said. Nor was a U.S. mediation role acceptable. But "if they wish to have closer ties of friendship with the triumphant democracy of Cuba? Magnificent!"[46] The meeting with Castro did not materialize, but U.S. officials maintained relations with the Santiago underground.

In another País letter to Castro, dated July 5, 1957, the Santiago rebel leader reported that the "very meritorious and valuable American embassy [he may have meant consulate] came to us and offered any kind of help in exchange for our ceasing to loot arms from their base [at Guantánamo]." A deal was apparently struck that gave the rebel leader Lester Rodríguez a two-year U.S. visa. País claimed that rebel maps and letters actually departed Cuba for the United States in an official diplomatic pouch at the same time with Rodríguez. "Good service," País remarked.[47]

In September, the CIA agent Robert Wiecha, undercover, took his post as a new vice consul in Santiago. He became not only a contact with but perhaps also a source of funds for the 26th of July Movement.[48] His appearance in Santiago, close to Castro's Sierra Maestra outpost, signaled a U.S. interest in monitoring and influencing a rebel force that had earned such attention by demonstrating its political clout. The CIA, throughout its history, has maintained contacts with different sides in civil wars. Such seems to have been the case in Cuba. If Batista won or lost, if Castro won or lost, the

CIA would be positioned to deal with either side, or with some third option in the middle.[49] Or so it seemed.

While these various interactions between U.S. officials and anti-Batista plotters evolved, Smith was telling Fulgencio Batista what the beleaguered Cuban leader wanted to hear and what Smith believed: The United States desired "to see the present government continue in office until succeeded by the administration to be chosen" in June 1958 elections.[50] At the same time, Smith accelerated his efforts to discredit Fidel Castro and the 26th of July Movement by suggesting that they were Communists. He later claimed in his memoir and in Senate committee testimony, in the red-baiting language and ill-supported accusations characteristic of McCarthyism, that *he* saw Communism in the rebellion before his embassy subordinates in Havana and his superiors in Washington—and that his warnings had been ignored.

Smith ordered CIA station chief Caldwell to review data on Communist party membership in Cuba. Smith thought the extant number of 10,000 card-carrying Communists too low. But the CIA officer, according to Smith, "had a closed mind," sympathized with the *fidelistas,* and displayed "the intellectual snobbism directed by the career officers against the political appointees."[51] The following month, the ambassador informed Washington that it was of the "utmost importance to determine promptly whether Castro [is] either collaborating with Communists or possibly [is] under their control. Information here does not confirm this."[52] Smith eagerly worked to confirm such Castro-Communist ties in order to save Batista, but the ambassador always came up short.

Wieland later explained that State Department officials did in fact consider the question of Castro's possible entanglements with Communists, but investigators found no evidence. "There were, of course, ample indications that he was unstable, increasing indications that he was unscrupulous, increasing indications that he was tyrannical in his leanings," all of which was demonstrated in Castro's scorn for the Miami Pact.[53] Wayne Smith, a low-level embassy officer, recalled that under Ambassador Smith "we were urged to leave no stone unturned in the effort to ascertain Castro's Marxist-Leninist proclivities—or lack of them." Like Wieland, Wayne Smith did not discover Communist sympathies in Castro, but rather "gargantuan ambitions, authoritarian tendencies, and not much in the way of an ideology of his own."[54]

Earl E. T. Smith would have been surprised had he read some of the rebels' correspondence at the time. In December 1957, for example, Che Guevara wrote to René Ramos Latour, País's replacement as underground leader, that Fidel Castro was a "leftist bourgeoisie," whereas Che himself was a Communist. Ramos remonstrated that what Che would eventually do—and apparently Fidel would not—was to replace Yankee domination with Soviet domination.[55] Throughout the rebellion, the *fidelistas,* hardly settled on a coherent ideology, heatedly debated the meanings of revolution and radicalism and the direction a new Cuba should take.

With no evidence that Communism had infested the 26th of July Movement, but with considerable information on Castro's tenacity and Batista's hardline resistance, both promising more disorder, Washington officials initiated a significant policy review in November. Prominent Cuban moderates who hated Batista and disliked Castro, and who wanted to cultivate friendly U.S.-Cuba relations, had been talking with Department of State officers for months. Their anti-Batista message impressed U.S. officials. Former Cuban National Bank president Felipe Pazos, Carlos Hevia of the Auténticos, and others built a case against U.S. arms deliveries to Batista's military. More, they gradually persuaded their American listeners that Cuban discontent with the Batista regime ran deep and wide in urban and in rural areas, cutting across economic classes and political parties.[56]

One moderate Cuban seemed to impress State Department officials a great deal. Dr. Joaquín Meyer of the Cuban Sugar Stabilization Institute visited Washington frequently. Although he worked for the Batista government he also criticized it. U.S. officials liked his temperate tone. The Cuban desk officer, Terrance G. Leonhardy of Wieland's office, informed Rubottom that Meyer's views are "most valuable" because Meyer was "pro-United States and is as neutral regarding the present Cuban political crisis as any official could be." Meyer told Leonhardy on November 14 that he reluctantly thought the time had come for the United States to intervene "in some tacit manner to save his country from chaos." If the United States thought Batista was likely "going to go, we [the United States] should tip the scales against him as soon as possible." If, on the other hand, U.S. officials believed that Batista could stay the course and would conduct elections, the United States should openly support the regime against the revolutionaries. At the same time, the United States would have to persuade Batista to create a favorable electoral climate; he would have to remove some of the "objectionable" members of the cabinet and security forces. Meyer himself, however, did not think that elections would end the Cuban crisis.

In a frank acknowledgment of the influential position of the United States in Cuba, Meyer advised that if Washington wanted to remove Batista, all it had to do was "get word to the right people in the Army and it would be all over in a hurry." Meyer then elaborated on why the United States had to intervene: Rebel sabotage of the sugar industry threatened the United States itself, for investors would lose money, consumers would pay higher sugar prices, and eventually the United States would have to supply foreign aid to help the island recover from the damage. "Act soon," Meyer implored.[57]

10

Taking Sides: Arms, Arrests, and Elections

A WEEK AFTER the meeting with Joaquín Meyer, the Office of Middle American Affairs delivered a long memorandum to Assistant Secretary Rubottom. This office, using information from the U.S. Embassy in Havana and the consulate in Santiago, and consulting with the Cuban desk officer Terrance G. Leonhardy, coordinated Cuba policy. William Wieland headed this unit of the State Department. The 1907-born Wieland knew Cuba well. He had worked there in the early 1930s as a newspaper correspondent, and he spoke Spanish. Like so many other Latin American specialists in the State Department, Wieland had served in many posts in the Western Hemisphere—Rio de Janeiro, Bogotá, San Salvador, and Quito—before settling down in Washington. He became director of the office on May 19, 1957, and the Cuban crisis soon preoccupied him.[1]

Written by Wieland and his deputy director C. Allan Stewart, the long memorandum, titled "Possible United States Courses of Action in Restoring Normalcy to Cuba," described "a state of near-rebellion" in Cuba that threatened U.S. citizens and large North American investments and provided "fertile ground" for Communists. The bedeviling question was: How could the United States restore order "short of direct intervention?"

Wieland and Stewart advised that the United States serve as a "discreet middleman" between the Batista government and its opposition in order to achieve free elections and end the killing and destruction. To realize this objective, the United States, perhaps through the Organization of American States, might have to bring "outside pressure" on Batista to ensure an honest presidential election. The State Department officers mentioned Meyer's cry for immediate action. In fact, because Batista was in such dire straits, he might welcome U.S. suggestions for an election plan that included an amnesty for political prisoners. One provision of such a plan was that Castro would be required to leave the country until after the balloting.

Wieland and Stewart spelled out other, apparently less appealing scenarios. The United States could adopy a "policy of aloofness from Cuban internal affairs." Batista would likely fall and "chaotic and tragic" consequences would follow. Or the United States could foment an armed forces coup by encouraging disaffected officers. "It is possible one of them might assassinate Batista." Such a coup could be stimulated by an announcement that unless Batista curtailed his brutalities the United States would end weapons sales and suspend MAP shipments. Not wanting to lose the U.S. assistance that gave them sustenance and stature, the military officers would supposedly oust Batista. As another option, if the opposition did not cooperate with an election plan, the United States could "adopt harsher measures to curtail their revolutionary activities here." The two advisers did not state a preference, but, given conditions and the stakes at the time, the option of "aloofness" was unacceptable.[2]

Intending to initiate debate, the State Department sent the memorandum to Ambassador Smith. Wieland even traveled to Havana to confer with the ambassador and his staff. The discussions proved "extremely helpful and constructive," thought Wieland. He reported that he and the embassy team believed that, "however small the odds," they had to try to effect a constitutional solution. He also noted a high degree of unanimity of opinion in the embassy. As for Smith, Wieland commented that their views were quite similar—the two diplomats were "back-stopping each other." This report certainly contrasted with Smith's later denunciations of Wieland and with the evidence of divisiveness in the embassy. Had Wieland misread the ambassador? Had the ambassador misled Wieland? The two officials certainly had some differences of opinion, but Smith's later harsh accusations were no doubt designed to shift blame from himself to others for the U.S. failure to block Castro.

While in Havana, Wieland also met with the veteran journalists Jules Dubois of the *Chicago Tribune* and Ruby Phillips of the *New York Times.* Then he talked with Cubans outside the government, most of whom said that Batista had to leave office before normalcy could return to the island.[3]

Ambassador Smith, Minister-Counselor Daniel M. Braddock, and First Secretary John L. Topping joined Wieland to draft a paper. Largely reflecting the earlier Wieland-Stewart memorandum, this new report of December 7 did reveal some differences—highlighting, for example, that Castro had socialistic leanings because he advocated the nationalization of public utilities. Ambassador Smith's touches also became evident in the statement that U.S. interests "would be best served by the continuation in office of the present government until the end of its elected term or at least until after elections." The report noted that Batista would become more cooperative (restoring constitutional guarantees and purging odious members of his government and armed forces) if the United States squeezed more tightly on Prío in Florida.

U.S. influence could be exerted on both Batista and oppositionists to endorse and arrange free elections, the report recommended. There was no mention of a military coup. The clear conclusion: Stick with Batista until the June elections eased him out. (The formal transfer of power was scheduled to take place February 24, 1959.)[4] As for Batista's successor, the report said nothing. Given Prío's sordid reputation among moderate and radical Cuban rebels alike, and given the U.S. desire to prosecute the former Cuban president in order to gain Batista's cooperation for a peaceful transition, Prío's name ranked low on any U.S. list of alternatives to Batista and Castro.

About two weeks later, the debate continued in a memorandum drafted by Wieland and Leonhardy which Washington sent to the Havana embassy for comment. More specific about procedure, this document identified several phases of U.S. action. The ambassador's task was to persuade Batista to create a positive "atmosphere for compromise" with the opposition. The ambassador would extend carrots to Batista to promote his cooperation: Cuban orders for weapons would be filled and the investigation of Prío would be advanced. After making these assurances, Smith would ask Batista to restore constitutional guarantees, declare an amnesty for political prisoners, denounce violence on both sides, and punish and remove police and military officers who committed unnecessary brutality. Smith was also supposed to appeal to Cuban judges to deal firmly with terrorists and to appeal to journalists to avoid incendiary editorials and stories.

Assuming that Batista acted along such lines, the embassy would begin to draw the government and oppositionists into talks. One condition had to be met at the start: Batista had to agree to retire from politics and give up his position as military chief. (The earlier, obviously unworkable proviso that Castro had to leave the island was not recommended again.) What if the militant opposition did not respond favorably to Batista's concessions and refused to participate in elections? The United States would likely offer "full and open support for the present regime." On the other hand, "*only* if" Batista did not cooperate, the United States, "short of intervention," would attempt "to hasten the ultimate fall of the Batista regime" while backing "responsible elements within the opposition" who would protect U.S. interests.[5] This document's message was unmistakable: Stand with Batista until after an election and block Castro in favor of moderates. The Officer in Charge of Caribbean Affairs, Edward S. Little, later explained that State Department officials were looking for a third alternative—a "middle group" between Batista and Castro.[6]

The election option became the pivotal part of the U.S. effort to quiet Cuba down. If the election alternative had prospered and worked to stem the bloodletting by providing a peaceful transition to a non-Batista government, today it might be ranked high in Cuba's history as an instrument of salvation. But, from the beginning, the option actually had little chance for success, as many Cubans consistently emphasized to U.S. officials. Cuban politics had simply become too polarized. Few people trusted Batista. Would he not rig the election as before?

Yet the election option was about all the United States had to work with because neither the Havana embassy nor the Department of State was ready to push Batista out. Nor were they willing to accept a government led by Fidel Castro. And a third force had not yet mustered political strength. Nor was the United States ready to intervene directly to force a settlement on all sides. Circumstances were not yet so crisis-ridden as to require a blatant display of hegemonic fiat. So U.S. officials decided to press Batista while hanging on with him, hoping that elections would solve the Cuban problem, hoping as well that U.S. interests would not collapse with him. "There is no alternative to dealing with the presently constituted government in Cuba, however much we may disapprove of certain acts of that regime," because large U.S. investments and the well-being of thousands of Americans living in Cuba were at stake.[7] The United States had to help Batista "weather the test," Wieland concluded.[8]

Shortly after Christmas, Ambassador Smith reported that he and the new memorandum "are not too far apart in our ideas," but his response was lukewarm at best. Smith disagreed with State's assessment that "the situation in Cuba" was "progressively deteriorating." Smith countered: "On the contrary, the present GOC [Government of Cuba] appears to be having progressively better control of the situation." Putting a good face on the Cuban crisis and counting on an orderly transfer of power through elections, Smith recoiled from any plan that in any way suggested Batista's ouster. He requested a meeting with State Department officers in Washington. The question came down to just how closely the United States should back Batista before the presidential election.[9]

A Cuban-American deal was in the making before Ambassador Smith departed for the United States: The Cuban government would restore constitutional guarantees and, in return, the United States would deliver military equipment that Cuba had already requested. In 1957 the Department of State had approved 11 Cuban requests for arms; at the close of the year, seven still awaited decision. Of the seven, argued Wieland, three deserved State Department approval: 100,000 rounds of 20 mm ammunition for the Cuban Navy, 10,000 hand grenades, and 3000 75 mm howitzer shells and two aiming devices. Not even making a nod toward the legal requirements that such equipment be used in hemispheric defense, Wieland left no doubt about the purposes of these weapons: "controlling movements of small boats which are suspected of carrying arms to rebel groups, including that of Fidel Castro" and, in Oriente Province, "combating attempts to burn the cane fields."[10] About this time, too, the Foreign Office in London learned that the State Department did not object in principle to British arms sales to Cuba. Anglo-Cuban weapons deals began to hatch.[11]

On January 11, Smith reported that he had obtained from Batista "definite assurances" that constitutional guarantees would be restored. And, the ambassador claimed, he had gained the concession "without any commitments from US." Smith pressed Washington especially to send the 20 armored vehicles Cuba had requested in mid-1957, for if Cuba "is forced to cancel this order because of our deferring delivery, it may seriously damage my overall objectives."[12]

In the morning of January 16, 1958, Smith met with Assistant Secretary Rubottom. The ambassador argued for the armored cars as a necessary part of any plan to prod the Cuban government toward an "orderly transition of power." Smith was authorized to inform Batista that, despite "strong" public and congressional opinion against the sale of arms to a "dictatorship," the armored vehicles would be delivered on schedule—March 4 at the earliest. But if Batista failed in the meantime to create conditions favorable to an "acceptable election," the United States might cancel the sale. The ambassador could carry good news back to Havana, then, for Secretary Dulles agreed that the delivery of cars and other arms would go forward, knowing, as Rubottom had warned, that the transaction would stir up negative public opinion in the United States. U.S. officials would simply have to "face up" to the criticism.[13]

At a press conference on the sixteenth, Ambassador Smith said little about the pro-Batista direction of U.S. policy, mentioned that the Cuban government would likely lift the ban on constitutional guarantees, and issued the obligatory but hardly persuasive statement pledging "non-intervention in the internal affairs" of Cuba.[14] He managed, however, to kick up another storm around himself. Thinking he had gone off the record, he remarked bluntly that the United States could not do business with Fidel Castro. Some of the reporters jotted down his words, and before long Castro himself knew about

Smith's comment.[15] When he returned to Cuba, Smith further revealed the tilt toward Batista when he said, "I wish that business in general in the United States [which was then experiencing a recession] was as flourishing as it is in Cuba."[16]

In early 1958, then, Castro had to wrestle with new dangers to his movement. The U.S. government, rather than pulling back from the Batista regime, was in fact shoring it up. Smith's public comments strongly suggested that the United States had decided to snub Castro and the 26th of July Movement. The State Department began to act on arms shipments to Batista forces. In early February, State approved delivery of the hand grenades, howitzer shells, and ammunition.[17] In mid-February a long-delayed shipment of U.S. ammunition and trucks was delivered to Batista's armed forces.[18] Batista was also placing orders for weapons in a number of other countries. In contrast, Castro repeatedly complained that his troops were underarmed.

U.S. authorities were even preventing Castro's sister Juanita from entering the United States, and they continued to interrupt anti-Batista activities. In Houston, Texas, for example, U.S. agents arrested nine persons and seized one million dollars' worth of weapons, clothing, and medical supplies destined for Cuba. In New York, the man Castro boosted as Cuba's next president, Judge Manuel Urrutia Lleó, sensed a hardening of U.S. policy toward the rebel cause. Like Castro and others, Urrutia lamented that "Cuba fights again, but tragically alone."[19]

Castro must have been distressed, too, that Jules Dubois rejected a 26th of July invitation to visit rebel forces in the Sierra and instead sent Castro questions about the subject that dominated the Cold War era: Where did he stand on Communism?[20] The *New York Times,* moreover, would not permit Herbert Matthews to go to Cuba again because the newspaper refused to assign a reporter to a story "in which there is even an appearance of a personal involvement."[21] Was Castro in danger of losing the propaganda war?

Fidel's squabbling with other dissidents, especially shrill in his dismissal of the unity pact, further split the opposition. When it seemed that Batista was buying carbines and napalm from Nicaragua and the Dominican Republic, rebel troops needed more arms, and Castro had to appeal for help from Costa Rican President José Figueres, who eventually obliged with a planeload of weapons. Castro's holier-than-thou attitude also persuaded some oppositionists to prefer a reformist military junta over Castro. The Directorio Revolucionario, moreover, expressed its independence from Castro by opening a new guerrilla front in the Escambray Mountains in central Cuba—much closer to Havana than Castro was. Faure Chomón, who led the expedition to the Escambray, told a Cuban journalist in confidence that once his forces dislodged Batista, he intended to "dispose of Castro next."[22] In this second front, a dishonorably discharged American veteran and soldier of fortune, William Morgan, served as a commander. (After Castro's victory, Morgan produced frog legs for the U.S. market. But when he apparently became a gunrunner for anti-Castro rebels, the revolutionary government executed him in March 1961.)[23]

Castro's objective of disabling the Cuban economy was also sputtering. The burning of cane fields and sugar mills caused damage, but mill owners and sugar planters began to take effective security measures in cooperation with the army, and the cane cutters, worried about losing their jobs, opposed the M-26-7's destructive campaign.[24] In fact, 1957 had been a good business year, with high sugar prices and continued foreign investment, and analysts like those at the United Fruit Company were predicting a profitable 1958 too.[25] After all, the rebellion was still largely confined to the southeastern end of the island.

Castro also recognized that Batista was taking the political initiative by promising elections, inviting parties to name their candidates, and restoring constitutional guarantees (done on February 25, except in Oriente Province). Even the ailing 75-year-old former President Ramón Grau San Martín accepted the Auténticos's nomination to run for the presidency. In a slap at Castro and other militants, Grau urged his compatriots to redeem the nation "by the pacific road of the ballot box."[26]

Meanwhile, the regime bore down on urban revolutionaries. The arrest of Armando Hart and other underground leaders on January 10 set back the Castro movement. "You can't imagine," Haydée Santamaría wrote from Havana to Celia Sánchez in the Sierra Maestra, "what the repression here in the plains is like! Its really frightening; I think we'll all end up prisoners, one by one, or something worse."[27]

Castro railed against these many misfortunes. In one particularly explosive letter in January to rebel leaders in Santiago, he lectured his comrades "to get things straight" about military orders he had issued. He was "at the end of my rope after a series of incidents that have occurred these last months." More, "I'm tired of having my feelings misinterpreted. I'm really not meanly ambitious. I do not believe I am the boss."[28] Despite such tortured disavowels, Castro frequently bossed, and the resulting discord he spread in oppositionist ranks weakened the anti-Batista rebellion. But Castro was not at all interested in ousting Batista for the benefit of old-line parties. He wanted the dictator out and Castro in.

At the same time, the troubles besetting Castro buoyed U.S. officials, for they had decided in January that Castro must be prevented from succeeding Batista. Later that month they asked U.S. embassies throughout Latin America to submit reports on Castro and the 26th of July Movement because "the information now available is contradictory, inconclusive, and of inadequate detail."[29] The following month, they talked about getting "the other side of the Castro story to the United States press as well as to Congress"—that is, the negative story.[30] They encouraged a prominent Cuban attorney to discuss with his U.S. clients how they might stem the adverse publicity the Batista government received in Congress and the press.[31]

Ambassador Smith optimistically remarked that Cuba was moving toward "normalcy" and that "Fidel Castro is losing prestige." Smith, who always emphasized the sunny side of Batista's Cuba, further informed Washington that the electoral process promised good results: Although the revolutionaries were splintered, other, more moderate opposition groups seemed to be rallying to present a united front in the forthcoming election.[32]

Although Castro and his M-26-7 insurrectionists detested Carlos Prío Socarrás, they fully grasped that his indictment on February 13 and subsequent arrest symbolized the stronger pro-Batista posture of the United States. They complained that a "deal" had been struck between the State Department and Batista.[33] A grand jury indicted Prío and eight other Cubans for conspiracy to violate U.S. neutrality laws by supplying money to military expeditions operating from U.S. territory and by sending arms to Cuba from the United States. The *Corinthia* and *Philomena III* incidents became central to the U.S. case against Prío. Assistant Attorney General William F. Tompkins remonstrated that Prío and the other conspirators had sent assassins to Cuba to gun down Batista.[34]

Prío voluntarily surrendered at the office of the U.S. marshal in Miami. Handcuffed like other defendants, Prío was then led down the street to the Dade County jail. Along the route, Cuban demonstrators denounced the U.S. government. When Prío learned that some of the other prisoners could not raise the bail of $3000 each, he elected to remain in jail with them.[35] Nationalistic Cubans of many persuasions decried the treatment of Prío, Cuba's last constitutionally elected president, as an insult to the nation, a humiliating public spectacle. The "man who slept in Churchill's bed at the White House, tonight sleeps in a cell in Miami."[36] Such public criticism did not deter Tompkins, who flew to Havana to meet with Batista to seek the help of the Cuban government in gathering further evidence and persuading reluctant witnesses to testify in court against Prío.[37]

The various setbacks M-26-7 suffered in early 1958, especially the Batista-U.S. bargain, prompted Castroites to counterattack. The Havana embassy observed an "increasing boldness and aggressiveness," and tended to think that the heightened activity derived from desperation more than from strength.[38] In mid-January Castro's forces staged a bloody attack on Pino del Agua, a well-defended sugar refinery in Oriente Province. The rebels killed some twenty while losing eight themselves. A few days later in New York, Cuban revolutionaries approached an executive of the huge Atlantic-Gulf sugar company and asked for a financial contribution. Should the company refuse and the revolution succeed, rebels warned, the firm "would be remembered."[39] To counter the Prío-financed Junta of Cuban Liberation in Miami, Castro created a new "Committee in Exile" in New York.[40] After a good deal of preparation in a mountain camp, Radio Rebelde went on the air February 24, further undercutting Batista's attempt to silence his critics. Plans went forward to send Raúl Castro with a special column to open a second front in Oriente Province (he reached his region of operations, the Sierra del Cristal, on March 10).

M-26-7 activists and other insurrectionists also reinvigorated terrorist activities. In Oriente, they torched warehouses (including a Sears, Roebuck facility) and destroyed buses. A 135,000-gallon Sinclair Oil Company storage tank and a chemical plant owned by Air Reduction Company of New York went up in flames.[41] In Havana, oppositionists detonated bombs time and again, and all over the island they assassinated policemen and soldiers.

On February 23, to claim headlines and to demonstrate that the regime could not protect visitors or maintain order, 26th of July members in Havana kidnapped the Argentine racing car driver Juan Manuel Fangio, who was scheduled to race in Cuba's Gran Premio, the nation's equivalent of the Indianapolis 500. His captors grabbed him in the Lincoln Hotel. The race itself came to a premature close when, in the sixth lap of a 90-lap race, a Ferrari spun out of control and plowed into the crowd, killing several people and maiming many others. After suspension of the race, Fangio was released unharmed. The race's outcome, the U.S. Embassy reported, "will add to the sense of tragedy and frustration which blankets the domestic political scene."[42] The very day Fangio was snatched, rebels attacked the Nicaro nickel plant. They cut communications lines, overpowered guards, and seized weapons they claimed had recently been shipped from the U.S. naval base at Guantánamo.[43]

Fidel Castro also once again cultivated the foreign press to carry his message to a wide, especially North American audience. Fathoming what Castro believed at any one time has long preoccupied historians. The rebel's early 1958 remarks to journalists illustrate the problem. Because he appeared conciliatory and moderate, critics later charged that he put on his "democratic mask" only for tactical reasons.[44] Was he simply playing to North Americans because his movement was in such trouble after the U.S. decision to back the *batistianos* and oppose the *fidelistas?* Or was the M-26-7 commander listening to reformers inside and outside his movement and rethinking programs—was he "sobering up" as he gained "political maturity"?[45] In any case, Castro did not sound very radical in the many statements he released in February.

Coronet magazine carried a "Why We Fight" article from the 26th of July leader. Castro promised to establish representative government, encourage political pluralism, and end corruption. He said his new government would overcome illiteracy and sponsor land reform (vaguely mentioned as the clarification of land titles and legal ownership). Need U.S. companies worry that he was dangerous? He disavowed earlier statements that his movement would nationalize the foreign-owned public utilities. "I personally have come to feel that nationalization is, at best, a cumbersome instrument" that would enfeeble private enterprise and hamper "the principal point of our economic program—industrialization."[46]

In early February, *Look* magazine published Castro's interview with Andrew St. George, a freelance photographer-reporter who had traveled through Cuba in 1957. A naturalized U.S. citizen, St. George was born in Hungary, joined the anti-German resistance during World War II, and worked from 1946 through 1952 for the U.S. Army Counter Intelligence Corps. After two years' study of comparative literature at Columbia University, he began to write popular magazine articles and serve as a special correspondent for the National Broadcasting Company.[47] In Cuba, St. George interviewed Che Guevara, among many others. Che tagged him an FBI or CIA agent, and, according to one source, St. George did in fact work for or with the CIA.[48]

Castro's *Look* exchange with St. George closely ressembled his *Coronet* message. He promised free elections and once again reassured the business community by distancing himself from nationalization. Business, in fact, should welcome the 26th of July Movement as a "boon—no more thieving tax collectors, no plundering army chieftains or bribe-hungry officials to bleed them white. Our revolution is as much a moral as a political one." When asked the familiar, annoying question about whether his movement was Communist-inspired, Castro responded that the Cuban Communists felt a closer kinship for Batista than for M-26-7. Castro directed his strongest words against U.S. policy, but he spoke in the language of traditional American ideals. Castro admonished Washington that "you kill the democratic spirit" in Latin America by supporting dictators like the Dominican Republic's Generalissimo Rafael Trujillo and Venezuela's Marcos Pérez Jiménez. "Do you think your tanks, your planes, the guns you Americans ship Batista in good faith are used in hemispheric defense?" asked Castro. No, Batista used U.S. weapons "to cow his own defenseless people."[49]

On the first anniversary of Herbert Matthews's scoop, the *New York Times* sent another distinguished correspondent to Cuba. Homer Bigart was no stranger to war or rebellion. He had already covered World War II, the civil war in Greece, the Korean War, and the beginnings of the civil rights movement in the United States. Irreverent and witty, the 50-year-old reporter had become famous among his colleagues for digging out the story.[50] In February 1958 Bigart and his cohorts easily evaded army patrols near Manzanillo and hiked through rough terrain to rebel headquarters. There they spent two weeks in the Sierra Maestra, where Bigart interviewed Fidel Castro. After his journey Bigart not only published several articles but also briefed the U.S. Embassy in considerable detail.[51] He found camp conditions primitive. Many rebel soldiers, dirty and sloppy in appearance, suffered from intestinal parasites. The food was monotonous—Bigart vowed never to eat another malanga, a starchy root. Castro told the reporter that he had 1000 rebel troops, and could order 400 of them into battle at one time.

Bigart observed that although a ten-person National Directorate supposedly governed M-26-7, "Castro spoke and acted as an absolute ruler, and appeared to be obeyed as such."[52] Castro admitted that the 26th of July Movement had not yet worked out a formal political and economic program, but he favored land reform and regulation of foreign investment to ensure Cuba a fair share of profits. The rebel excoriated the U.S.-owned public utilities and bitterly protested the flow of U.S. arms to Batista, which, he insisted, kept the dictator in power.

With the help of an interpreter, Bigart also held several conversations with Che Guevara. The frail, asthma-suffering rebel who held a doctorate in medicine from the University of Buenos Aires described himself as a leftist, not a Communist. Che said his strong anti-imperialist, anti-U.S. views stemmed from the 1954 CIA intervention in Guatemala that toppled the reformist Arbenz government. Che had been in Guatemala at that time. Bigart sensed uneasiness among some 26th of July leaders about Che's politics and about charges that the movement might be Communist-influenced.[53]

Castro rejected any solution to the Cuban crisis that included a military junta's coming to power, and he outlined a plan designed to take the political high road away from Batista and the United States: If government troops pulled out of Oriente Province and Batista granted a general amnesty, the 26th of July Movement would run a candidate (Manuel Urrutia Lleó) in the presidential election. The Organization of American States, however, must supervise the election. The OAS option actually did not originate with Castro but with Ambassador Smith, who had presented it to Batista on February 20.[54]

Castro also talked at length with the Spanish journalist Enrique Meneses, who tramped into the Sierra Maestra Mountains in December 1957 to report for *Paris Match,* which published his first article in March after it had been smuggled out of Cuba to Miami. To the Batista regime's dismay, *Bohemia* in Havana republished his work. Meneses, like all journalists who interviewed Castro, wondered whether Castro was "deliberately leading me astray." But "four months of living together, by day and by night, in peace and in war, talking to all of them, have convinced me that it would have been impossible to give a false impression of their characters and their ideas over such a long period without giving themselves away." As for the perennial question about links between Communism and M-26-7, Castro insisted that "I hate Soviet imperialism as much as Yankee imperialism! I'm not breaking my neck fighting one dictatorship to fall into the hands of another." Castro eagerly sought information from Meneses about Nasser's land reform program in Egypt, from where Meneses had once reported. Castro himself leaned toward a program that took uncultivated land owned by the government and distributed it to the *guajiros,* who could, if they wished, work the land collectively.[55]

The more Senator Mike Mansfield, a member of the Foreign Relations Committee, read in the press about Cuba, the more he wondered whether the United States was in fact neutral in the island conflict. At one hearing in early March, the Montana Democrat pestered Assistant Secretary Rubottom about the meaning of the Prío arrest and the continued shipment of arms to Batista. Mansfield captured the essence of the case many Cubans were making against the United States: "In other words, a government which came into power in Cuba by usurpation, and which maintains a military dictatorship, can buy arms in the United States, but a constitutionally elected President is put in jail for trying to assist in the overthrow of that Government?"

Rubottom tried to deflect any hint of preferential U.S. decisions. Mansfield would not let him ignore the query: "Do you think that is the kind of policy that is likely to help this country keep its reputation of devotion to freedom among the people of Latin America?" Rubottom weakly allowed that the problem was "difficult." Senator Wayne Morse of Oregon followed up, charging that U.S. military aid was keeping Batista in power. Rubottom retorted that Batista had been elected president in 1954. The Oregon senator shot back: "They have elections over there in Russia, too." During these heated exchanges, Rubottom also made one of the strongest U.S. public statements yet against Castro. If the Cuban rebel came to power, Rubottom predicted, he would simply impose another dictatorship on the island.[56] The next day, ignoring the last comment, Radio Rebelde broadcast the news of the Mansfield-Morse questioning of Rubottom on arms for Batista.[57]

Despite growing public criticism in the United States of their Cuba policy, U.S. officials had decided to oppose Castro and to ride out the rest of Batista's presidency. As their discussions and reports of late 1957 and early 1958 demonstrate, most U.S. officials, in Washington and Havana, agreed that Fidel Castro and his militant 26th of July Movement had to be prevented from taking power. Several worries pushed them to this anti-Castro judgment. Overall, the violent acts of the rebellion endangered U.S. economic and strategic interests. U.S.-owned property on the island was being destroyed and American lives jeopardized. Sugar production—so central to U.S. trade with Cuba—had become unsettled. Should he win, moreover, Castro intended to reduce the immense U.S. presence on the island by nationalizing key industries and distributing lands to poor farmers—restructuring Cuban society at the expense of North American interests. His ties with avowedly radical rebels such as his brother Raúl and Che Guevara raised doubts that he would ever respect U.S. interests. Fidel's headstrong personality and drive for power also caught the attention of U.S. officials. He stoutly refused to tolerate compromise or to fold up the rebellion in favor of a peaceful, electoral solution. Castro loomed too as a threat to U.S. hegemony in Latin America, for he endorsed revolutions against dictators, some of whom stood fast with the United States as allies. U.S. officials correctly read Castro's anti-imperialism as anti-Americanism.

Knowing full well that they could not control Castro, fearful that he would end Cuban if not Latin American dependency upon the United States, sensing an ill-formed yet radical danger arising from a volatile, bearded mountain rebel of folk-hero status, and influenced by anti-Batista, anti-Castro Cuban moderates who said they would sustain friendly relations with the United States in a post-Batista government, U.S. policymakers took their stand: Stop Castro.

To contain Castro, U.S. leaders had to have Batista's cooperation. It was denied them. Batista refused to create the conditions necessary for an orderly electoral process. He refused to follow U.S. instructions. He harbored his own brand of Cuban nationalism. Indeed, he too resented outsiders who told him how to run his nation's affairs. Batista probably even bugged U.S. Embassy telephones.[58] Although he coveted American arms, he chafed at any U.S. pressure that would diminish his power. Batista balked, and thus the plan put together in January in Washington soon unraveled.

III

Riding the Tiger to Defeat

11

Batista's Self-destruction and the Suspension of Arms

BATISTA'S ANSWER TO the spreading war was to get even tougher. He closed the public secondary schools. He discouraged a much-publicized effort by Roman Catholic bishops in Cuba to form a national unity government. This attempt at national harmony formally died when Castro also rejected it. Batista red-baited his many critics with undocumented charges that the insurrection followed Communist dictates. Stiff-arming the U.S. request to punish police and military officials known for excessive brutalities, the Cuban dictator appointed a new chief of national police and a new SIM Chief, whose missions appeared to be more crackdowns. So repressive was the environment that a judge had to flee Cuba after he handed down an indictment against a navy officer and police officer for their part in the torture and disappearance of a prisoner after the Cienfuegos uprising.

Looking for ways to influence Batista to moderate regime violence and thus to quiet the public furor that fed Castro's rebellion, and sensitive to criticism from members of Congress and newspaper editors about Batista's use of U.S.-supplied weapons, the State Department decided to enforce the long disregarded terms of the 1952 Mutual Defense Assistance Agreement. Arms from the United States would now have to be used for hemispheric defense. Cuba could, if it wished, request permission to use the weapons for internal security. MAP-trained infantrymen carrying MAP-grant equipment and operating in the Sierra Maestra against rebel forces became the test. Stewart and Wieland prepared the case for this turnaround, and they cleared the key State telegram of February 28, 1958, with Deputy Assistant Secretary of State William P. Snow, Under Secretary of State C. Douglas Dillon, and other officials in the State and Defense departments.[1] Dulles and Rubottom were apparently away at the time.

Ambassador Smith recognized the import of this change and warned Washington that if news of it became public, Batista might fall. State officials rejected Smith's appeal for delay. On March 3 they summoned the Cuban Ambassador and told him that the United States could not tolerate protracted use of *grant* assistance in violation of the agreement. (Arms *bought* by Cuba were not affected.) The U.S. position, remarked Ambassador Campa in grand understatement, "will certainly not be of any help to the Government of Cuba."[2]

Ambassador Smith had dutifully carried out the January plan to prepare Cuba for moving Batista out before Castro could come in. In his many meetings with President Batista and other high-level officials, Smith had offered U.S. military equipment while he pressed for decisions that would ensure free elections. Even as the outburst of death-dealing terrorism and government countermeasures staggered Cuba, Ambassador Smith remained optimistic. He believed Batista when the Cuban president and other officials asserted that M-26-7 and the Partido Socialista Popular were fighting hand-in-hand and when the regime repeatedly promised free elections and welcomed United Nations (rather than OAS) observers.[3] As late as March 8, after a meeting with Batista, Smith found the president "conciliatory."[4]

Other U.S. observers waxed far less sanguine than Smith. The legal attaché in the Havana embassy, for example, informed J. Edgar Hoover that Batista's "control" of Cuban politics was "weakening."[5] A report from the office of the Assistant Chief of Staff for Army Intelligence in Panama perceptively concluded that "the time may be passing when a fairly stable economy, popular political apathy, and opposition dissensions allowed him [Batista], with the help of the military, to gain time in which to effect a more or less orderly transition of power to a hand-picked successor." The report quoted Carlos Márquez Sterling, the Partido del Pueblo Libre presidential candidate who had broken away from the Ortodoxos. He explained why Batista could hold no enthusiasm for an election: "This is Batista's dilemma: if the election is honest, he will have to yield the Government to the political opposition; if it is a fraud, he will have to yield the Government to the rebels."[6] The Office of Naval Intelligence, moreover, reported that "there is no evidence of definite ties between the Communists and the current revolutionary movement led by Fidel Castro."[7]

Washington officials also doubted whether Batista would ever cooperate to create a fair electoral environment. He would not even declare a partial political amnesty, and arrests of critics continued. U.S. officials learned that many moderate oppositionists were certain that Batista would never permit honest elections.[8] Diplomatic officers in Washington, in other words, increasingly questioned Smith's judgment. Some probably shared the opinion of one embassy officer who thought that a "tough" career ambassador could have persuaded Batista to hold an honest election and then eased the way for the dictator's retirement in Florida.[9]

On March 12, Batista profoundly poisoned the political environment and his relationship with the United States. Once again he suspended constitutional guarantees (freedom of expression, right of assembly, freedom of movement) and imposed press censorship. He took this extreme step, sure to distress U.S. officials and inflame U.S. public opinion, because it seemed the only way to preserve his power. Muzzling the Cuban press became necessary to keep the critical reports of foreign journalists out of Cuban newspapers and magazines. Batista also suspended constitutional guarantees to protect his police and military officers; their murder and torture cases could now be transfered from civil to military courts. Batista too was taking precautions to counter Raúl Castro's opening of a second M-26-7 front in the northern part of Oriente on March 10. With his decision of the twelfth, Batista crushed prospects for national reconciliation and doomed the already flagging elections option.

Soon a number of cabinet members resigned, throwing the Cuban government into disarray. Fidel Castro issued a manifesto which declared "total war against the tyranny" and demanded that the "morally bankrupt" Batista regime step down or face unparalleled bloodshed. The "final blow" would be administered through a general strike backed by military action. Castro's March 12 missive also informed Army soldiers that if they surrendered to M-26-7, they would not be punished—but they must bring their weapons with them.[10] Adding to the regime's woes, the head of the pro-government Confederation of Cuban Workers warned Batista that members were now less controllable and that many favored a general political strike. Sensing trouble, sponsors of the St. Petersburg-to-Havana yacht race canceled the event.[11]

In mid-March the Committee of Cuban Institutions, composed of more than 40 national religious, fraternal, professional, civic, and cultural associations, demanded that Batista resign. Leaders of the National Association of Pharmacists, the Knights of Columbus, and the Cuba Council of Protestant Churches were among the many who endangered themselves by speaking out so frankly. Just before the petition became public, Ambassador Smith, watching his hopes for elections fade, had failed to dissuade Dr. Raúl de Velasco, president of the Medical Association, from endorsing the petition. The doctor was taken aback by Smith's failure to grasp the extremism of Batista's repression and by the ambassador's naïve faith in elections. "I have to defend the life and interests of United States business and the United States subjects here," Smith reminded the physician. "The life and interests of the Cuban subjects are also at stake," Velasco answered. "Nobody's life is worthwhile in Cuba today. There is no way in which you can convince the Cuban people that any [election] offer by Batista will be carried out."[12]

Batista had never before been so besieged from so many quarters. "A plague of stinging buzzing mosquitoes seemed to swarm about the dictator's head, none consequential in itself, but collectively infuriating, maddening, and—like mosquitoes—potentially capable of starting a fatal infection," remembered the journalist Robert Taber.[13]

As these events unfolded, Herbert Matthews went back to Cuba, although the censored newspapers were not permitted to print a word about his arrival on March 14. The *Times* writer began to take notes on horrid and heroic stories in Havana, including the torture by police of Esther Lina Milanés Dantín, a teacher and DR member. She ended up in a hospital, where the attending doctor took photographs of her battered body to record the evidence. An iron rod had been thrust into her vagina. The police also tortured a young man caught at the same time: Officers tore off an ear, smashed a foot, and crushed his testicles. When he would not talk, they kicked his pregnant wife in the stomach. Then he talked. SIM officers next entered the hospital to claim Ms. Milanés. Ten nuns defiantly formed a wall in front of her bed, refusing to let the intruders pass. After the SIM men left, doctors moved her by ambulance to the Argentine Embassy for protection. She filed charges in court against her torturers. The news spread across the city and abroad.[14]

As this and other accounts of Batista-regime brutalities swirled around Cuba, the Havana Hilton prepared to begin business. Wondering whether to open on schedule because of volatile island politics, Hilton executives had consulted U.S. officials, who advised that they proceed. The gala inaugural ceremony, heavily guarded, went forward on March 23 as planned.[15] For some, including hundreds of North American guests in the magnificent facility, there simply was no civil war. At most, it existed far away from the capital city. The opening of the Havana Hilton reflected another side of Cuba—a cautiously optimistic side apparent in the North American community, a side buoyed by the presumption that Cuba would ultimately survive its current political turmoil and that the anti-revolutionary United States would help it do so.

Despite rebel sabotage and political chaos, then, some business people could look forward to a good 1958. Sugar prices were up, as were sales of automobiles, trucks, and electrical appliances. Plans for the expansion of the nickel-cobalt operation at Moa Bay went forward, while the U.S.-owned

Manatí Sugar Company was building a new plant to convert bagasse into a fiber board for the production of furniture, doors, and other construction materials. Business-watchers predicted a ready and profitable market for the sugar byproduct throughout the Caribbean and Central America. At the same time, Allied Chemical purchased half-interest in two Cuban companies.[16] The First National City Bank of New York advised its customers that the overall trend of business in Cuba was "satisfactory," with the building trades and department stores particularly active.[17] "Considering the political situation," exclaimed another analyst of the Cuban economy, "it is amazing that business is as good as it is."[18]

Positive business news hardly lifted the pessimism that now permeated the State Department. Could Batista survive the chaos tormenting his country? Rubottom now thought not. The Batista regime, the Assistant Secretary telegrammed Smith, "has utterly failed to convince Cuban people and certainly U.S. public of its intention carry out free elections." While some groups, like M-26-7, altogether dismissed the promised elections as a sham, other oppositionists asked for a postponement of the elections until November. The State Department, having few other options, endorsed the later date to allow "tempers to cool." Looking ahead with trepidation, Rubottom asked the Havana embassy to give its best estimate of which group or individual would head the Cuban government if Batista were "removed from the scene."[19]

In the midst of the crisis, Ambassador Smith met with Batista on the evening of March 13. Busts of Abraham Lincoln surrounded them. During the insurrectionary period and later, Cubans across the political spectrum revealed their admiration for the sixteenth President of the United States. After Fidel Castro won power, for example, he created a Commission for the Revision and Modification of Historical Monuments. The USS *Maine* monument in Havana became its first assignment. Installing new plaques that disparaged the "voracity of imperialism" and celebrated Cuban independence, and replacing the eagle with a dove, Cuban authorities also took down the busts of President William McKinley and General Leonard Wood. A bust of Abraham Lincoln replaced them.[20]

During the rebellion against Batista, some touted Fidel Castro as a new Lincoln determined to lead his people out of bondage. Years after his victory, Castro was asked why he did not openly espouse radicalism before 1959. Drawing a lesson from the history of the U.S. Civil War, and revealing more certainty about his political bent than he actually had before his triumph over Batista, Castro claimed that President Lincoln had wanted to free the slaves from the beginning, but he had hesitated to take such a controversial step so as not to strengthen his opposition. "In our case," Castro explained in hindsight, "to have stated a radical program at that moment would have resulted in aligning against the Revolution all the most reactionary forces, which were then divided. It would have caused the formation of a solid front among the North American imperialists, Batista, and the ruling classes."[21] In March 1958, Batista's "solid front" was indeed cracking, including his U.S. support. Like Lincoln—at least the Lincoln in Castro's mind—the clever Movimiento 26 de Julio leader savored the disunity among his opponents.

Fulgencio Batista too admired Lincoln, but surely for different reasons—strength under fire, firm leadership, a willingness to use force against violent rebels? Vice President Richard M. Nixon's 1955 toast in Havana could not have pleased the Cuban president more: Batista and Lincoln shared a humble background and Batista had achieved the stature of the renowned American leader.[22]

"I could not help but notice that he showed no signs of wear and tear," Ambassador Smith observed of the Cuban president three years later.[23] Because Batista's primary response to accelerating civil war in March 1958 was to increase the size of the army, he asked Smith "whether he could expect delivery of arms already negotiated for with US." The ambassador told the Cuban president that U.S. policy had not changed: Cuba would receive its deliveries of weapons but had to observe their proper use.[24]

On the fourteenth of March, Smith, who continued to believe that Batista sincerely desired honest elections, informed Washington that Batista would survive the current crisis. Should he fail to hold power, any new government would face the "serious problem of controlling Fidel Castro." But Smith, in another misjudgment that revealed his deep distrust of M-26-7 and his shallow understanding of the polarization besetting Cuba, did not think that Castro, although an important figure, would dominate the nation's politics if Batista were removed. The ambassador expected that Batista's departure would sap the 26th of July of its cohesiveness and momentum.[25]

Just a little more than an hour after Smith despatched these conclusions to Washington, the State Department shocked the ambassador with a terse telegram: "Department has requested Customs to suspend export license for 1,950 M-1 rifles for Cuban Army." These Garand rifles, for which an export license had been granted, awaited loading on a ship docked in New York City. The March 14 telegram went on: The United States was suspending action on "all Cuban arms requests and shipments" (apparently both MAP grants and private sales) because of the Cuban government's failure to create favorable conditions for fair elections.[26]

The history of the decisionmaking for this critical shift in policy remains baffling. Secretary of State John Foster Dulles was in Asia from March 7 to 18 to attend a meeting of the Southeast Asia Treaty Organization in Manila and a meeting of the U.S. Far East Chiefs of Mission in Taiwan.[27] Christian A. Herter was Acting Secretary when the decision for suspension was made.

Assistant Secretary Rubottom opposed the aid suspension, but he too was away from the department. He had a speaking engagement in the American Southwest and took the opportunity to take a vacation with his family until late March. Although he had participated in the ongoing review of the question and opposed suspension because he thought it would deliver a blow to an already reeling regime, he was not consulted about the arms suspension decision. Driving along in Louisiana, Rubottom heard about it on CBS Radio. He stopped at the first opportunity, found a telephone, and called William P. Snow to confirm the news.[28]

William P. Snow, Rubottom's deputy, had handled the issue during Rubottom's absence. The Maine-born, 51-year-old career Foreign Service Officer, using drafts prepared by Officer in Charge of Cuban Affairs Terrance G. Leonhardy, despatched the key telegrams to Havana and communicated with Acting Secretary Herter. Wieland, who favored the cut-off, testified later that the decision for the full suspension did not originate in his Office of Middle American Affairs, although the temporary suspension of the shipment of rifles did.[29]

Was Snow authorized from the start to declare a *complete* suspension? His important telegram of March 14 to Smith mentions the suspension of the rifles shipment *and all other arms.* Yet Snow's informational memorandum of the same day to Herter, submitted *after* the stop order on the rifles had already gone out and the telegram to Havana had been sent, included not a word on a suspension of *all* arms.[30] Nor did Snow ask Herter for his approval of the rifles suspension. Did Herter know that Snow intended, in the telegram to Havana, to curb all arms, and that that was the message given to the Cuban government? Was it evident to all appropriate officials in Washington at the start that the suspension was comprehensive?

Not until March 26, long after Snow's initial telegram to Havana, did Secretary Dulles initial his approval of a March 24 recommendation from Snow that "further shipments of combat arms to Cuba not be authorized until Cuban conditions improve to the point where arms furnished will be dependably used for hemispheric defense and not used up [in] internal strife."[31] Did Snow and his associates make what turned out to be momentous decisions on their own about both the rifles and the full suspension *before* seeking approval from either the absent Assistant Secretary of State and the Secretary of State? Snow's March 24 memorandum for the secretary emphasized the "temporary" nature of the suspensions. Yet his March 14 telegram to Havana never used the word; on the contrary, it spoke of "long range US interests" that would be served by suspending "all arms requests and shipments."[32] Dulles got one message from Snow, Batista got another. Had Snow's telegram to Havana highlighted the temporary status of all suspensions, perhaps the "devastating psychological effect" on the Cuban government that Ambassador Smith observed would never have occurred.[33]

Cloudy as the decisionmaking appears, why the suspension became U.S. policy is clear. Cubans themselves forced the issue. Batista and the Cuban oppositionists left the United States little other choice. U.S. officials worred as much or more about *Cuban* public opinion as about opinion at home; they worried about "increased resentment" toward the United States from Cubans who associated Washington with Batista's "strong arm methods."[34] Cuban criticism of the U.S. military missions and arms shipments cut across parties and organizations. Almost every group championed an end to military aid for Batista. Even Cubans willing to participate in Batista-conducted elections, like the Auténticos' Grau, denounced the U.S. military linkage. "If this continues," a prominent Cuban attorney soberly remarked, "the United States will have but one friend in Cuba—Fulgenico Batista."[35] The shooting opposition, moreover, was doing everything it could to sabotage U.S. plans to shore up Batista.

The Cuban president himself had jettisoned the January agreement with the United States by issuing his March 12 decree. Since he had basically made honest elections impossible, the U.S. carrot in the bargain—arms—had to be pulled back. U.S. officials in Washington had believed in Batista, had made a deal with him, had placed their hopes in the fair elections he repeatedly promised. They felt jilted and betrayed. Smith put it too strongly in his memoirs, but he nonetheless captured Washington's sullen mood when he wrote that after March 12 he could no longer "engender any support in the State Department for the Batista government."[36] Washington officials had suffered a "severe setback," another official observed.[37]

This was not simply a matter of saving face. Department of State policymakers hoped that Batista would bow to U.S. pressure and create a favorable electoral environment. They wanted to block a Castro triumph achieved through arms. U.S. officials still clung to the elections option. Although this may seem rather remarkable given the inhospitable conditions in Cuba, it is explicable: No group other than Castro's M-26-7 had yet positioned itself to take power should Batista, an ally of the United States, leave office precipitously. The United States could find no satisfactory third alternative to Batista and Castro at the time. Elections seemed the only course available at the time.

There was also a compelling reason for acting immediately. M-26-7 leaders in the United States had learned about the shipment of the 1,950 M-1 rifles, which were obviously not going to be used for hemispheric defense, the longtime fictional rationale for aid to Batista. They even knew the serial numbers of the weapons, and they intended to release the information. On March 14, pickets showed up to protest at the New York pier where the rifles awaited loading.[38] The 26th of July Movement received the details of arms transactions with Batista from a sympathetic informant in the Cuban Embassy in Washington. Angel Saavedra, a uniformed secretary to the military attaché, spied for the movement and passed intelligence data to Ernesto Betancourt. Smith soon warned Batista of the infiltration and suggested that the entire staff of the Cuban Embassy be replaced to rid it of Castro's agents.[39] But Smith's intervention came too late.

Although they knew that their decisions would unsettle the regime, U.S. officials did not believe that the suspensions would cause Batista's imminent collapse or deny him the capacity to fight. The decisions were supposed to be kept quiet, out of the public eye, and they were expected to be temporary—enforced only until Batista mended his ways. From the beginning, Department of State officials labored to assure the Cuban government that the suspension did not represent an overall change in U.S. policy. They also signaled Cuba that the suspension did not constitute recognition of a state of belligerency, a decision which conceivably would have activated international law to the extent that the United States would have had to demonstrate true neutrality and stand back altogether from Batista.[40]

Ambassador Smith, of course, who on March 15 informed the Cuban government of the rifles suspension, endorsed the Batista regime's protest against the stoppage of the M-1 rifles shipment and added that "continued suspension of arms shipments" would "encourage the rebels" and "weaken the Cuban Government and possibly lead to its downfall, probably with attendant violence and risk to American lives and property."[41] Smith explained to Foreign Minister Gonzalo Güell not that the United States intended by its decision to push Batista toward reform, but rather that the United States had been put in an "embarrassing position" by "overwhelming criticism from Congress and press."[42] Smith misled the Cuban government, for U.S. officials had hoped that Batista would retreat from the excesses of his rule.

Although keenly sensitive to the consequences of his decisions, William Snow obviously did not believe that they imperiled Batista. Batista's own intransigence threatened the regime. The "temporary" suspension, Snow calculated, would finally prod the Cuban president to reform his nation's politics.[43] Snow's prognostication proved wrong. If anything, political life for Cubans became more uncertain and more hazardous. On March 20, the government postponed elections until November 3, and soon the police were hunting down the 44 members of the Bar Association who had dared sign a letter demanding that Batista give way to a unity government. The president of the Havana Bar Association hid in a church before finding his way into exile in the United States.[44]

Eager to keep the suspension under wraps to prevent rebel exploitation, the Department of State did not issue a press release on any aspect of the arms issue.[45] If asked, State officers planned to explain that a shipment of rifles was being held up temporarily according to the standard U.S. practice of studying arms deliveries to any tense foreign setting.[46] When Manuel Urrutia visited the State Department on March 17 to protest U.S. arms shipments to Batista and the continued presence of the U.S. military missions in Cuba, Wieland and Stewart never uttered a word about the suspension.[47]

Not until March 27, and only after being asked and then playing down the story, did State Department officials explain publicly that a shipment of rifles was being suspended to permit an opportunity to consult further with the Cuban government. No hint of a total ban appeared in the report in the *New York Times,* on page nine. The page-one story on Cuba was actually the seizure of *El Orión* by the U.S. Coast Guard. The 60-ton vessel, five miles at sea from Brownsville, Texas, cradled a substantial cargo of weapons and ammunition worth $20,000 and 35 Cubans dressed in M-26-7 fatigues. Their expedition to Cuba foiled, the arrested insurrectionists soon staged a hunger strike that, unlike the U.S. suspension of arms to Batista, captured headlines. For weeks after this episode anti-Americanism swelled among Cuban exiles at rallies where they berated U.S. authorities for persecuting the anti-Batista movement in the United States. Most of the Brownsville gunrunners were ultimately found guilty; they received suspended sentences and three-year probations.[48]

About the only news that cheered U.S. officials was the Partido Socialista Popular's announcement that it endorsed Castro's call for all-out war. Fearful of being left behind as the insurrection advanced, the PSP had belatedly decided to take a step toward the rebel movement it had once denounced. The State Department determined to publicize this apparent linkage between the Communists and 26th of July in order to tarnish Castro. "Please handle fast," Rubottom instructed his staff.[49]

On March 31, Representative Charles O. Porter of Oregon issued a brief statement on the arms suspension that the State Department had given him.[50] Still, at John Foster Dulles's April 1 press conference, reporters spared the Secretary of State any questions on the subject of the blocked arms. Seldom in Cuban-American relations or in U.S. foreign-relations history has such a pivotal decision been camouflaged for so long.

State Department officials could also believe that Batista would not fall fast from his palace because he was getting arms elsewhere—from Trujillo's Dominican Republic and Somoza's Nicaragua. The Nicaraguan dictator sold to Cuba 30 British Staghound armored cars equipped with 37 mm guns that he had bought from Israel.[51] Thus Batista did not face a shortage of weapons; he had enough guns and bullets to fight his enemies—if his army used them well. Wieland even hinted that the suspension would hurt the Cuban revolutionaries, into whose eager hands so many U.S. weapons had fallen.[52]

The U.S. tactic with Batista was common enough: Explain to angry foreign government leaders that they had to initiate reform policies to satisfy American public opinion, which in a democracy ruled supreme. If the foreign government would only shift course, went this familiar litany, the opinion would become more understanding and tolerant, thus permitting policymakers to set things right again. *After March 14,* State Department officials probably even welcomed some public criticism to buttress their message to Batista.[53] Batista would not bend.

For weeks, as U.S. public pronouncements of neutrality rang hollow, critics in Congress and the media had argued that arms deliveries to Batista signaled that the United States was taking sides in the Cuban civil war by blatantly propping up a dictator. But the pressure on the State Department should not be exaggerated; it was hardly immense. No Gallup poll sounded public opinion on the issue and most Americans remained ignorant about Cuba's troubles. Congress never voted a resolution to halt arms to Batista. Nor did Congress trim Cuba's fiscal year 1959 allocation in the Mutual Security Act of 1958; in fact, Cuba received an increase. Although President Eisenhower had been briefed on the suspension before his press conference of April 9, not one journalist asked him to comment on any aspect of the Cuban crisis, then entering the period of the general strike. Recession, Defense Department reorganization, and atomic tests were the topics the President handled that day.[54]

Some members of Congress kept Cuban issues alive, if not on the front pages. Democrats controlled both the Senate and the House of Representatives. On the Senate Foreign Relations Committee sat liberals such as Mike Mansfield of Montana, Hubert H. Humphrey of Minnesota, and the always outspoken Wayne Morse of Oregon, who chaired the subcommittee on Latin America. Although no liberal on domestic issues, J. William Fulbright of Arkansas joined the others in doubting the Eisenhower Administration's wisdom in sending U.S. aid to dictatorships in Latin America. As a committee member, Republican George D. Aiken of Vermont also raised questions.

Echoing calls from Latin Americans themselves, many of these members of Congress pressed for aid for economic development rather than for military establishments. They argued that the best method for stemming radicalism and Communism throughout the Third World was not army repression but economic growth. Paraguay's Alfredo Stroessner, Nicaragua's Somoza, the Dominican Republic's Trujillo, *and* Cuba's Batista were but a few of the Latin American tyrants whose iron fists rallied rather than crushed radicals.[55]

In January 1958 a popular movement had overthrown the Venezuelan dictator Marcos Pérez Jiménez. His ouster and controversial resettlement in the United States had brought the issue of arms and dictators to the forefront. Later, after Vice President Richard Nixon's riot-wracked trip to South America in May, and after the fall congressional elections, which the Republicans lost badly, criticism of U.S. links with unpopular dictators heightened.

One member of Congress whom anti-Batista Cubans found sympathetic to their cries was Representative Adam Clayton Powell. The African-American Baptist minister from New York City had long championed the underdog, civil rights, and Third World nationalism. But he was hardly the ideal ally, for his reputation included an ostentatious life-style of dapper clothes and beautiful women—and expensive overseas junkets where he displayed both. Through a member of a rebel club in New York City, Mario Llerena and Manuel Urrutia made contact with Powell. In a meeting at his Harlem church, Powell promised that he would raise his voice against arms for Batista. So did Representative Porter of Oregon, who had been condemning military aid for several months.[56]

Apparently unaware that the M-1 rifles were being held back, Powell took the House floor on March 20 to denounce U.S.-Cuba military ties. Using classified documents from Llerena, who had received them from Saavedra in the Cuban Embassy, Powell read lists of weapons the Cuban military had secured under contract in the United States over the preceding two years—grenades, rifles, communications equipment, and more. He demanded an end to future cargoes and a withdrawal of the military missions. Porter followed in a few days with his own chastisement of U.S. policy.[57] Senator Morse soon pounded Rubottom once again about U.S. assistance to "totalitarian" governments in general and the proposed increase in Mutual Security aid for Cuba in particular. Eager himself to win the Cold War, Morse boomed: "You just feed Russian propaganda when you pour money into a vicious dictator state such as Batista's state."[56]

Foreign Minister Güell would not hear any of it; he blamed Cuba's troubles on "Communistic" rebels who would not cease their violence until they had become the government of Cuba. Batista vowed to hang on to power and to launch a military offensive against rebel forces in the mountains. Then he could deal with political questions from a position of strength. "We must weaken Castro to make him play ball," declared Güell. A sympathetic Ambassador Smith reported the regime's message to Washington, and added that "it is relatively easy to change dictators but very hard to get rid of dictatorship." The people of Cuba were "riding the tiger," and "exchanging tigers is no solution."[59] Smith would stick with Batista. He preferred the "dictator Batista" to the "would-be dictator Fidel Castro."[60]

Smith began a campaign to roll back the suspension. He found allies in the U.S. military. When the State Department asked Smith in early April to provide information on the activities of a MAP-trained unit to determine if the army company might be in violation of the U.S.-Cuba agreement, Smith quickly replied that the Chief of the U.S. Army Mission believed "if internal security is a part of hemisphere defense, use of MAP trained units and equipment against rebel forces would not be improper." On the other hand, "if internal security was not part of hemisphere defense, use of such assistance would be improper without prior US consent." The MAAG chiefs questioned State Department policy, arguing that the Defense Department had revised MAP instructions to give "high priority" to internal security within the hemisphere defense program.[61]

The Joint Chiefs of Staff also filed a protest against the arms suspension. Chief of Naval Operations Admiral Arleigh Burke, speaking for the JCS, criticized the State Department for suspending arms without consulting the military brass. This unfriendly act undercut Cuba's ability to maintain internal security. The decision had to be reversed. The suspension, moreover, violated national policy, because it encouraged Cuba to buy weapons from sources other than the United States and might prompt other Latin American nations to procure their military equipment outside the United States.[62] U.S. policy had long sought to exclude European weapons sales from the hemisphere, to rivet Latin American militaries to U.S. equipment and thus to U.S. influence.[63]

Pressure from Ambassador Smith and the military, and Rubottom's return to the department, actually began to open the door for Cuba again. In the first week of April 1958, the Department of State began approving requests for "non-combat equipment." Department officers reviewed a list of export licenses already validated for arms shipments to Cuba and another list of pending export license applications. The officer who supervised the earlier arms suspension, Deputy Assistant Secretary Snow, could not find many "non-combat" items on the lists, which included the Garand rifles suspended on March 14, as well as pistols, helmets, machine gun tripods, tank spare parts, shells, and aircraft engines. Now willing to send something to Batista, Snow speculated that depth sounders and helmets might qualify. "I think we should consider releasing a selection of these items without much delay," he told Wieland. In mid-April, in what must have been some stretching of definition, the State Department granted export licenses for two C-47 airplane engines purchased by the Cuban Air Force.[64]

Yet, alert to controversy, Snow's boss Rubottom persuaded Cuba's new Ambassador, Nicolás Arroyo, to withdraw an order for "unarmed" light naval craft in the sub-chaser category. The definition of "non-combat" could only be pushed so far, after all. The Assistant Secretary at the same time expressed his hope that improved conditions in Cuba would permit the United States to reconsider the question of arms shipments. It is telling that Rubottom did not press Cuba for political reform as a quid pro quo, perhaps because in April Cuba had canceled orders for combat weapons.[65] By early May, the Department of State had eased its arms policy even more: Sales of combat equipment would be approved "selectively."[66]

Oppositionists—including Fidel Castro, noting that the U.S. military missions still functioned and that U.S. military equipment still flowed to Batista—thought the United States duplicitous. Was this neutrality? they asked. Although U.S. officials denied the charge, anti-Batista leaders also protested that the United States conspired to help Batista purchase arms in Nicaragua and the Dominican Republic.[67]

Particularly searing to a wide spectrum of Batista's opponents was another showy U.S. affair, this time to honor General Francisco Tabernilla. Criticism had thundered upon U.S. officials in November 1957 when his son Carlos received the U.S. Legion of Merit. Now, on March 28, 1958, officers of the U.S. military missions sponsored a testimonial luncheon to celebrate the elderly father's promotion to Chief of the General Staff of the Cuban Armed Forces and General-in-Chief with five stars. After the much-publicized event, Rubottom advised Ambassador Smith to avoid giving "undue warmth" to Cuban military figures in such "touchy" times. "The opposition groups in the United States have a very effective publicity machine at work," Rubottom complained, and members of Congress were listening.[68] Smith gave little ground, defending the gathering as standard MAAG practice. He consented, however, to instruct his military mission chiefs to clear with him all future functions "involving publicity."[69] Cuban critics of the regime fumed. The Tabernilla luncheon, Manuel Antonio Varona sternly informed Senator Fulbright, boosted "a despised, crumbling military dictatorship."[70]

During the March crises, Ambassador Smith never forgot that one of his primary assignments was the protection of threatened U.S. interests. At the end of the month he sent embassy civilian and military officers to Oriente Province for an inspection of the Nicaro and Moa Bay operations. "Let revolutionaries as well as GOC [Government of Cuba] see US watchful concern for American business interests," he telegrammed Washington.[71]

12

Terrible Mood: Castro and the General Strike

Anxieties quickened in Cuba in late March and early April.[1] Everybody braced for a showdown. Greater numbers of Latin American, European, and U.S. reporters tramped around the island, witnessing police and rebel violence. Ruling by emergency decree, Batista prepared his forces to meet a rebel-planned general strike. His gunmen chased down dissidents of all kinds, often killing M-26-7 suspects on the spot. Revolutionaries replied in kind. DR terrorists assassinated police and *batistianos.* Rebel units also ravaged Oriente Province, skirmishing with Cuban forces, sabotaging, halting traffic. The alarmed U.S. Embassy activated the early stages of an evacuation plan for U.S. citizens. Stickers with American flags were readied for distribution to Americans to post in their house windows, and $25,000 was squirreled away in the embassy safe should endangered Americans require immediate relief.[2]

In those frantic times, Castro entered one of his irritable, imperious moods. "This is a complete fuck-up," he complained to Che Guevara. The M-26-7 *jefe*'s stock of ammunition was low because precious supplies had not been distributed properly.[3] Sam Halper of *Time* magazine, visiting Castro in the Sierra Maestra, bore witness to the rebel arms shortages. This reporter, contrary to later criticism that the U.S. press was invariably fawning toward Castro, thought Castro acted like an "autocrat" among his *barbudos* and had an unquenchable thirst for power."[4] Mario Llerena, soon to feel the 26th of July leader's wrath, also observed Castro's "ever expanding ego."[5] By late April Castro was writing Celia Sánchez about disorganization in the movement: "It's enough to drive me mad." He went on: "It is obvious we are overwhelmed with problems, and the work is exhausting."[6]

Batista on the other hand seemed confident. On Easter Sunday, April 6, he entertained several American journalists at his Kuquine estate. Delivered by limousine, Roy Brennan of the *Chicago Sun-Times,* Milt Sosin of the *Miami News,* and others eyed Batista's prominently displayed medals. A bust of Abraham Lincoln stood close to a copy of the Emancipation Proclamation. "I am a great admirer of Mr. Lincoln," Batista allowed, "and I think that we stand for the same things." But what about the revolution, the reporters wanted to know. The revolution "has been crushed," he answered. "There never really was a revolution, anyway; just a lot of fuss by a bunch of communists." What about Fidel Castro? "We must kill him. . . . He cannot be allowed to live," Batista replied.[7]

With Batista determined to smash his opponents, the nerve-wracking problem for Cubans across the island became safety and survival. In Santiago, General Alberto Del Río Chaviano summoned civic, business, and labor leaders to warn them that he intended to meet revolutionary activities with "extreme repression." The general and SIM also foiled a coup plot by junior army officers.[8] Besides suspending constitutional guarantees and imposing press censorship, the Batista regime issued emergency decrees which sapped the judicial system of any semblance of independence, required dismissal of workers who participated in any general strike, and ordered the imprisonment of employers who closed their business doors during a strike. Increasing numbers of Cubans tried to flee the country. A State Department intelligence report measured the accelerating political polarization: "Moderate middle elements have either withdrawn from political activity or aligned themselves with the opposition, and [a] showdown involving violence and bloodshed appears inevitable." The only "possible third alternative" to Batista and Castro seemed to be a military-civilian junta.[9]

With a general strike approaching and bombs detonating every day, North Americans joined Cubans in worrying about their lives and property. Tourists began leaving Havana, and hotel reservations plummeted. Jules Dubois of the *Chicago Tribune,* cut off by Cuban censors when he tried to file stories, located tourists heading back to the United States and persuaded them to carry his copy out of Cuba. Branches of New York's First National City Bank and Chase Manhattan suffered a run on their resources from depositors. Cuban business slowed considerably and the sabotage of properties increased. Telephone poles were sawed off, electric cables cut, and Texaco gasoline tanks torched. Alarmed about potential violence to their baseball players, International League officials asked Ambassador Smith if they should go ahead with plans for the season opener in Havana in mid-April. It was their call, answered Smith. The president of the Texas-based Zapata Off-Shore Company, George Bush, contacted the State Department because he feared one of his oil drilling rigs near Cay Sal Bank might be molested.

North Americans in Oriente Province became particularly edgy. Baptists began to evacuate their missions as rebels moved freely through the region. U.S. companies in the province—Texaco, Moa Bay Mining, and Cuban Nickel, among them—evacuated employees' dependents. A tanker lay off the coast of Santiago de Cuba to evacuate Texaco employees should such drastic action become necessary. Rumors spread that the rebels were targeting the Nicaro facility because it was U.S.-owned. Plant officials installed a two-way radio transmitter to connect with the Guantánamo Naval Base and the Santiago consulate. North Americans living in the vicinity of Guantánamo City moved their automobiles for safekeeping to the naval reservation. Back from his trip to Cuba, Matthews told a State Department officer that greater hatred for North Americans and greater bloodshed for Cuban youth lay ahead.[10]

In Santiago, a couple of days before the strike, Roy Brennan, Robert Taber, and other American reporters sought out Andrew St. George, the freelance photographer-journalist who had been in the mountains with Fidel Castro and was now on assignment for *Life* magazine. As they talked with the adventurer in a restaurant, a boy sympathetic to the rebel cause rushed in to tell St. George that the Casa Grande, the hotel where the American journalists were staying, was swarming with armed men. St. George hurriedly led his colleagues out the restaurant's back door. While St. George went into hiding in the city, Brennan and others crept back in the darkness to the hotel to retrieve their typewriters and luggage. Soldiers wielding Tommy guns promptly arrested them and jailed them at Fort Moncada. Guards "festooned with hand grenades like pineapples on trees" glowered at the prisoners.[11] Meanwhile, St. George notified the American Consulate. U.S. officials intervened to gain the release of the journalists, who then flew back to Havana in time to watch the general strike unfold.[12]

In the United States itself, rebel clubs cheered their compatriots. In New York, in March, Mario Llerena released the first issue of *Sierra Maestra.* On March 29, members of one of the M-26-7 organizations in Chicago marched in front of the Cuban Consulate in Chicago and handed out leaflets denouncing U.S. military aid to Cuba. Two days later, the rival Club Patriótico 26 de Julio de Chicago demonstrated in front of the Federal Court House and staged a hunger strike. On the eve of the general strike, 250 people packed the grand ballroom of the Belvedere Hotel on West 48th Street in New York. Raúl Chibás, Manuel Urrutia, and others spoke vigorously from the dais under a 8' × 15' portrait of Fidel Castro (smaller reproductions sold for $1.00

each). They asked for money, pled for unity among oppositionists, and denounced the sale of U.S. arms to the Batista regime. An FBI special agent attended the gathering and dutifully reported the event to his superiors.[13]

As the destruction of property in Cuba mounted through early 1958, few North American business executives spoke publicly about the rebellion or the forthcoming strike. They did not want to cross the rebels. Some optimistically downplayed the anti-Americanism of the rebellion. One hotel manager remarked that foreign business should always expect trouble, from inflation to revolutions to hurricanes. "These things are just part of the risk of doing business in Latin America."[14] In contrast, a very frank chairman of the board of Warner-Lambert Pharmaceuticals, Elmer Bobst, defended Batista, blasted Castro, and berated the U.S. press for exaggerated stories. Only "teenagers, criminals, and Communist hirelings" followed Castro. As for regime-perpetrated violence, Bobst snapped: "I understand policemen in the U.S. are not too gentle with cop killers, either."[15]

The Batista regime was certainly not gentle on April 9 when the general strike that Fidel Castro had promised in his March manifesto finally began. Yet another setback in a series of disasters for the steely insurrectionist—the Moncada attack, prison, arrest in Mexico, the *Granma* voyage and landing, bare survival in the Sierra Maestra, severe disunity in the opposition, shortages of military equipment, and a number of near misses had all gone before—the much ballyhooed general strike fizzled not so much because Batista crushed it but because the 26th of July Movement flubbed it.

In the early March planning stages leading to his manifesto, Castro seemed to believe that "the plains" (primarily the urban wing of the movement) might advance the revolution that "the mountains" had been sustaining. But he did not trust followers outside the mountains; he feared them as rivals and disliked them as moderates, and he could not control them. It seems that he harbored doubts about whether a strike could succeed. "If he [Batista] succeeds in crushing the strike, nothing would be resolved," Castro wrote in a pre-strike letter. "We would continue to struggle, and within six months his situation would be worse."[16] But there were grounds for rebel optimism. Much was going the rebels' way: the U.S. arms suspension and the widespread public disaffection from the regime evident in the petition of the Civic Institutions.

Still, even M-26-7 insurrectionists in Havana hesitated at first to endorse a general strike. Chief of Militias René Ramos Latour warned Castro that except for Oriente the movement suffered weaknesses. He predicted heavy casualties in the cities. As always, weapons shortages plagued the rebels. The capture of the *Orión,* the seizure by U.S. authorities of an arms supply in a Miami hotel on March 31, and a National Police raid on a rebel storage site near Havana that netted two truckloads of weapons had pinched the movement.[17] A C-46 transport plane full of machine guns, automatic rifles, and bazookas flew from Costa Rica to Mexico and then to the Sierra in late March, promising some relief.

Havana became the primary target of the strike. Success in the capital city, it was assumed, would ignite the rest of the island against Batista and permit Castro's mountain troops to break out and meet head-on Batista's dispirited and confused forces. The Civic Resistance Movement, students, workers, and others were supposed to be organized and armed by hardcore 26th of July members who readied more than 40 houses in Havana for meetings and the storage of weapons. At prearranged sites, Molotov cocktails would be distributed to the youthful insurrectionists. Rebel violence throughout the city would frighten workers and they would stay home from their jobs. So it was anticipated.

The strike had originally been set for March 31, but not enough arms had been accumulated in the cities by that date. M-26-7 activists in Havana grew resentful, grumbling that Castro was probably hoarding guns for his own mountain troops. There was something to this complaint. In a post-strike letter, Castro mentioned that there had been a tug-of-war over the arms delivered from Mexico, which he believed should remain in his war zone.[18]

Some conspirators wondered if the strike ought not to be shelved, but Castro's designated strike leader Faustino Pérez insisted that it begin on April 9. The evening before, representatives of several opposition groups met with M-26-7 conspirators to plot the next day's attacks. Members of DR, Prío's Auténtico party, Triple A, and other organizations bristled at what the 26th of July Movement plotters ordered them to do, for weapons had still not arrived, and Molotov cocktails and demolition charges essential to disabling Batista's armored cars and tanks had not been adequately stocked. "One of my jobs is to blow up this tunnel [under the Río Almendares in Havana]," remarked the engineer and 26th of July activist Augustin Capó. "But I haven't got the explosives I need yet. What can I do with only 32 lbs of dynamite? And the general strike's tomorrow!"[19]

Other handicaps worried the urban warriors. Word about the strike was already on the streets and no contingency plan had been drafted in case the strike failed. The Auténticos and Triple A denounced the operation, refusing to participate. The evidence is conflicting on whether the DR pledged support. M-26-7 activists kept the Communists of the PSP at arm's length, thus reducing even more the number of fighters that could be hurled against the regime. The PSP underground publication noted on the eve of the strike that "the forces of disunity remain present."[20] Once again a divided opposition worked to Batista's favor.

Poised for a noon start, M-26-7 commandoes were surprised at 10:00 a.m. to hear the crackle over their clandestine radios: The strike would begin at 11:00. Pérez had inexplicably changed the schedule. SIM and the police seemed ready for any hour. Havana's banks were closed and stores were nearly empty. As revolutionaries appeared suddenly in building windows, on rooftops, and at intersections, regime forces gunned them down. Few Cubans could escape police traps. Reporters listened to chilling exchanges on National Police radio. "We have a man who says he is a lawyer. He has a gun in the glove compartment of his car and a permit to carry it," came one call into headquarters. "Kill him!" came the response. "We have ten prisoners," went another. "No prisoners. Kill them," answered the despatcher.[21] Casualties climbed into the hundreds.

M-26-7 militias stimulated panic but not public support. Cautioned by the pro-Batista leaders of the Cuban Confederation of Workers to stay at their jobs, workers were too scared to leave their workplaces for Havana's unsafe streets, now alive with gunfire and strewn with disabled vehicles. A bomb blast put the electric company out of operation, but damage to businesses was surprisingly light. In other Cuban cities the militias failed to destroy key facilities or unleash a popular uprising. Property damage was greatest in Santiago.

The Communists, perhaps welcoming catastrophe for their political rivals, may have cooperated with the police to identify M-26-7 militants. "Well, you didn't see any Communists get killed yesterday," remarked one PSPer. "We keep our heads."[22] "Those bloody Communists!" cried a young rebel commando whose assignment had been to seize the studios of CMQ-TV, proclaim the strike before the cameras, and then blow up the building, "They stopped us doing it," and soon the police stormed in, "probably tipped off by the Communists. It's a miracle we're still alive."[23]

Before April 9 was out, the regime had triumphed over the insurgents. *Time* magazine reflected widespread opinion in the United States: The rebels had been hurt, but Batista had escaped "with hardly a bruise."[24] Indeed, "I expected a Waterloo but what happened was a Dunkirk," said an elated Batista.[25]

As for the anticipated break-out attacks from the Castroite forces in Oriente Province, they never took place. Had Castro set up the urban revolutionaries for disaster in order to stymie political rivals? Was he cool toward the strike project? Castro certainly ranked urban underground insurrectionists as rivals, and he feared that they would cut a deal with disaffected military officers to create a junta that would elbow him aside. Castro had maneuvered to defeat his rivals before, and Batista had often assisted him by killing them. Was the disaster of the general strike calculated as a similar case?[26]

Although the indictment is plausible, would Castro take the risk of losing thousands of urban loyalists who had been sending him valuable money and arms? Could he have known just how poorly Pérez and his compatriots had planned the strike, and how ill-armed they were? Was he so sure of himself and his mountain forces that he was willing to sacrifice his allies and supplies to Batista's marauders? A hint is there, but the evidence is not, and Cubans to this day tiptoe around the delicate subject. In any case, after the foiled strike, with many urban political moderates discredited or dead, Fidel Castro and his guerrillas gained advantage among the revolutionaries.[27]

Castro told his secretary and companion Celia Sánchez that the revolution had suffered a "moral rout." Now, as never before, salvation "rests in our hands." He said he would never trust "the organization" again. By that he meant the urban-based National Directorate of the M-26-7. He had to bear responsibility for the "stupidity of others" at this moment, he complained, but "in the future we ourselves [the mountain guerrillas] will resolve our own problems." Castro also chided rivals who accused him of "caudilloism."[28]

He denounced Prío for "living in luxury in Miami" while the rebels ate roots.[29] To Mario Llerena and Raúl Chibás in the United States he sent a blistering letter denouncing them and other M-26-7 exiles for failing to send enough supplies. He itemized their sins: egotism, trickery, incompetence, negligence, and disloyalty. In his familiar hyperbolic language, Castro charged that he had not received so much as a bullet from abroad (he had earlier lectured the urban rebels in Cuba that they too had not even sent one weapon).[30] Castro also complained about being short of medicine—especially for the severe toothache that tormented him.[31]

Rebel leaders knew that they themselves had failed to plan and execute carefully, but some also blamed the United States for their troubles: U.S. authorities on the eve of the strike had interrupted arms destined for the insurrectionists. Had the United States not backed the regime through arms supplies and the military missions, moreover, Batista could not have blunted the effort so handily. Castro sneered that the United States, the supplier of weapons to dictators, had canceled the sale of arms to Batista, but little had actually changed. "The United States sells arms to Somoza and Trujillo. Somoza and Trujillo sell them to Batista."[32] He told the Argentine journalist Jorge Ricardo Massetti that U.S. arms "massacred the Cuban people" and kept Batista in power.[33]

From the beginning of the strike, Ambassador Smith kept in touch by telephone with Washington. On the morning of April 9, Smith reported that Havana was quiet. "The war," he snidely remarked to Wieland, "is being fought in the United States."[34] Shortly after noon, Smith called Wieland to report some explosions and shootings, but little else. In the late afternoon they talked again. Smith could report good news: busses were running, stores were open, banks were operating, telephones were working. The power outage darkened only Old Havana and Vedado.

But stories like that of Neal Wilkinson could not have helped the already troubled hotel business. A freelance writer/correspondent from New York, Wilkinson was sitting at a terrace cafe not far from his hotel close to the Presidential Palace. A gasoline bomb, hurled from a speeding car, exploded. Batista guards stormed the cafe. Although Wilkinson shouted "*Americano!*" they whipped him and other victims. He eventually got to the U.S. Embassy; after a doctor treated his wounds, embassy staff whisked the correspondent out of the country.[35]

The International League fared better than Wilkinson. League authorities had scheduled their baseball season to open in Havana on April 16. But because of the unsettled political conditions surrounding the general strike, the Buffalo Bisons team balked. League headquarters warned the club that it would have to forfeit games if it did not show. Smith intervened to assure all concerned that the violence was in faraway Oriente Province. "If children can play in the streets of the city," the Ambassador wondered, "why can't grown up men play in the stadium?"[36] Nonetheless, special precautions were taken for the opening on April 17, when 10,000 fans poured into the stadium. Two military detectives sat on the Bisons' bench, while police escorts provided protection for the players as they moved to and from the ballpark. With a three-run rally in the ninth inning, Havana beat Buffalo 6–5 that night. This game and three others were played without political incident.[37]

After the strike, U.S. Embassy officials analyzed the reasons why Batista had won the day. Apparently using information from rebel contacts, they suggested that a military coup had been planned to coincide with the strike, but, given the brevity of the strike turmoil, the young conspiring officers never moved. Although embassy officers seemed surprised by the opposition's ineptitude, they measured well the factionalism in the anti-Batista movement. U.S. officials took some pains to discount the regime's assertion that Communists and insurrectionists were one and the same. Quite the contrary, the 26th of July Movement and other revolutionaries had consistently rebuffed Communist overtures for alliance.[38] At the National Security Council meeting of April 14, CIA Director Allen Dulles agreed. Dulles told President Eisenhower, moreover, that Batista had re-established order after the "fizzled out" strike. "Castro will now have to start on a new tack."[39]

While Ambassador Smith and the U.S. military continued to press for a resumption of arms shipments to Cuba, U.S. journalists and U.S. officials alike speculated over whether Castro should be counted out after the strike debacle.[40] Impressed by the failure of the strike, some officials identified a new phase in the rebellion, a new opportunity to reach a political solution. New that the rebels had proven that they could not capture the cities, Smith too thought the time ripe for moving again toward a "peaceful solution."[41] The solution, argued one State Department official, could come only from the "better elements" and the "*responsible*" groups—in short, not from Batista, Castro, Prío, Grau, and others whose sordid pasts disqualified them.[42] Once again, in the search for the third force, the elections option took on new life. It would prove elusive.

In the last couple of weeks in April, Smith helped shape a more optimistic U.S. official mood through reports to Washington. Cuba had returned to "normalcy." By "normal" Smith and his colleagues meant largely that business was reviving after the March slump. Tourist traffic was increasing and the sugar harvest was on schedule. The Moa Bay Mining Company began to bring back dependents of its employees.[43] Certainly there was nothing "normal" about Cuba's political world, unless repression and violence fit the definition. New Batista decrees clamped controls on the political freedoms of workers and employers in public-utilities industries. As for the public schools, Batista abolished governing boards in all of the nation's municipal districts. On April 25 the regime extended the suspension of constitutional guarantees for another 45 days.[44]

Far from Havana in Oriente Province, Consul Park F. Wollam reminded U.S. officials that "Santiago is not normal no matter what conditions prevail in the rest of the island." Indeed, terrorist bombs still exploded, rebel and army units still clashed, and the Nicaro nickel enterprise still lost equipment to theft. Tourists may be returning to Havana, Wollam observed, but he hoped that they "do not come here."[45]

In the United States itself there seemed no respite from Cuban entanglements. Rodolfo Masferrer, pro-Batista member of the Cuban House of Representatives and brother of the notorious Rolando Masferrer, whose private army of "tigers" beat, tortured, and killed regime opponents with government approval, visited Miami on April 19. Anti-Batista Cubans roughed him up at the airport. The Cuban House of Representatives declared Miami the home of "gangster acts as frequent as they are reprehensible," and the mayor of Havana, whose own son had endured a similar beating earlier, censured Miami as an unsafe city. Cubans spent $400 million in Miami in 1957, the mayor announced; in the future they would spend their money elsewhere. Governor Leroy Collins of Florida apologized for the Masferrer incident and pledged to apprehend the assailants.[46]

From Fidel Castro's perspective, U.S. territory should figure even more prominently in the war against Batista. Distressed but uncowed by the strike failure, Castro on May 3 convened a Sierra Maestra meeting of M-26-7 to assess the damage and gain the offensive. As Celia Sánchez, Che Guevara, and others witnessed, Castro stormed against the plains rebels, including Faustino Pérez. In a farmhouse at Los Altos de Mompié, tempers, shouts, and accusing fingers punctuated the discussion. The division between the *llano* (plains) and the *sierra* (mountains) widened. Jealousies had long festered, but a fundamental issue was at stake: Should the rebels primarily fight an urban or rural guerrilla war?

Fidel Castro reorganized the 26th of July Movement to place decisionmaking firmly in his hands. He named himself commander-in-chief of the armed forces, including the urban militias, and Secretary General of the National Directorate. Primacy went to the guerrilla war in the countryside. From this "decisive reunion" of 26th of July leaders at Mompié, Che remembered, Fidel Castro emerged with supreme authority.[47] Some Havana underground activists resented Castro's criticisms and charged him with caudillism, but others dutifully followed him.[48] One biographer of Fidel Castro has claimed that the strike failure and the Mompié meeting also marked Castro's self-conscious shift to a Marxist-Leninist organizational structure of centralized leadership.[49] If one follows this line of argument, it points logically to the conclusion that, although Castro had little use for the PSP, he did welcome Communist models on how to wield power. Still, the available record does not provide evidence for Castro's self-conscious adoption of Marxist-Leninist principles in early 1958.

One advantage Castro sought from the Mompié meeting was a more active and more productive role for the 26th of July Movement in the United States. He ordered Haydée Santamaría to Miami to coordinate the work of the exiles in collecting funds, shipping guns and ammunition, and creating unity among the feuding factions.[50] Mario Llerena was called to Miami to meet with her. There he faced what he called "a little kangaroo court." Blamed for ineffectiveness as chairman of the Committee in Exile, Llerena withstood "a barrage of invective." He quickly resigned. Santamaría and other accusers gave him no credit for generating the anti-Batista publicity represented in Powell's speech or for his patient lobbying with the State Department.

Llerena has claimed in his memoirs that after the failed general strike, middle-class professionals like himself, in exile and in Cuba, were pushed to the fringes of the rebellion.[51] Still, never comfortable with M-26-7's championing of violence and its rejection of politics as a vehicle to remove Batista, and long suspicious of Castro's personal ambition, some moderates moved to the sidelines by choice. Their dilemma was real, for they could in no way abide the usurping *bastistianos* nor wholly embrace the radical *fidelistas,* by mid-1958 the two primary protagonists in the Cuban conflict. Llerena broke with Castro; others, with no other place to go, went along with the master of the Sierra Maestra. Politics became more polarized.

Castro did insist on *his* small mountain army as the vanguard of the revolution, require adherence to *his* commands, act nothing like a democrat, badger his critics, and judge anyone other than his loyal *barbudos* as challengers to be isolated. This transformation of the rebellion may have been based more on Castro's fixation with sheer survival and on his quest for power than on a newfound ideological commitment to class warfare or a Marxist-Leninist scheme for decisionmaking. That Castro by mid-1958 wanted Batista and some U.S. interests pushed out of Cuba is clear. What he wanted to put in, other than his own leadership, is not clear. In any case, just a few days after the Sierra Maestra summit meeting at Mompié, Castro, in a self-described "terrible mood," once again denounced other rebels for their "stupid mistakes" and "imbecility." "What these people have for brains is pure s---."[52]

13

Operation Fin de Fidel *and U.S. Weapons: Anti-Americanism Ascendant*

"It is nearly impossible not to intervene in a country as closely associated with us as Cuba," Earl E. T. Smith wrote a few years after his ambassadorship.[1] Smith, of course, intervened to preserve the Batista regime and to prevent a Castro victory. The ambassador failed. He and Eisenhower Administration officials failed because Batista's repressive behavior further alienated Cubans and pushed politics to the extremes, and because U.S. interventionist decisions and actions aroused profound hostility among diverse anti-Batista rebels. In this process, President Dwight D. Eisenhower seemed removed from the key decisions. He left policymaking to his subordinates, apparently comfortable with a management style that often kept him ignorant of details and that presented him with problems usually only when they reached a point of severe crisis. At least such was the case on Cuba policy.

President Eisenhower seemed only vaguely aware that, in the latter half of 1958, anti-Americanism heightened among Cubans opposed to the regime. Many more came to see the United States as the enemy, responsible for sustaining an unpopular dictatorship and for prolonging Cuba's agony. The revolutionary government's strident anti-Yankeeism of 1959 and after, which seemed to surprise so many observers, derived from long smoldering Cuban resentments against twentieth-century U.S. imperialism. But these discontents drew especially upon the immediate life-and-death experiences of 1958, when many anti-Batista Cubans who hoped the United States would abandon the dictator grew disillusioned. U.S. military links with the Batista regime rested at the core of those embittering experiences.

Ambassador Smith went to the State Department during the second week of May to press for a renewal of arms shipments to Cuba. Washington officials had heard his pitch before: The ambassador could never persuade the Cuban president to create conditions for an orderly, electoral politics unless the United States removed the stick of suspension and used the carrot of shipment. Wieland challenged Smith, explaining that the ambassador simply failed to understand the unpopularity in the U.S. press, public, and Congress of arms deliveries. If criticism of the Cuban case did not abate, Washington officials worried, Congress might call into question U.S. ties with other Latin American military dictatorships—if not the entire U.S. military assistance program in the hemisphere.[2] Smith remained dissatisfied with the renewed shipments of so-called "non-combat" equipment. Batista wanted more, and so did Smith.

Assistant Secretary Rubottom, facing hostile congressional opinion every time he testified before committees, struggled on several fronts to manage the Cuban problem. He tried to mollify Smith until such time as U.S. policy on arms might be able to shift. Rubottom began to discourage other nations from shipping arms to Cuba.[3] He worked to quiet American public criticism of U.S. connections with the Batista regime and to assure Congress that the United States was honoring neutrality. Rubottom had to keep the U.S. military alerted that it should avoid incidents that might be read as U.S. support for the Cuban government. He pressed the Batista government to move toward truly free elections. Finally, Rubottom shaped U.S. policy to block a 26th of July Movement victory. He failed in all cases. Before he, Smith, and other U.S. officials could grasp the enormity of the failure, Fidel Castro and anti-Americanism had advanced beyond the point the United States considered acceptable.

Vice President Richard Nixon's tour of South America in May 1958 exposed currents of anti-Yankeeism and "a strong wave of anti-dictatorial feeling" in the hemisphere.[4] José Figueres, former president of Costa Rica, regretted, he said, that Nixon was spat upon. "But people cannot spit on a foreign policy which is what they meant to do."[5] The Nixon debacle also drew comment from Cubans and turned the spotlight once again on the issue of U.S. support for Latin American dictators. Rebels hastened to point out that the fierce hostility Nixon faced in Argentina, Venezuela, and elsewhere mirrored that in Cuba—resentment against U.S.-endorsed dictatorships. Cubans also remembered that just a few years earlier Nixon had toasted Batista. "The time may come when Nixon may receive a cold reception in Cuba," predicted one oppositionist.[6]

In Caracas, Venezuela, Nixon faced angry stone-throwing and club-wielding demonstrators who smashed the safety glass of the vice president's surrounded car. When President Eisenhower ordered the U.S. military to prepare to extricate Nixon, the Guantánamo Naval Base readied itself for a rescue mission tagged "Operation Poor Richard." One thousand U.S. troops hurried to the base. GTMO-based ships headed for Venezuela with instructions not to go within sight of land so that their movements would not be detected. Few secrets were kept and Venezuelans who expected an invasion cried foul. Cuban oppositionists criticized the island's role as a launching pad for operations against other Latin Americans. The Nixon episode rekindled Cuban sentiment against the prominent U.S. military on their island.

Pro-government Cubans, for their part, took the occasion of the Nixon affair to remind U.S. leaders that Communism threatened Latin America, including Cuba. They admonished Washington that it had not been as tough toward insurrectionists as it should have been. As a gesture toward the United States, however, Batista censors banned newspapers from printing any stories of Nixon's troubles in Lima, Peru, where university students stoned him.[7]

At the same time, Batista's chances for survival seemed to improve, as did U.S.-Cuban relations. After the general strike, terrorist attacks declined. Political parties stepped up their campaigns for the November election and Batista invited United Nations observers to witness the balloting. To someone like Ambassador Smith, who believed Batista's pledge of honest elections, Cuba seemed to be moving toward a political solution. Through stern measures Batista also seemed to be regaining control. School administrators, judges, and labor unionists unfriendly to the regime were being purged. Under tight censorship, the press could not criticize the government or report rebel activities. The suspension of constitutional guarantees was extended. The pro-government labor confederation held a May Day rally that U.S. Embassy officers applauded as "orderly and enthusiastic."[8]

U.S. officials also measured progress in other areas. After a meeting in Washington, a reluctant Cuban Sugar Stabilization Institute bowed to U.S. wishes to increase the sugar crop so as to ease an anticipated shortage of sugar in the United States because of production shortfalls in Hawaii and Puerto Rico.[9]

The regime's military prospects also seemed brighter as it geared up for a major offensive against the rebel army in Oriente Province. Government forces formed a blockade around the Sierra Maestra to halt the flow of men and matériel to 26th of July forces. Operation FF (*Fin de Fidel*) anticipated trapping Castro himself high in the mountains. Batista assured Ambassador Smith that in two months "the mop-up should be completed."[10]

The offensive began in May, and military skirmishes with rebel units multiplied. On May 7 government airplanes strafed rebel positions around

Santiago, pounding San Juan Hill for an hour. The armed forces received and tested two helicopters it had bought from Britain. The government also offered a qualified amnesty to insurrectionists who had committed no crimes; if they surrendered, they could return to their homes. The Army General Staff soon claimed that hundreds of rebels turned themselves in.[11] Actually the opposite was probably happening. The people of Oriente Province, reported Ruby Phillips, "say the reason for the growing strength of the rebels is the savage repressive methods of the armed forces" and their "ruthless killing" of hundreds, which has "caused hundreds to join Castro and the other groups, since life with Castro is far safer than living defenseless in the lowlands under control of the armed forces."[12]

An anxious Castro spurred his mountain legions to get ready for Batista's summer offensive, and he implored the urban underground to send weapons and more weapons for the "decisive moment."[13] The mountain insurgents readied mines and grenades to interrupt the army's march "at each bend in the road."[14] They strung telephone wires to improve communications; they manufactured bombs in their workshop, built bunkers, and dug trenches. They waited to snipe at, to ambush, to harass, to terrify the 14 battalions of some 10,000 troops amassed against them.

As he prepared to defend his mountain bastion, Castro also kept his propaganda machine going. Fewer journalists were hiking into the mountains, so he answered questions through radio transmissions. Jules Dubois, following the Nixon caravan in South America, initiated one exchange from Venezuela in mid-May that annoyed Castro. Privately Castro complained to Che that Dubois asked "all the well-known and tendentious queries."[15] Dubois's first question: "Are you or have you ever been a Communist?" An impatient Castro answered "No," but explained, "I understand that this is a question that every North American newspaperman feels compelled to ask." Only Batista, the rebel said, branded the 26th of July Movement "Communist" in order to win arms from the United States. Castro sharply criticized the United States for arming a bloody dictatorship.[16]

Batista's public relations firm continued to counter rebel propaganda with its *Report on Cuba*. Attempting to exploit anti-Communist sentiment in the United States, the newsletter declared the existence of a "hemisphere-wide Communist plot." The evidence? The Castro rebellion and the attacks upon Nixon. Americans should travel to Havana for fun and night clubs, the publication implored, because "the 'Revolution' seems to be a thing of the past; after all, it's 700 miles away" from Havana.[17]

Castro's followers and rivals in the United States also remained active. Several exiled labor leaders joined Prío at a New York rally attended by 1500 people. Prío, out of jail on bail, continued to agitate against the regime. The former president's case became tied up in the courts after he filed a petition for change of venue from Miami to New York. State and Justice Department officials grew frustrated as judges took their time to make a decision.[18] Prío's case languished while he seemed less and less a central figure in the insurrection, although his money continued to fund it.

Other Cuban exiles lobbied the State Department, wrote letters to members of Congress, or spoke at public gatherings. José Miró Cardona, president of the Havana Bar Association and a leader of the pro-Castro Civic Resistance, detailed for a gathering of Miami lawyers Batista's destructive campaign against the Cuban judiciary. The Federación Estudiantil Universitaria staged demonstrations in front of the United Nations in New York and demanded that the Committee on Human Rights take notice of Batista's crimes against students. Cuban exiles also assaulted Cubana Airlines pilots at the Miami International Airport.[19]

In Chicago, the Cuban Liberation Movement of July 26, receiving free legal advice from the attorney Constantine N. Kangles of Kangles, Getto, and Bunge, raised money for the Castro forces and propagandized against Batista.[20] Rebel followers stirred publicity for the cause in different ways. At Penn Plaza Park in Newark, New Jersey, at midnight on July 26, they flew the flag of their movement, a red and black banner with a white star and the number "26" at center. Sixteen hours later the police confiscated the flag and took four Cubans into custody. Not sure what charges could apply, the police soon released them.[21]

Some M-26-7 rebels did not fare as well. While detectives were searching a West Side apartment in New York City for narcotics, they accidentally came across an arms cache. The submachine guns, pistols, M-1 Garand rifles, and carbines had been dismantled and wrapped for shipping. Two men were booked.[22] In Lynwood, California, near Los Angeles, U.S. Treasury agents stormed a garage that was serving as a factory for the manufacture of machine guns for Castro's army. "The biggest arsenal since the days of Dillinger," claimed a Treasury agent. Five people were arrested. "When word reaches Cuba about our arrests," said one of the captives, "our families will suffer greatly."[23]

Carlos Franqui, appointed to head Radio Rebelde, proved more successful. In exile in Miami, he conspired to load a plane with electric fuses for mines, Italian carbines, and 20,000 bullets for his return to Cuba. Falsifying documents—declaring that the aircraft was destined for Jamaica—Franqui took off from the United States in a Cessna piloted by Pedro Díaz Lanz. The plane made a rough landing on a small airstrip in the Sierra Maestra. After offloading the craft, the radical journalist hiked to Castro's La Plata camp, where Fidel heartily welcomed him. Franqui sensed that he received such *abrazos* not only because he had delivered weapons but also because he had come from Miami, not Havana. The mountain forces still blamed all troubles on the urban underground. "There was a total lack of mutual understanding. It was a wall," Franqui noted.[24]

Franqui's cargo arrived just in time. Batista forces began their offensive in late May. The rebels gradually gave ground, moving deeper into the rugged mountains while punishing government troops through selective attacks, sniper fire, and field mines. Batista's ill-trained recruits often broke ranks under rebel hit-and-run assaults. From his La Plata headquarters, with Celia Sánchez at his side, Castro schemed to defeat the advancing Batista forces, isolating one platoon after another. The underarmed rebels eagerly collected the army's abandoned weapons, but there never seemed to be enough, and ammunition ran low or did not fit the rifles in hand. The rebels always worried about quality too. Two young revolutionaries preparing an inventory listed ten captured San Cristobal rifles that the Cuban army had no doubt acquired from the Dominican Republic. "More Cristobal rifles? Junk!" snapped one rebel. The other agreed, and quipped: "I think we made a mistake when we got the United States to stop shipping arms to Batista. Now we can never capture decent rifles."[25]

With every battle the count of dead and wounded climbed. From the air, government planes strafed and bombed. For ten hours on May 30, B-26s and other aircraft raked rebel positions. Radio Rebelde competed with official government reports to claim victories. The State Department complained that the radio broadcasts were always berating the United States, sounding too much like Communist slogans.[26] Meanwhile, rebel-placed bombs went off in the cities, and the DR front in Las Villas Province became more active.

Despite the excalating violence and signs of mixed economic performance, North American business in Cuba still expressed some optimism about the future. Perhaps the press censorship worked to hide the story of continued political turmoil or perhaps Batista's propaganda was working. More generally, representatives of U.S. interests, nurturing the hegemonic presumption, probably believed that the government offensive would succeed or that Castro, however troublesome, would not antagonize them. He had said as much in his February 1958 interviews. In any case, analysts pointed to some economic improvements: the construction industry was growing; Cuban Electric and Cuban Telephone were expanding; General Electric put up a new factory to construct switchboards and other electrical equipment; the new $5 million U.S. Rubber tire factory near Havana was beginning operations; and the Havana tunnel opened. Robert Kleberg of the

King Ranch continued "very bullish" about the island as he prepared to buy more land in Cuba. And the sugar crop, hardly interrupted by the rebels, promised a good return for the year.[27]

In early June the Cuban army cranked up the war against Raúl Castro's forces. "They rape, murder, and rob," Fidel's brother protested. "Their planes hurl firebombs, delivered to them by the Yankees" at the Guantánamo Naval Base. "These monstrous crimes must be revealed to the entire world."[28]

At the same time, Fidel inspected the damage done to a sympathizer's house in the Sierra Maestra. In one of his most trenchant and prophetic anti-U.S. comments during the insurrection, penned in a brief letter to Celia Sánchez on June 5, he asserted: "When I saw rockets firing at Mario's house, I swore to myself that the North Americans were going to pay dearly for what they are doing. When this war is over, a much wider and bigger war will commence for me: the war that I am going to wage against them. I am aware that this is my true destiny."[29]

That war lay ahead of him. The war before him turned nasty everywhere. On June 15, Batista police tortured, probably raped, and then killed Maria and Cristina Giral in the city of Cienfuegos. Both young women were members of the Civic Resistance. Their employer was José Ferrer, a businessman, who, at CIA agent William Caldwell's urging, visited Ambassador Smith to relate the tragedy. "What do you want us to do, send down the Marines?" barked Smith. "No," answered Ferrer, "but I hope you realize what is happening here." Smith asked, "Have you gone to the police to file a complaint?" Incredulous, Ferrer replied: "What police? The same police who killed the girls?" Smith, obviously not getting the point: "Well, have you gone to the courts to complain?" Ferrer: "What courts and what judges?"[30] The ambassador never seemed to fathom the depths of Batista repression or the accelerating Cuban disaffection from the regime.

By June 19 the Cuban army was moving dangerously close to the crest of the Sierra Maestra and Fidel Castro's La Plata command post and airfield. "We're running the risk of losing not only the territory but the hospital too, the radio station, the bullets, mines, food, etc.," Castro wrote Che Guevara, whom he ordered to send reinforcements at once.[31] As mortar shells fell on his base, Castro eyed enemy soldiers through his binoculars. Controlling the roads and paths, M-26-7 sharpshooters forced the intruders back. La Plata and Castro survived. Batista's big push was being blunted.

The bloodletting continued for weeks. A turning point was reached on July 21, when a government battalion surrendered after the prolonged battle of El Jigüe. The rebel army slaughtered cattle to feed the more than 220 prisoners and then turned the captives over to the International Red Cross.

The commander of the government troops, Major José Quevedo, shifted sides and became a *fidelista.*

As the government offensive weakened in July, Guantánamo and its precious water supply embroiled the United States and the 26th of July Movement in dispute. Earlier in the year, as a contingency for an emergency, Washington had approved the stationing of U.S. guards on Cuban soil at the Yateras waterworks and aqueduct that supplied the naval installation. On July 1 the Cuban government requested that the United States relieve the Cuban military at Yateras. Rear Admiral Robert B. Ellis, the base commandant, hesitated because the rebels had never threatened the water supply and Raúl Castro had apparently issued orders that the Yateras facility was not to be molested.

On July 25 the Cuban army notified U.S. officials that it was removing its troops from guard duty at Yateras. Although the State Department believed that the "action of the Cuban Army seemed unnecessary" and the positioning of U.S. forces outside the base "might well cause strong resentment in Cuba," Ambassador Smith urged the cautious Ellis to act.[32] On July 28 the base commander sent 15 Marines to Yateras. At the time, Assistant Secretary of State Rubottom was traveling with the president's brother Milton Eisenhower in Central America.

Was Batista trying to provoke an incident between U.S. soldiers and the rebels so that the United States would intervene, saving his stumbling regime? Many of Batista's foes thought so. In New York, Raúl Chibás led a demonstration outside the United Nations. Angry exiles in Washington and Miami bombarded the State Department with telephone calls. Cuban throughout Latin America protested. Fidel Castro announced that the stationing of North American forces on Cuban soil constituted "aggression."[33] But, always fearful of U.S. intervention to save Batista, Castro tempered the response: Withdraw your Marines, he told Washington, and the rebels would not interfere with the base's water supply. In Washington, Admiral Burke reluctantly bowed to State Department "policy considerations of a high order" and ordered Rear Admiral Ellis to move the Marines back to the base. They returned to GTMO on August 1. As Batista soldiers resumed guard duty at the water facility, Ambassador Smith charged that the United States had capitulated to rebel demands, undercutting the regime.[34]

During the summer fighting, 750 out of 800 soldiers of a MAP-armed and -supplied unit went into battle in Oriente Province. In June, U.S. newspapers and senators cried foul, and the State Department worried that all aid to Cuba would be cut off unless Batista pulled the unit out. But this was no easy task, because the MAP soldiers had been dispersed throughout the infantry forces fighting against the mountain guerrillas. But worse, argued Ambassador Smith, these were Cuba's best troops; if they could not fight, Batista might lose. The Cuban desk officer Leonhardy would not be dissuaded: "A treaty violation had occurred and could not be overlooked," he reminded Smith. The unit had to withdraw.

Although Cuba agreed to "sterilize" the MAP battalion, U.S. officials expected more than this vague answer. They demanded compliance with the Mutual Defense Assistance Agreement: U.S. aid must be used for hemispheric defense. Smith was directed to tell the Cuban government that all military men trained specifically under MAP and all MAP-supplied equipment had to be disengaged from combat. To monitor compliance, the MAP army unit had to be reassembled and stationed in the Havana area. A "bitter pill" for Cuba, to be sure, but required by law and by unrelenting critics.[35]

An angry Smith fought back on June 16. The MAP issue, he informed his State Department superiors, strained U.S.-Cuban relations, and the "only beneficiaries of this policy in the end may turn out to be the Communists." He understood why the regime could not comply with requests for sending MAP-trained armed forces to their barracks: 75 percent of the pilots in the Cuban Air Force, for example, had received some kind of MAP training. Smith asked for permission to help Cuba fudge the question; that is, let Cuba argue that the war against Castro contributed to hemispheric defense because the rebels were "in league with communism." Smith, having never supplied evidence of Communist links with M-26-7, sounded now more than ever like a mouthpiece for the Batista government. State Department officials firmly rejected his plea and disputed his farfetched analysis.[36]

By late summer, however, stymied by Batista's unwillingness to comply with Washington's demands, U.S. officials quietly let the issue fade. The MAP remained in Oriente in a vain attempt to hold back the rebel advance.[37] Later, State officials explained that they had not forced a confrontation with Batista over the use of MAP-trained personnel because the Castro cause would have received "a substantial boost we did not want it to receive" at a time when Washington was worrying about Communist influence in the rebel camp.[38] U.S. officials also appreciated "the unqualified support which the Cuban Government was giving our Government in the United Nations on our anti-Communist stand."[39]

Smith got his way too after he protested in mid-June against the State Department's refusal to clear an export license for ten T-28 trainer planes. State officials reconsidered and approved shipment of the planes, which could carry rockets, bombs, and guns. The T-28s, it was decided, could be tagged non-combat equipment.[40]

During this intra-government debate on aid to Batista, the Joint Chiefs of Staff again reinforced Smith by pressing the State Department to restore arms shipments to Cuba and to relax MAP restrictions. At a Department of State-JCS meeting on June 27, Admiral Burke echoed Ambassador Smith's charge that the Cuban rebels were "allied with Communism," and he took aim at State officers who were interpreting the 1952 agreement "too rigidly" and "trying to tell a sovereign nation what to do." But no resolution came from the meeting.[41] As before, President Eisenhower and Secretary Dulles do not seem to have busied themselves with the issues swirling around Cuba policy.

The debate within the U.S. government over MAP aid was soon eclipsed by one of the most dangerous episodes in U.S.-Cuban relations during the insurrection. In early May, Fidel Castro had received some disquieting news from his spies in the United States: The United States was about to ship 300 "aerial rockets" to Batista forces through Guantánamo Naval Base. Castro hesitated to "start a row" until he could confirm the report.[42] By the end of the month he had the evidence in hand: photographs of U.S. equipment being loaded into a Cuban transport plane (two planes, a C-54 and C-47, actually picked up the cargo) and copies of the paperwork generated by the U.S. military establishment for the delivery of rocket heads.[43] The U.S. Consul in Santiago reported that oppositionists were "disillusioned" by the episode, while Castro denounced the United States for lending itself once again to Batista's purposes. "One of the rockets has fallen into our possession," he told Venezuelan journalists. "It is numbered 462594—it is of North American manufacture."[44]

The Department of Defense, in apparently routine manner, was correcting an earlier error. The wrong (non-explosive) rocket heads had been shipped before the U.S. suspension of arms, and now the correct (explosive) ones were being sent in exchange. In late May, lower echelon State officials, knowing that the rebels had learned of the swap, had hastened to stop it, but they intervened too late. "Our friends in the Pentagon" had presented an "unfortunate *fait accompli*," complained one State Department officer.[45]

Not until July 3 did the Department of State manage to issue a press release to explain the exchange as the rectification of a mistake. State officials denied that the United States had lifted the arms ban or was permitting Cuban military forces to use the naval base. Anti-Batista leaders, on the other hand, dismissed the explanation as irrelevant: The new rocket heads were going to kill Cubans, many of them innocent people caught in war zones.[46] Embittered toward the United States, pressed by the Cuban Army, and harassed from the air by government warplanes during the summer offensive, the commander of the "Frank País" Second Front decided to retaliate against the United States. Raúl Castro had had enough of U.S. "neutrality."

14

Rocket Heads, Kidnappers, and the Castros

AT DUSK ON June 26, Raúl Castro's Second Front rebels attacked the Moa Bay Mining Company, a subsidiary of the U.S.-owned Freeport Sulphur Company.[1] The commandoes overpowered guards and captured ten American and two Canadian engineers. Several rebels and government soldiers died in the skirmish. "They'll be treated well and returned in a few days," a rebel assured the wife of one of the engineers.[2] An American employee of the company ran to a vessel in the harbor and radioed news of the raid to the Freeport office in New Orleans, which in turn informed the New York office, which called the company office in Havana. In this roundabout way Ambassador Smith then got word. Some 40 miles to the south of Moa Bay, rebels also seized another American, Desmond Elmore of the Ermita sugar mill.

U.S. officials soon began a multitrack effort to gain the release of the captives. Through the local Moa Bay manager, Consul Wollam in Santiago learned that Ambassador Smith had ordered him to investigate. Downed telephone lines between Havana and Santiago necessitated this indirect means of communication. In Washington, Terrance Leonhardy decided to publicize the incident through the U.S. press in order to generate hostile opinion that might persuade the rebels that their brazen act would damage the 26th of July cause. Anyway, the story would not remain quiet for long.

While Smith contacted the Cuban government to ask for its cooperation, Leonhardy approached Ernesto Betancourt. The U.S. diplomat chastised the "irresponsible elements within the Movement" who had staged the kidnapping. Betancourt "seemed unpleasantly surprised" and admitted that if anything untoward happened to the captives, the anti-Batista campaign would suffer a setback in U.S. public opinion. Ambassador Smith's first-blush response was that the incident was a "publicity stunt" similar to the earlier abduction of the racing car driver Fangio.[3]

Wollam flew to Moa Bay in a company plane with instructions from Smith *not* to enter into negotiations with the Castro forces. Officials in Washington agreed that "it was not possible at this time to deal directly with the rebel elements," probably because such contacts could be interpreted as official recognition. "Word" would be transmitted to the rebels in Cuba "in a discreet manner."[4]

On June 27, Raúl Castro's forces struck again. In the evening, at gunpoint, the *barbudos* halted a civilian bus traveling from Guantánamo City to the naval base. The vehicle carried 28 U.S. Navy sailors and Marines returning from liberty. The bus had been escorted to the outskirts of the city by Cuban Army soldiers. The U.S. Navy Shore Patrol followed by ten minutes and saw nothing unusual en route to the base. But the bus never reached its destination. Raúl's guerrillas seized it and quickly transferred the passengers to waiting trucks at a sugar plantation. The new kidnap victims were also taken deep into the hills, probably near the town of Palenque in Sierra del Purial.[5] One rebel warrior recalled the captives as "detestable men" because they "looked down on us Cubans as savages. Even as prisoners, they looked on us with contempt."[6] At the same time, M-26-7 activists in Havana and in the United States released copies of the requisition forms for the rocket-head deliveries. That day, too, Friday, June 27, rebel forces grabbed Richard Sargent, the Canadian manager of the Isabel sugar mill near Guantánamo.

Because the crisis was ballooning, Wollam's instructions changed: He could now contact the rebels directly in order to win the captives' freedom. How he was to gain their release he was not told. A graduate of Occidental College (B.A., 1938) and the University of California (M.A., 1940) who joined the Foreign Service during World War II and then served in Latin America and Italy, the Indiana-born Park F. Wollam could never have imaginged the rigorous assignment he was about to undertake.[7] To prepare for a journey into the mountains, Wollam talked with General Alberto Del Río Chiviano of the government forces and then with the Cuban Air Force pilot of a Beaver aircraft active in the area. The U.S. Consul received their assurances that they would halt air operations while the kidnapped Americans remained in rebel hands.

At 7:00 a.m. on Sunday, June 29, Wollam joined a Moa Bay Company employee and a Bethlehem Steel Company employee, both of whom had ties to the rebels, for a journey by jeep into the hills. The Bethlehem man, a

Cuban, served as guide. "We don't know where you're going, Park, how long you'll be gone, and whether they've added you to their kidnap list," remarked Balfour Darnell, the Moa Bay manager. "How long do we wait for some word from you?" Wollam advised that if Balfour had not received a message by Wednesday, "think of something else to do."[8]

After 45 minutes on the road, they took on a *guajiro* guide to direct them through the countryside. The four picked up hitch-hikers for short distances. At one point, two young men armed with shotguns hopped on the jeep. Shortly after 9:30, with the heat becoming oppressive, Wollam's vehicle parked next to a small palm-roofed house, or *bohío,* used as a rebel outpost. A friendly sergeant served them sweetened coffee while a message was sent forward that the consul was on his way. Wollam now worried less about an ambush. But at the next outpost Wollam faced two charging armed men with shotguns ready. They had apparently not gotten the correct message. Wollam's fellow passengers shouted off the rebels.

Before long the jeep passed a small wooden church, a Baptist mission with a service in progress, and then stopped at a country hut in the small village. Wollam spotted two camouflaged vehicles from the Moa Bay Company, one of them a pick-up truck loaded with medical supplies. A gruff lieutenant expressed rebel outrage over the U.S. delivery of the rockets (the rebels made no distinction between the heads and the rockets) and the slaughter of civilians by government airplanes. Wollam managed to see two of the Moa Bay captives, Hal Kristjansen from Winnipeg, Canada, and Bert Ross from Houston. Kristjansen related that he and another Moa Bay man had actually escaped for a time from their captors. When they ran out of food and water, they approached a *guajiro* for help. The peasant informed the rebels, who recaptured the two with a warning that they would be shot if they tried to escape again.

Wollam asked the rebel lieutenant to let him see all of the kidnap victims, and he requested their release. But they were scattered in small groups throughout the hills. Wollam was told he could visit most of them on his way to meet Commander Raúl Castro, the only person with the authority to free the men.

Wollam knew that the kidnappings were no stunt. He scratched a note on lined yellow paper: "We have arrived at the first stage and arrangements are being made to place us in contact with the principal officers of authority." He advised: "Keep the names of persons helping us very quiet and the less comment the better." Rebel contacts carried the note to Moa Bay.[9]

The jeep started off again, retracing ground toward the Baptist church. When a plane suddenly appeared, everyone jumped for cover. Wollam ended up in a pigsty under a shed containing a camouflaged Moa Bay tractor.

People of all ages scampered from the church into the trees and underbrush. Wollam recognized the aircraft as a Cuban Air Force Beaver. He had General del Río Chaviano's pledge, so he was not particularly alarmed. But soon the government plane swept down and dropped incendiaries, while its clattering machine-gun sprayed the area. In his official report a few days later, Wollam explained how "furious" he was after this strafing and bombing.

At the time, wet with the smelly mud of hogs, he was frantic, scared, breathless. He scratched on his paper pad: "Call them off!!!! Keep other planes out of area." More: "These were some near misses. If we get hurt—it is this!" At the bottom he scribbled: "They also tossed a bomb at a church. For the safety all any [sic] Americans, if nothing else—get them stopped?"[10] (This handwritten note and the earlier one found their way to Moa Bay by 4:45 p.m. Moa Bay manager Darnell and then Ambassador Smith quickly filed protests with the Cuban government.) Wollam began to believe oft-repeated rebel stories about the strafing of civilians or "anything that moved." Kristjansen also penned a brief note that reached Darnell: "We hope to get out of here soon because Batista and his [indecipherable letters] will finish everybody off if not."[11]

At noon Wollam's jeep arrived at another command post, a farmhouse called Los Pinos. There a rebel captain held the Moa Bay engineers Edwin Cordes of Fanwood, New Jersey, and James D. Best of New Orleans. They told Wollam that they had witnessed a funeral for the six rebels killed during the Moa Bay assault. Passionate anti-American speeches had dominated the somber gathering. Still, the hostages were well treated.

Wollam's odyssey continued in another jeep ride, this time toward Naranjo, deeper in the mountains. At about 4:30 p.m., after a rough trip over mule trails, streams, and a host of obstacles in the tropical forest, he arrived at a large country house notable for its inside plumbing. He saw four more kidnapped Americans: A. A. Chamberlain of Coral Gables, Florida; J. K. Schlissler from California; Roman Cecilia, formerly of New York; and Professor E. P. Pfleider, head of the Engineering Department at the University of Minnesota, a consultant to the Moa Bay Company. Wollam heard more rebel complaints against U.S. responsibility for Batista's air attacks. Dr. Lucas Moran and Major Belarmino Castillo Aníbal watched over the captives and pressed the rebel perspective upon them. Raúl Castro was nowhere in sight.

As Wollam headed for Naranjo on June 29, four jets and a rescue craft from the Guantánamo Naval Base flew 20 sorties in a vain search to locate the captured U.S. military personnel. At 2:45 that day, vice consul Robert Wiecha, the CIA agent at Santiago, departed the base by jeep to try to make contact with the rebels holding the naval personnel.[12]

In the morning of June 30, rebel Captain "Oriente" Fernández escorted

Wollam to a coffee *finca* near Calabazas, a village of some 100 people. It was raining, and it took three and a half hours to travel ten miles. At one point a deep mud hole swallowed the jeep. An ox was found to pull out the stranded vehicle. Along the way Fernández pointed to a flock of birds—the "rebel air force," he joked.

Taking residence in a four-room frame house with a corrugated roof that rang with noise every time it rained, Wollam waited for Raúl and saw more captives: William Koster of Akron, Ohio; Edward Cannon of Ontario, Canada; Howard A. Roach of Watertown, Maine; and Henry Salmanson of Portland, Oregon. All of the Moa Bay prisoners were now accounted for; tired and dirty, none had been harmed. Indeed, they were well fed. "On the house," a rebel remarked as he provided prisoners with a snack of beans and coffee. "The rebels took us swimming and on jeep tours," said one captive American.[13] The tours served a political purpose: The rebels wanted the Americans to see the damage Batista's airplanes had caused. Some military captives said they had become rebel sympathizers. When Rear Admiral Ellis at Guantánamo heard that, he snapped that U.S. military prisoners ought not to be "enjoying life" and "drinking beer." Thomas R. Mosness of Ames, Iowa, admitted that he might have gotten a little "buddy buddy" with the rebels.[14] Raúl Castro, relishing the thought of converting a U.S. serviceman to the cause, asked the young Navy airman to "come back in a month or so and stay with us." Mosness knew the limits: "I couldn't do that. That would be desertion."[15]

On the thirtieth, Raúl Castro's soldiers captured six more Americans, A. E. Smith, J. P. Stephens, H. F. Sparks, and J. G. Ford of the United Fruit Company at a sugar mill near Preston, and Sherman A. White and J. Andrew Poll, two managers of the U.S. nickel facility at Nicaro.[16] Now the rebels held 50 prisoners. The commander of the Guantánamo Naval Base reported that Ambassador Smith intended to issue an "ultimatum" the morning of July 1: If the captives were not returned within 48 hours, he would request permission from the Cuban government to take "whatever means were practicable" to free them. It is not clear whether Smith meant direct military intervention, a threat to use force, or something else.[17] In any case, Washington scotched any ideas of overt threats or military intervention.

The State Department nodded in the ambassador's direction by asking the embassy to make a special study of possible Communist influence in the 26th of July Movement. "The recent kidnapping of U.S. citizens in Cuba by rebel forces, the anti-U.S. tone of recent Radio Rebelde broadcasts, and the increasing disregard by the Castro forces for U.S. property," concluded the State Department "Instruction," suggested that the Communists "have made considerable progress in their attempt to influence the pro-Castro groups."[18]

But Raúl Castro did not grab the Americans because of Communist influence. Anti-Americanism, accentuated by recent evidence of U.S. military support for Batista, which deeply embittered M-26-7 and other rebel groups, is the most consistent theme in the episode. The rebels under Raúl's command and the local farmers in his district had been frequently bombed and strafed. The insurrectionists blamed the aerial attacks on the United States, Batista's military supplier, and they wanted to draw world attention to the problem. Reflecting widespread Cuban opinion, Rufo López-Fresquet defended the kidnappings as "a last resort" protest against the U.S. military presence in Cuba. The U.S. Army Mission "is instructing the Batista forces in better ways to kill the Cuban rebels (if they do not teach better military tactics what in the hell are they doing here?)."[19]

Raúl Castro also complained that the bombings were turning the *campesinos* as much against the rebels as against the regime; the aerial raids came only because of the presence of the insurgents. More, Americans scattered throughout the hills offered a protective shield, for, the rebels assumed, Batista's military would not risk killing Americans. Hence the Second Front rebels tagged their project "*Operación Antiaerea.*" In exchange for the captured Americans, moreover, the rebels might gain some U.S. recognition and force U.S. concessions. In the least, the kidnappings would embarrass the Batista regime and gain publicity.[20]

In the United States, the publicity was uniformly bad. The *New York Times,* strongly anti-Batista under Herbert Matthews's guidance, editorialized that the kidnappings—"juvenile escapades"—would only harm the Castro cause to oust the dictator. The editorial sent a blunt message to the rebel leadership, always eager to shape a favorable U.S. public opinion: Batista "stays in power because of a failure of this opposition over the past eighteen months to get together and act together. It is not the United States that keeps General Batista in office, nor is it the business of the United States to overthrow Latin-American dictatorships."[21]

Although most newspapers urged patient diplomacy, one clamored for "our firepower at the Guantánamo base to show him [Castro] what we mean."[22] Many Americans also urged "some Yankee guts." Tell the "sniveling cowards," telegrammed an Indianapolis man to the president: "Our citizens back alive and well or Castro dead."[23] Conservative Republican Senator Styles Bridges of New Hampshire grumbled that the United States could expect kidnappings as long as it displayed weakness in the world. Without saying that he would send in the Marines, Bridges asserted, "I'd get them out."[24]

President Eisenhower, in his very first public utterance on the Cuban rebellion, told a press conference that "we are trying to get live Americans back." The United States had to avoid doing "anything reckless." He denied that his administration had given "improper support" to the Cuban government, and he lectured the "dissident portion of the [Cuban] nation" that it was wrong to seize innocent people. The president revealed little understanding of the Cuban crisis.[25]

From his Cuban *finca,* Ernest Hemingway saw some humor in the crisis. He telephoned the Havana embassy and asked "when they [the rebels] were going to start picking up the F.B.I." He quipped that Fidel Castro would entertain more Americans on July 4 than Ambassador Smith himself would.[26]

Family members of the hostages began to worry, especially because the U.S. government had failed to notify them. On June 30, William Koster's wife sent an urgent telegram of complaint to President Eisenhower. "Certainly the families of these men deserve the courtesy of being informed," she wrote. "My husband is 62 years old and I am very much concerned about his health under these conditions."[27] Not until the evening of July 2 did Leonhardy give her a reassuring call. She mentioned with chagrin that some members of Congress were demanding military intervention that would only endanger her husband.[28]

In Oriente Province at Calabazas, meanwhile, Park Wollam was awakened at 1:30 a.m. on July 1 by shouts. A courier handed the consul a letter from Raúl Castro, who welcomed the U.S. envoy to the "Free Territory of Cuba." All day Wollam waited. With his soiled shirt in the wash for the day, he rested in a T-shirt. In the late afternoon, four jeeps stormed into the *finca* and parked next to the farmhouse where Wollam slumbered. Out jumped Raúl Castro, Vilma Espín (a young engineer who had studied at the Massachusetts Institute of Technology and a founding member of M-26-7 who later married Raúl), and several armed *barbudos.* The U.S. Consul immediately asked Raúl for the release of the prisoners. He added that he could not "negotiate," because he was only a consular officer. "This will not be so easy," Castro answered, for he required "certain guarantees."

The 27-year-old Raúl Castro, a thin man with a pencil mustache and shoulder-length hair pulled back in a small bun, launched into M-26-7's indictment of the United States: Batista's use of North American weapons to bomb the Second Front; U.S. military aid to Batista; the transfer of the rocket heads at the naval base and the use of the base to refuel and arm Cuban planes; Batista's concessions to U.S. investors (such as Moa Bay); the decoration of General Tabernilla. Wollam answered each point, explaining the mix-up on the rocket heads and categorically denying that the Guantánamo base was used to service Cuban military planes. The Cubans became "emotional about the women and children being bombed and strafed in their territory," Wollam reported later. Four hours of intense discussion resolved nothing.

After dinner, under lantern light, they met again in the farmhouse. Raúl said he would release five captives if Wollam would deliver to Washington demands that the United States halt aid to Batista and offer some form of recognition to the rebels. Wollam promised to pass the rebel message on to Washington. "How do I know you will keep your word?" asked Raúl. "I want some high official to come to Calabazas and give me solid guarantees you will keep hands off our war."[29] He asked for Ambassador Smith. Impossible, answered Wollam. Would the United States treat the rebels as co-belligerents? "We must be recognized." Impossible, Wollam replied. Raúl settled on two fundamental demands: The United States must not let the Guantánamo Naval Base serve Batista's military and the United States must stop further arms shipments to the regime. After this conversation, Wollam sent a request to the Santiago consulate for a U.S. helicopter to lift the captives from the mountains. The message reached Santiago ten hours later.

While Wollam debated Raúl Castro on July 1, the Buffalo Bisons beat the Havana Sugar Kings 5–2 and the brother of Batista's presidential candidate Andrés Rivero Agüero was assassinated. That day, too, Secretary of State John Foster Dulles held a press conference. A reporter noted that Americans were also captive in China, East Germany, and the Soviet Union. Was there a pattern? "I think it is impossible to treat these different cases as though they were all alike," Dulles responded. But, in all instances, the United States would not be blackmailed—the United States would not make political concessions.[30]

That very day the CIA agent Vice Consul Robert Wiecha made contact with some rebels, who promptly arrested him. "This is very serious [arresting a North American official]," Wiecha lectured. "Do you know what you are doing?" A bottle of Bacardi rum soon warmed the environment.[31]

By July 2 worried embassy officers in Havana had not yet heard from Wollam, to whom they had sent three messages and medicine by courier.[32] But, on that day, to the amazement of entertained local inhabitants, the helicopter the naval base had dispatched at Wollam's request landed safely in a pasture two miles from Calabazas. As he waited to see off five kidnap victims, Wollam was handed a last-minute letter with the two rebel demands he had heard the day before, and a third: a U.S. "*delegado*" attached to the Second Front to observe Batista's use of U.S. military aid.[33] Eager to rescue at least some prisoners and hampered by poor communications and transportation in and out of the Segundo Frente, Wollam tucked the paper in his pocket and did not argue. The helicopter lifted off with the five prisoners; it returned 90 minutes later to pick up Wollam, who relished a hot shower.

The Navy helicopter flew without incident to the Guantánamo base. Wollam contacted Washington and Ambassador Smith to explain Raúl Castro's terms for the release of the remaining captives. From the perspective of the Havana embassy the first two demands (no more arms for Batista and no use of the naval base for Cuban military operations) were already U.S. policy and thus acceptable, but the third (an observer) was not, "because it would imply recognition."[34] In Washington, the president and the National Security Council heard a brief report on the kidnapping crisis from CIA Director Allen Dulles, who sounded optimistic.[35]

Smith telegrammed Washington that "Wollam states that these people are fanatics with blind spots and determined. Wollam says he is sure there are Communists in the group but mixed up in thinking."[36] Actually, Wollam said little about rebel-Communist connections. After conferring with Smith in Havana, Wollam flew out of the capital city on July 4 to prepare for another trek into rebel territory. Smith reported that Wollam judged Raúl Castro's rebel group "isolated, young, irresponsible, emotionally and intellectually unstable and liable to sudden changes of decision."[37]

Meanwhile, on the Fourth of July in the mountains, the kidnappees enjoyed a roast pork meal while rebels shot off rifles to simulate firecrackers. After the Independence Day feast they played baseball. The game had to be called in the seventh inning when Batista planes flew overhead, although the score by that time left no doubt about which side was going to win: Rebels, 10; Invited Guests, 4.[38]

What did Fidel think of his brother's actions? Raúl had acted on his own in his own war zone. On June 22, after stewing for days about the delivery of the rocket heads to Guantánamo and without seeking approval from his brother, Raúl had issued Military Order No. 30 for the roundup of North Americans. Fidel, battling Batista's forces in the Sierra Maestra, probably first learned of the capture of the Moa Bay employees on June 28 via a Cuban government announcement. First he was "unbelieving" and then "furious."[39] Radio Rebelde soon explained that the Sierra Maestra command, because of poor communications, had not heard directly from the Second Front and could not confirm the kidnappings.[40] On July 3 Fidel sent a firm message to Raúl through a rebel radio broadcast: Release the North Americans. Their government, not these citizens, deserved blame for the shipment of weapons to Batista and the "inhuman bombings of Cuban civilians."[41]

On July 5 Wollam returned to Calabazas, where he soon complained that he had to conduct his delicate business in "a goldfish bowl" of reporters and photographers who had arrived—Lee Hall and George Skaddings of *Life,* the freelancer Andrew St. George, and Robert Taber and Wendell Hoffman of CBS, among others from several nations. St. George and Taber had flown in a chartered plane from Miami. Their aircraft ran out of gas and made a daring dead-stick landing on a mined airstrip. Like others who exploited their rebel underground connections to get to Calabazas, Jules Dubois left the Guantánamo Naval Base in a sputtering taxi and through various means of transportation finally reached the "Free Territory of Cuba," but only after he suffered injuries from a jeep accident that dislocated his shoulder and rearranged his nose.[42] The CIA's weary Wiecha also arrived after having been held prisoner at rebel gunpoint for three days; he had failed to see the captured Marines and Navy men.

On the 5th of July a helicopter landed at Calabazas and took out three of the United Fruit people. The helicopter returned to take more, but a "big powwow" got under way and the aircraft was sent back to the naval base empty.[43] The next day Raúl Castro, seemingly unmoved by his brother's order, arrived at Wollam's small dwelling to negotiate. Raúl seemed to relish the world press's witnessing of the event and the cameras' clicking away at every opportunity. Wollam handed him copies of recent statements by Ambassador Smith and Secretary Dulles on the U.S. policies of not sending arms to Batista and not servicing his war planes at Guantánamo and on the rockethead transfer. The young Castro wanted more than press releases; he insisted on negotiating. Thereupon, Wollam beckoned Hall, Skaddings, St. George, and Taber. The newsmen frankly told the assembled rebels that they were making a major mistake, because public opinion in the United States had turned sour. The meeting was adjourned, with nobody happy but perhaps the reporters, puffed up by being called forward to participate in critical talks.

The next day, July 7, Raúl Castro lectured Wollam and Wiecha that the U.S. government had neither offered adequate guarantees nor taken steps toward recognizing the "Free Territory of Cuba." Still, the Second Front commander assured U.S. officials that more prisoners were being moved to Calabazas for possible release. Wollam was shown a 750-pound fire bomb—"a large red can type thing"—that purportedly fell on rebel territory. Told that the United States was the source, Wollam retorted that Batista could buy weapons anywhere on the world market. The rebels insisted that the United States could "control" such arms flows if it wanted to. In any case, the Americans and Canadians were now "international witnesses" to the bombing of civilians.[44]

By this time, "Operation Anti-Aircraft" was under way. That is, the rebels intended to hold American prisoners, especially the military men, for as long as possible to deter further government aerial attacks, which had relented during the crisis. Wollam intended to stand "firm" with the rebels, even though he found the experience "wearing on the nerves."[45] He managed to send from Calabazas to the naval base some long, frank letters intended for the Havana embassy, some of whose officers were now assigned to the base. CIA Havana station chief William Caldwell was also at the base, coordinating Wiecha's activities.

Raúl Castro departed Calabazas, but Wollam stayed there for two more days to supervise the evacuation by helicopter of more civilian captives. On the seventh, three were flown to freedom. Two days later, five more were airlifted to the naval base.

Ambassador Smith railed against the slow pace of the releases and pressed the State Department for more vigorous action: Issue a statement that if the rebels did not free the hostages in 72 hours, the United States would consult with the Cuban government on joint measures. The State Department rejected an ultimatum because it would endanger the lives of the captives. State officers urged instead that Wollam arrange a meeting with Fidel. Smith disapproved, replying that such a conference would irk the Cuban government.[46] The following day Wollam reported to the U.S. Embassy in Havana, where he collected radio equipment and a CIA radio operator, Sam Boki.[47] Raúl approved Wollam's use of Boki's shortwave radio to improve communications with the embassy. On the twelfth the last American and Canadian civilians held captive were set free.

State Department officials not only had to conduct business with the rebels and Ambassador Smith, but also with irritated U.S. military brass. On July 3, Rear Admiral Robert J. Stroh had upbraided William Wieland over the State Department's decision, taken the day before, to suspend delivery of the T-28 planes to Cuba. Wieland coolly asked the naval officer "if he wanted to see the Americans in the hands of the rebels killed."[48] Delivery of the aircraft at that moment in the kidnapping crisis could only have inflamed it. Grudgingly accepting the decision but not giving up, Ambassador Smith informed the State Department that Batista had "a personal interest" in the T-28 aircraft, that the United States must meet its commitment to ship the planes after the captives were freed, and that not a word of the suspension should be leaked to the press.[49]

In Washington, on July 10, State Department and Pentagon officials met to discuss the crisis. Some of the military brass stepped back from intervention. Admiral Jerauld Wright, Atlantic Fleet Commander, pointed out that operations in Oriente Province "would be similar to those at the outset of the Korean conflict in that the rebel groups would undoubtedly disintegrate and reappear as peaceful farmers." The war would be difficult and prolonged. The U.S. military might have to tramp through dense forests and swamps for two to three months. Few officers who remembered the impasse in Korea wanted any part of such a war.[50] State Department officials, including Robert Murphy and Rubottom, agreed that getting into Cuba would be easy, but getting out would be hard. If the United States intervened militarily, Murphy asserted, "we might with luck find their [the captives'] corpses."[51]

But the hawkish Admiral Arleigh Burke, believing that the rebels served Communist designs and upset that U.S. prestige was being damaged, persuaded the Joint Chiefs of Staff to demand immediate action: (1) remove restrictions on military assistance to the Cuban government; (2) give the Cuban government 72 hours to gain the release of the captives; (3) offer Cuba whatever assistance was needed to effect the release; (4) notify the Castro rebels that they had 72 hours to set the hostages free before U.S. intervention. The Joint Chiefs insisted that a Marine Regimental Landing Team be ordered to the Guantánamo Naval Base.[52] "Of course people would get hurt" if the United States took military action, Burke acknowledged.[53]

The U.S. government wisely chose diplomacy over military muscle. Still, and remarkably, the U.S. military in Cuba, with Ambassador Smith's approval, continued to demonstrate publicly U.S. support for the Batista regime. On July 12, in the midst of this very tense kidnapping crisis, the U.S. Naval Attaché and officers of the U.S. Navy Mission joined President Batista at the dedication of a new building for merchant marine cadets in Mariel.[54] It was just the sort of public display that before had angered the rebels and fueled anti-Americanism.

Wollam returned to the mountains on the twelfth with Boki in an attempt to arrange a place for the evacuation of the Guantánamo enlisted men. More discussions and more delays followed until the fourteenth, when Wollam flew by U.S. helicopter to a new, well-hidden site Wollam called Caujerí. Raúl had informed him by cordial letter that the new rendezvous, not on the maps, was marked by four white sheets in the form of a rectangle. The helicopter could pick up the Marines and sailors there.[55] But when Wollam arrived, the rebel commander at the isolated spot balked. He informed a protesting Wollam that the Americans would be released in small groups over several days. Wollam lost his temper, charging deceit. On July 15 five sailors, two Marines, and a Cuban bus driver lifted off for the air base. The rest remained prisoners. The rebels were gambling that the United States would not intervene militarily.

Wollam then settled down for a long wait at Caujerí, all the while pleading for the release of more hostages. During this time he met Charles W. Bartlett, Jr., a 20-year-old machinist's mate of the U.S. Navy. Sympathetic to the 26th of July Movement after he witnessed Batista soldiers beating up civilians, he had fled the ammunition ship USS *Diamond Head* at Guantánamo on July 4. Wollam and Wiecha failed to persuade the deserter to return voluntarily. (Months later, in October, Bartlett showed up at the naval base and was later court-martialed.)[56]

On July 18, while the Buró Represión de las Actividades Comunistas was arresting seven "Communists" in Havana, the rebels in Oriente Province freed all remaining captive U.S. military personnel. Fidel probably ordered the release; four of his men had traveled for eight days through the jagged terrain to instruct Raúl to obey orders. Raúl Castro's command announced that the Americans were now needed elsewhere, presumably in Lebanon, which the United States had just invaded during another Mideast crisis.[57] Jules Dubois claimed credit for persuading Raúl to use this gesture "to get off the hook."[58]

Finally, on July 19, Wiecha, Boki, and Taber were lifted out by helicopter. Wollam and the injured Dubois were flown out later, but not before more trouble. Ambassador Smith had managed to get a message to Wollam telling him to protest the rebels' disarming of a Marine guard along the Guantánamo base fence. Wollam's rebel handlers held up his departure while they penned a denial. (Later, M-26-7 apologized for the incident and returned the gun.)

His kidnapping operation over, Raúl Castro entered $1400 in his account book for "expenses for the Americans"—beer, soft drinks, and more.[59] Park Wollam received the State Department's Superior Service Award. In February 1959, curious to know who Raúl's allies were, Wollam attended the rebel's wedding to Vilma Espín.[60] After the Castro brothers took prisoners at the foiled Bay of Pigs operation in 1961, Department of State researchers dug into their official files looking for helpful leads on how Raúl's captives had gained release in 1958.[61]

15

Frankenstein, Texaco, Nicaro, and a Toughened Attitude

Although Fidel Castro frowned upon his brother's brazen, unauthorized kidnapping of North Americans, a bouyant Raúl Castro saw the prolonged escapade quite differently. His rebel forces earned a needed rest from ground combat and air attacks—time to reorganize and build in anticipation of showdown battles. As long as the North Americans remained captive in the hills, Batista's armed forces and planes stood down. "One American is worth an antiaircraft battery," a rebel lieutenant declared.[1]

The Batista regime, a mere bystander while U.S. officials tramped and helicoptered in and out of rebel territory, once again appeared weak, if not helpless. The Cuban government, it was demonstrated, could not protect foreigners. Nor could Havana help but permit the United States to negotiate with the regime's enemies. The kidnapping crisis drew attention to U.S. deliveries of armaments, damaging U.S. pretensions of neutrality. Although Raúl Castro probably did not know it at the time, the hostage crisis also prompted Washington to suspend delivery of the T-28 airplanes.[2]

During the crisis, the *fidelistas* sought to mend political fences by signing the Caracas Pact on July 20, forming the Frente Cívico Revolucionario. Castro dictated the language of the unity agreement over a radio hook-up to Manuel Antonio Varona in the Venezuelan capital. Castro, the only major anti-Batista insurrectionist on Cuban territory, was in a position both to lead and to conciliate. M-26-7 and eight other opposition groups, including DR, Prío's Auténticos, Civic Resistance, and Montecristi, but not the PSP, agreed to coordinate their anti-regime efforts and to back Castro's choice for president, Judge Manuel Urrutia. In conspicuous contrast to his earlier denunciations of other rebels, Castro waxed magnanimous in the pact statement, praising fighters outside the mountains. The United States, on the other hand, drew criticism for its continued aid to "the dictator."[3]

As many similar statements and the kidnapping episode revealed, anti-Americanism was surging through the rebellion against Batista, intensifying M-26-7 resolve. And the rebels' "anti-American line" rankled officials in Washington and Havana.[4] They too had felt helpless during the crisis—forced to negotiate with insurgents who vowed to overthrow a government that the United States recognized and supported, forced to follow the rebel timetable, forced to shelve a military response for fear of harming the captives, forced to hear ever popular repudiations of their claims of U.S. even-handedness in the civil war. Anti-Americanism in Cuba fed anti-Castroism in the United States. Both accelerated from late summer through fall 1958.

As U.S. officials pondered their limited options in Cuba, their hostility toward the Castro movement deepened. Two days after the last Americans were released, the single-minded Ambassador Smith renewed his pleas to deliver the T-28 trainers. "For US to do otherwise," read his telegram to Washington, "could be interpreted in Cuba that US had been intimidated by kidnappers."[5] At the same time, the diplomat's allies in Havana, the members of the U.S. military missions, also pressed for the T-28s and shipment of non-combat equipment. Soon after the release of the kidnap victims, in fact, the Departments of State and Defense expedited Cuban requests for non-combat equipment: signal flares, communications gear, tools, radar sets, and demolition machinery.[6] U.S. officials also decided to maintain the military missions in Cuba, despite intensified pressure from a broad range of anti-Batista groups that their very presence in Cuba prejudiced U.S. neutrality.

C. Allan Stewart, Wieland's deputy director, took a fact-finding trip to Cuba in July. He talked with Smith and the heads of the military missions. Stewart discovered a flourishing Castro movement because a majority of Cubans—especially the young—stood against Batista. The unpopular "iron hand" Cuban president "has permitted the development of a Frankenstein which he is powerless to crush today." Stewart expressed alarm that Castro's forces were growing day by day, especially outside the Sierra Maestra. The U.S. envoy also reported that moderate dissidents who supported Castro's movement in the capital city had become uneasy about the young rebel leader. "They say they have created a Frankenstein which may, with the passage of time, take over the control of the country." One dissident leader urged Stewart to back "moderate opposition elements" in order to "head off the growing strength of Castro."

If the United States shipped airplanes to Cuba, Stewart wrote, rebels would likely direct reprisals against Americans and force a mass evacuation. The military mission officers, however, sided with Batista and downplayed the regime's repression and killings. Stewart regretted that these military men were "loathe to consider the political aspects" of the problem. Military aid to Batista, Stewart concluded, "would only prolong the eventual showdown." The United States had to find alternatives before Castro became so strong that he could "dictate" Cuba's political future. Castro might very well become "as much of a dictator as Batista."[7]

Both the Havana embassy and the State Department began to write policy papers on Cuba. Rambling and hesitant, the papers recommended little more than hanging on with Batista through the November elections while maintaining the arms suspension.[8] U.S. officials did not like either Batista or Castro, but, given the mood of the Cuban people, Batista's self-destruction, and M-26-7's momentum and control at the local level, the United States had no other options—unless it was willing to launch a military intervention to save Batista or sponsor a military coup to topple him. Although a military ouster of Batista followed by a military-civilian junta always received a hearing in U.S. policymaking circles, the obstacles to it stood high: Batista would not budge; Castro steadfastly opposed this alternative; the military leaders who might stage a coup, like Colonel Ramón Barquín, were in prison and could not easily be freed; even if military renegades managed to depose Batista they still had to face an entrenched and expanding M-26-7.

Most critical, as the U.S. Army's Assistant Chief of Staff for Intelligence observed: "We are unable to identify any prospective leadership in the Cuban Army competent to overthrow the Batista regime. . . ."[9] Ambassador Smith for once had it right—a military coup was "too risky to attempt." Thus the United States seemed trapped, he argued, in a "drifting policy."[10]

President Eisenhower presided over the National Security Council on August 7, but Cuba did not receive even a mention. As before, Cuba had obviously not pressed itself upon him as a burning issue. Better put, his advisers had not pressed Cuba upon him. At that meeting, when the CIA Director surveyed the world, he pointed to the Taiwan Strait as a Cold War flashpoint. Chinese Communist forces were building up opposite Quemoy (Jinmen) and other offshore islands and Chiang Kai-shek (Jiang Jieshi) was becoming "edgy." At the same time, in the tense Middle East, a U.S. invasion force in Lebanon continued to back the beseiged government. NSC discussion also focused on long-range ballistic missile programs—Jupiter, Thor, Atlas, Titan, and Polaris; U.S. policy toward Korea, where large numbers of

U.S. soldiers stood sentinel; and sub-Saharan Africa, where decolonization was producing a host of new, independent nations in a strategic area. Silence on Cuba.[11]

Although the Cuban crisis paled next to events elsewhere, at the departmental, assistant secretary, and office levels of decisionmaking in Washington, and certainly in the Havana embassy and Santiago consulate, the Cuban question commanded day-by-day and often hour-by-hour attention. The CIA was closely watching, too. But with Batista's well-armed forces still intact and not yet clearly in collapse, with Fidel Castro still confined to the mountains at the southeastern end of the island, with elections still holding out some hope for a peaceful transition of power, with no evidence of a Communist threat, and with the expectation that non-Batista, non-Castro "moderate" leaders would emerge, U.S. officials had not decided by early August to advise the president that a client state was in danger of being lost.

Eisenhower Administration officials were also not moved by a powerful public opinion or domestic political needs that might have elevated Cuba to high-crisis status. Although some members of Congress followed the events of the rebellion, most did not. Nor did the general public, which, although it could become passionately vocal on individual events like the kidnapping of Americans, seldom expressed itself in any coherent way on Cuba policy. The managers of the Gallup poll did not count Cuban affairs among the topics worth asking about, for they did not conduct a single poll on Cuba in the period of the rebellion. The U.S. press, more anti-Batista than pro-Castro, never spoke with one voice.[12] Anyway, the U.S. media had traditionally given Latin America comparatively scant attention, and in the 1950s Cuba always had to compete with dramatic events in Europe and the Middle East and the Cold War peril of the nuclear arms race.[13]

Cuba may not have reached a point of urgency in the United States, but to U.S. policymakers it looked less and less manageable. The rebels advanced while Batista faltered, his military no longer a reliable pillar of his power after the failed summer offensive. Mounting guerrilla tangles with police took many more lives on both sides as the terror of bombs, assassinations, torture, and sabotage rocked Havana and other cities. Although some new factories opened in 1958 and the sugar crop was good, the Cuban economy deteriorated in the fall: Sugar prices dropped, decreasing revenues compared with 1957; the transportation and communication systems functioned only intermittently; passenger car and truck sales dropped by nearly half from 1957 figures; tourist hotels attracted fewer customers; the coffee crop dropped below forecasts; cautious businesses reduced their inventories; and unemployment rose.[14]

The Cuban presidential election campaign went forward, keeping alive the hopes of some Americans that the outcome would quiet the tempest, but the absence of constitutional guarantees poisoned the political atmosphere. The Batista regime still imposed limitations on the right of assembly and freedom of speech. Arrestees were denied pleas of habeas corpus. The highly charged issues of U.S. neutrality, the arms embargo, Cuban use of MAP aid, and the military missions would not go away. As FBI agents watched, Prío's Auténticos, DR, and M-26-7 held rallies in the United States to collect money, recruit insurrectionists, and build up pressure on U.S. officials to break with Batista. Spanish-language newspapers such as *El Diario de Nueva York* regularly reported the stories of Cubans who had fled the island.[15]

Heightened anti-Americanism in Cuba manifested itself in accelerating attacks on foreign interests and demands that U.S.-owned and other foreign companies pay taxes to the insurgency. The ongoing Cold War, too, required U.S. officials to worry about the 26th of July's possible linkage with Communists. For Dulles, Herter, Rubottom, Snow, Wieland, Stewart, Leonhardy, and others in the State Department, the late summer and fall offered no respite from Cuban troubles—and no certain policy path.

By the middle of August, Batista's offensive had collapsed, his forces unable to dislodge the insurgents from their mountain stronghold. A small rebel army had held off more than 10,000 government troops. Batista was now thrown on the defensive. Rebels of all camps gained converts and new donations, and Fidel Castro's mystique as the unconquerable Cuban deepened. Early that month, seeing a dim future for the regime, high-ranking Cuban military officers secretly approached an unreceptive Castro with the suggestion that he endorse a military junta to replace Batista. The Cuban military, once the mainstay of the regime, had bent in battle and was now flirting with treason. Ambassador Smith might have gotten the arms embargo lifted, Batista admitted, "if the Army had shown any effective demonstration of power by winning a decisive battle." But after the failed summer offensive, "the active units could not win even a skirmish."[16]

In late August, launching a 26th of July offensive, a rebel force of several hundred men under the command of Che Guevara and Camilo Cienfuegos began to move out of the Sierra Maestra through Camagüey Province toward the central province of Las Villas. The M-26-7 columns made it to the Sierra de Escambray only after weeks under trying conditions—food ran short, mud and swamps impeded travel, ambushes took their toll, informers gave their whereabouts away, troops got lost. That Batista's army could not stop them revealed just how unreliable the government's military had become. In the Escambray mountains, the *fidelistas* encountered staunch rivals, for DR, Communist, Prío, and other units already operated in the territory. Despite the existence of the Caracas Pact, in the Escambray the feuding began immediately over tactics, strategy, politics, and ideology. Even M-26-7 leaders there argued among themselves about the revolution's future programs. Fidel Castro bemoaned the "quarrels and division."[17]

Back in Oriente Province, the columns led by Raúl and Fidel Castro went on an offensive designed to clear the eastern province and seize Santiago. In August, too, the rebel commander-in-chief issued an order requiring all sugar proprietors in Oriente to make a "contribution" to the rebel army for each sack of sugar produced. If they did not pay, he vowed to attack their properties.

Fidel Castro had apparently feared that the kidnapping crisis might cause the U.S. arms embargo to be lifted. But by September he sensed it would hold. A reversal on the stoppage would surely unleash an uproar in both Cuba and the United States and demolish altogether U.S. pledges of neutrality. Confident that Batista had to struggle to acquire arms while rebel forces increased theirs, aware of discontent among Batista army officers, and fearful that a military coup could pre-empt rebel success, Castro communicated secretly with some army officers to spur their disaffection from Batista.[18] As he kept wary eyes on his Cuban rivals, on Batista's military, and on the United States, Fidel Castro seemed more than ever before in a position to set his nation's course and to defy U.S. interests.

As violence engulfed the island in late 1958, rebel confrontations with foreign and U.S. interests and the destruction of U.S.-owned property became commonplace. State Department and Havana embassy officials came to believe that American companies were being singled out and that the attacks on property and demands for taxes revealed a vigorous strain of anti-Americanism in the Castro movement.[19] Would the threat to U.S. interests become so great that the United States would feel compelled to intervene to protect them? As William Wieland once warned Ernesto Betancourt, if the extortion and sabotage against American companies continued, "I did not know how heavy the pressures would be brought to bear on the United States Government nor could I predict the consequences."[20]

Eager as U.S. officials were to protect U.S. businesses in Cuba, they could do little short of direct armed intervention in Oriente because the Batista government was fast losing control of the province and the rebel military seemed able to move at will. At a time when some members of Congress and the press claimed that the United States was losing the Cold War in the Third World because it had aligned itself with dictators, U.S. military intervention seemed unlikely. The case that might have justified armed intervention—that the *fidelistas* were aligned with the Communists and Soviets—could not be made persuasively because there was no proof whatsoever of such a linkage. And even if U.S. officials had contemplated the extreme step of a military expedition, they were restrained by yet another obstacle: No other anti-Batista group or leader was ready or strong enough to challenge Fidel Castro for leadership of a post-Batista government.

By early October, Wieland could recommend no more than leaking a report on property damages to journalists in order to deter the Castro brothers, who, sensitive to U.S. public opinion, might relent. Rubottom scratched on Wieland's memorandum: "I do not expect the U.S. press will find this information very newsworthy."[21] State Department officers were pleased to hear, however, that Urrutia, Varona, Miró Cardona, and other exiles had pleaded with Castro to cease the "blackmail" of North American firms in eastern Cuba because of the negative publicity it aroused in the United States.[22] But the rebels on the ground in Cuba were not listening. Local Cuban conditions dictated U.S. caution. U.S. companies would have to take their hits.

Losses sustained by U.S. firms in Cuba for the first nine months of 1958 totaled more than $2.25 million. How did U.S.-owned businesses handle the rebel pressure and sabotage? They asked the Cuban government for protection, but found the Cuban Army ineffectual. Some companies evacuated dependents. Some flatly refused to turn over any money to the 26th of July Movement. When the New York-based Francisco Sugar Company rejected demands for payment of 15 cents a sugar bag, rebels dismantled track, pushed cars into the water, and then tried to wreck the company's new port.[23] Despite Ambassador Smith's strong advice to "pay not one cent of tribute," some firms fearful of sustaining losses paid taxes to the rebels.[24] On the other hand, the Holston Trading Company, in El Cristo, Oriente, fed up with rebel raids, decided to close its mine altogether. The dismissed mine workers must have made a strong case to M-26-7, because the local rebel commander soon promised to return a stolen jeep if the mine stayed open.[25]

Business executives also met regularly with U.S. officials to ask for assistance. Washington and the Havana embassy criticized rebel attacks on American interests and "other manifestations [of an] anti-U.S. point of view," including requests for financial contributions.[26] Yet when company executives asked State officers what "protection" the U.S. government would give them if they resisted paying "tribute," the reply was unsatisfying. Washington could help only in regions where the Cuban government had jurisdiction, "but in this particular case large areas of Eastern Cuba are under control of rebels who have no status recognized by this country."[27]

The State Department's growing antagonism toward M-26-7 gained momentum with each rebel assault on U.S. interests. In August, rebels burned 10,000 bags of sugar at a United Fruit Company warehouse in Central Boston and levied a "tax" of $186,000 on the firm; Radio Rebelde asked Cubans to boycott Pepsi Cola and Miller's High Life because they contributed to pro-Batista radio; and a bomb blew out windows at the Havana Hilton.[28] In September, fire in a Goodrich rubber plant destroyed 60,000 tires; Cuban Electric reported seven cases of sabotage in Oriente and five in Santa Clara; rebels also demanded money from Bethlehem Steel. The Batista Air Force bombed the grounds around the Nicaro plant, where rebels also burned a Nicaro tractor and "helped themselves to anything they wanted."[29] United Fruit's vice president complained that "hardly a day passes when either horses, mules, saddles, material and equipment are not taken from the various farms" at Preston. United Fruit estimated late in the month that it had lost $50,000 worth of property to the rebels.[30]

In October the rebel war further disrupted the Cuban economy and threatened U.S. interests. An unexploded bomb was discovered at Havana's Coca-Cola bottling plant. Rebels hijacked an airplane flying between Guantánamo and the United Fruit plantation at Preston, while rebel agents asked the First National City Bank of New York in Cuba to contribute $100,000 to M-26-7 for schools and hospitals. And a coastal schooner loading gasoline for the Moa Bay Mining Company blew up. On the twenty-third of the month, rebels kidnapped two United Fruit employees, both of whom had been grabbed before in June. They were held for only a day near Preston and released unharmed.[31]

Rebels particularly harassed the Texaco refinery and the Nicaro plant, intensifying the hostility of U.S. officials toward the *fidelistas.* Texaco had long before become entangled in the Cuban civil war, the company's property a prime target in Oriente. Rebels stole equipment from the pump house that delivered the refinery's water supply, hijacked tank trucks, confiscated jeeps, demanded thousands of pesos, and frequently interrupted traffic heading for the Texaco facility from Santiago, four miles away. Although the company built a high fence around its property and Cuban Army and Navy detachments patroled the area, the depredations continued. Thirty-seven U.S. citizens worked at the plant, and they worried about the safety of their dependents in Santiago. On October 1 rebels assaulted the company's mail truck as it traveled from Santiago to the refinery; they shot the Cuban driver and stole the vehicle.

Park Wollam once more met with rebel leaders. He, the Texaco refinery manager, and a rebel representative conferred on the second of October, but nothing was resolved. Back in Washington, Terrance Leonhardy suggested to the Texaco vice president that the manager remind his Cuban employees that the plant might have to shut down, throwing them out of work. This seemed the only means available to persuade the rebels to halt their harassment. When Texaco asked the State Department if Wollam could be present at a future Texaco meeting with the rebels, should such an event become

ABOVE: Street violence in Havana, January 1954. (*Miami News Collection, Historical Association of Southern Florida, Miami*)
BELOW: President Fulgencio Batista (*left*) and Vice President Richard M. Nixon (*right*) in Havana, February 1955. (*The Richard Nixon Library & Birthplace, Yorba Linda, California*)

Fidel Castro in Central Park, New York City, October 1955. (*Reprinted with the permission of Macmillan Publishing Company from* Fidel Castro *by Jules Dubois. Copyright © 1959 by Jules Dubois. Originally published by the Bobbs-Merrill Company, Inc.*)

TOP RIGHT: Fulgencio Batista (*left*) and President Dwight D. Eisenhower (*right*), July 1956. (*National Park Service Photograph, Dwight D. Eisenhower Library, Abilene, Kansas*)
BOTTOM RIGHT: U.S. Embassy, Havana. (*U.S. Department of State Photograph, National Archives, Washington, D.C.*)

CESEN LOS ASESINATOS
DE NUESTROS HIJOS
MADRES CUBANAS

Riviera Hotel Brochure. (*Historical Association of Southern Florida*)

TOP LEFT: Riviera Hotel, Havana. (*Reprinted with permission from Vicente Báez, ed.,* La Enciclopedia de Cuba, *San Juan, Puerto Rico: Enciclopedia y Clásicos Cubanos, Inc.*)
BOTTOM LEFT: Protest march in Santiago de Cuba, July 1957, during Ambassador Earl E. T. Smith's visit to the city. The banner reads: "STOP THE ASSASSINATIONS OF OUR SONS. CUBAN MOTHERS." (*Reprinted with permission from Victor Báez, ed.,* La Enciclopedia de Cuba. *San Juan, Puerto Rico: Enciclopedia y Clásicos Cubanos, Inc.*)

U.S. products in Cuba. (*Reprinted from United States Cuban Sugar Council,* Sugar: Facts and Figures . . . 1952. *Washington, D.C.: United States Cuban Sugar Council, 1952*)

Ambassador Arthur Gardner (*left*) and General Francisco Tabernilla (*right*), Army Chief of Staff, Cuba. (*Reprinted with the permission of Macmillan Publishing Company from* Fidel Castro *by Jules Dubois. Copyright © 1959 by Jules Dubois. Originally published by The Bobbs-Merrill Company, Inc.*)

Ambassador Earl E. T. Smith. (*Library of Congress, Washington, D.C.*)

Roy R. Rubottom, Jr., Assistant Secretary of State for Inter-American Affairs. (*U.S. Department of State Photograph, National Archives*)

William A. Wieland, Director, Office of Middle American Affairs, Department of State. (*AP/ Wide World Photos*)

ABOVE: Park F. Wollam (*striped shirt, center*), U.S. Consul, Santiago de Cuba, July 1958, during the kidnapping crisis. (*AP/World Wide Photos*)
BELOW: Raúl Castro (*second from left*) shakes hands with Vice Consul and CIA agent Robert Wiecha (*right*), July 1958. (*Courtesy of U.S. Department of State*)

PARK
WE WANT PRIO FREE
BATISTA MEANS CORRUPTION AND MURDER
WE WANT PRIO FREE

Fidel Castro in the mountains. (*Cuban Collection, Yale University Library*)

Top left: Carlos Prío Socarrás, former President of Cuba. (*Miami Herald*)
Bottom left: Demonstration in Miami, February 1958. (*Miami Herald*)

Celia Sánchez (*right*), Fidel Castro's assistant. (*Cuban Collection, Yale University Library*)

William P. Snow, Deputy Assistant Secretary of State for Inter-American Affairs. (*Courtesy of U.S. Department of State*)

Raúl Castro during the rebellion. (*Cuban Collection, Yale University Library*)

Old Havana church and Pepsi-Cola sign. (*Reprinted with permission from Victor Báez, ed.,* La Enciclopedia de Cuba. *San Juan, Puerto Rico: Enciclopedia y Clásicos Cubanos, Inc.*)

President Dwight D. Eisenhower (*left*), Secretary of State John Foster Dulles (*center*), and Under Secretary of State Christian A. Herter (*right*). (*National Park Service Photograph, Dwight D. Eisenhower Library*)

William D. Pawley, U.S. emissary to Batista. (*AP/Wide World Photos*)

ABOVE: Reunion of Fidel Castro (*left*) and Herbert Matthews (*right*) in Cuba, January 1959. (*New York Times*)
BELOW: Che Guevera (*third from left*), President Manuel Urrutia Lleó (*fifth from left with glasses*), and Camilo Cienfuegos, rebel commander (*right*), January 1959. (*AP/Wide World Photos*)

ABOVE: Fidel Castro after triumph. (*Cuban Collection, Yale University Library*)
BELOW: Northeast gate at the border between Cuba and the U.S. Naval Base, Guantánamo Bay. Senator Frank Church of Idaho, who in the mid-1970s helped expose CIA plots to assassinate Fidel Castro, visited the base in 1961. (*U.S. Navy Photograph, Frank Church Collection, Boise State University*)

necessary, State officers pulled back. Wollam's presence at a meeting with rebels might lead the insurrectionists to think that they could press demands on other U.S. firms and would surely spoil the consul's relations with local Cuban government authorities. On October 17, rebels sabotaged the company's water well.

Three days later insurrectionists under Raúl Castro's command seized two Americans and seven Cubans traveling between the oil refinery and Santiago. U.S. officials in Washington publicly denounced the kidnappings, while Smith urged Wollam to tell his rebel contacts that if the Americans were not released immediately, the ambassador would recommend the sale of U.S. arms to Batista. The State Department, once again disciplining its chief diplomat in Cuba, killed the ultimatum. The captives were set free three days after their abduction. Charles R. Bennett, a construction chief, and Kenneth D. Drews, a machinery foreman, recounted that they had been picked up by a rebel squad and taken to a camp, where the commander ordered their release. Fidel Castro soon explained that the men had been seized because they happened to drive into an ambush Raúl's warriors had set for Batista forces. Fidel "boiled with resentment" against State Department officials for criticizing his troops.[32] The Texaco men should have known the area was dangerous at night; they had to be detained to protect the rebel ambush and themselves. No harm was done to the captives, he said. In fact, the rebels had never killed a North American. Why didn't the State Department protest *Batista's* brutalities? Cuban lives were as important as North American lives.

After the rebels let their prisoners go, they asked Texaco for $500,000 in cash or $300,000 in arms as a down payment on the taxes the company would have to pay under a future revolutionary government. Delaying a response, Texaco began to evacuate its dependents. On November 3, the day of the Cuban presidential election, a force of 30 rebels stole Texaco's last two jeeps, as well as 300 gallons of gasoline. Refinery manager Charles Cutbirth threatened to close the plant. The rebels, used to exploiting Texaco as a transportation and fuel supplier and worried about the unemployment of several hundred workers, soon promised to refrain from molesting the refinery.[33]

Throughout the war in Oriente Province, the rebels also preyed on the U.S.-owned nickel plant at Nicaro. They repeatedly attacked Cuban Army guards and moved freely in and out of the plant site at night. Many Nicaro workers sympathized with the insurgency. Just days after Raúl's forces released the American hostages in late July, a rebel band overpowered two guards at the Nicaro plant and stole a road grader and 1000 gallons of gasoline.[34] At one point in the summer, Batista tried to use rebel assaults on North American property to pry open Washington's arms door. He informed Ambassador Smith that he would station 1000 soldiers at Nicaro if the United

States would supply 2000 rifles and other supporting equipment. The State Department killed any such deal.[35]

Raiders attacked Nicaro once again on September 29. They killed an employee and kidnapped two company guards.[36] By early October the rebels had hauled away Nicaro equipment worth $200,000. Late in the month, after Batista troops withdrew from Nicaro, the rebels occupied the facility, established a garrison, and promised not to interrupt plant operations. During the early morning hours of October 24, the USS *Kleinsmith,* a transport ship from Guantánamo, arrived at Nicaro and began to evacuate more than 50 dependents of Nicaro employees. Later in the day, Washington ordered the aircraft carrier *Franklin D. Roosevelt* to the scene to provide helicopter airlift if needed. President Eisenhower was kept informed throughout the crisis.

Just as the evacuation began, Cuban warships appeared, landed government troops, and commenced to fire into the plant area. After fleeing the launches sent to ferry them to the *Kleinsmith,* the evacuees hid in a warehouse. At noon the insurrectionists began to leave Nicaro, but Cuban employees at the plant blocked the departure of the evacuees until they received Batista Army assurances that they, the workers, would not be punished for their rebel sympathies. Although their exit was delayed, the U.S. citizens finally made it to the Guantánamo base, from where they were flown to Miami.[37]

On October 25, Fidel Castro lambasted the re-entry of Cuban forces as a "maneuver by Batista and Mr. Smith, along with various officials of the U.S. Department of State, to bring about a landing of North American forces."[38] Castro began to worry anew that Batista sought to Americanize the Cuban civil war. As for Nicaro, the Cuban Army reoccupied it. Plant managers decided nonetheless to suspend operations. In Washington, Assistant Secretary Rubottom remarked that a "showdown" was approaching and that U.S. officials "had a considerably toughened attitude toward the rebels in the last few days because they seem to have come almost to a point of no return. . . ."[39]

16

Burning Up the Wires: The Quest for Communists and Arms

As U.S. OFFICIALS protested rebel malevolence against the Cuban economy—and hence against North Americans and U.S.-owned property—they increasingly wondered about Communist ties with the 26th of July Movement. They also struggled with the persistent question of the U.S. military linkage with the Cuban regime. The three issues—attacks on American business interests, possible Communist influence in the insurrection, and continued U.S. military connections with a faltering dictatorship—became intertwined in the American mind. Did the anti-Americanism evident in the kidnapping of U.S. citizens and attacks on American-owned properties indicate that Communists had gained a foothold in the 26th of July Movement? Should that be the case, should not the United States sustain the Batista regime through military assistance in order to forestall a victory for international Communism?

Ambassador Smith more than any other American official thought in worst-case terms. His stubborn hammering of the spread of Cuban anti-Yankee sentiment, the threat of Communism, and Batista's need for U.S. help to combat a Communist-infested Castro movement did not persuade Washington to lift the weapons suspension. But the ambassador's unrelenting criticism of Fidel Castro and the *barbudos* helped intensify U.S. hostility toward the rebels and nurture a receptivity to some form of U.S. intervention.

Anxiety about the role of Communists in the Castro camp no doubt contributed to the U.S. "toughened attitude" that Assistant Secretary Rubottom mentioned. From the time the *Granma* landed in Oriente, U.S. officials had been probing as much for any Communist proclivities in the 26th of July Movement as for the wellsprings of profound Cuban disaffection toward Fulgencio Batista and his long-time patron, the United States. Even so, the anti-Americanism expressed by the ardent Castroites of the movement, as much as their alleged Communist inclinations, perhaps explains the growing official U.S. hostility toward them. As Rubottom said, "It didn't make any difference whether Castro was a Communist or not. He was so obsessed in his hatred for the United States. . . ."[1] But perhaps the two categories cannot be separated. Like many Cold Warriors of the day, Ambassador Smith believed that people who were anti-American must also be pro-Communist.[2]

Despite Ambassador Smith's claims that the PSP and M-26-7 had become close partners in the war against Batista, U.S. officials, try as they did, could not find evidence of a Castro-Communist pact, important Communist infiltration, a Communist takeover, Soviet ties, or a commitment by Fidel Castro to Communist dogma. Heard were the usual comments that Che Guevara preached Marxism and that Raúl Castro had once visited Eastern Europe. Familiar were the charges by U.S. business executives with interests in Cuba that both Castro brothers had become Communists.[3] A cautious statement by Minister-Counselor and No. 2 officer of the U.S. Embassy, Daniel M. Braddock, revealed his awareness of such sentiment: "It appears certain that he [Guevara] is an extreme leftist, ultra-nationalistic, and strongly anti-American. In addition, the charge has repeatedly been made that he is a Communist. Whether or not that is true, he is certainly an obvious channel for the Communists to try to use in their attempts to infiltrate and influence the Cuban revolutionary movement, in which he is becoming an increasingly prominent figure."[4]

U.S. officials also knew well both Batista's and Smith's repeated remarks that because the PSP and M-26-7 both sought to oust Batista and their messages sounded alike, they were following the same line—the Communist line. "The recent kidnappings have brought to light the Communist influence and penetration in the Raúl Castro group," the ambassador concluded in one of his many exaggerations and inaccuracies. "The Communists and the 26th of July Movement have the same prime objective—the overthrow of Batista."[5] Using numbers produced by *batistianos* and faithfully recording every Batista official's claim that Communists lay behind the island's troubles, Smith urged U.S. officials to quadruple their estimate of Communist party strength in Cuba to more than 100,000.[6] In October 1958, to cite yet another example of unsubstantiated reporting, Smith informed Washington that one of the candidates for the presidency told him that 80 to 85 percent of the 26th of July Movement was Communist. And, "when Raúl was attending Havana University there was talk that he was a homosexual."[7]

Always lacking proof of an alliance between Communists and *fidelistas* or Communist control of the rebel movement, Smith once asked the U.S. Embassy in Mexico City for information on Communists among the three Cuban exile groups there. The answer: No evidence existed that Communists dominated the exiles.[8] FBI investigators, moreover, had discovered "no communist direction, control or influence of the July 26 Movement in the U.S."[9] Smith nonetheless pushed on, often retreating to a Cold War truism that revealed little about the sources of the anti-Batista insurrection or about M-26-7's political goals: The clever Communists hovered, just waiting for the right moment to exploit the chaos churned up by the rebels.

Most U.S. officials judged Castro's ideas fuzzy and undeveloped—nationalistic and anti-American to be sure, but not communistic or pro-Soviet.[10] Even when anti-Batista leaders such as former Prime Minister Varona confided to State Department officers that they sought to hem in Fidel Castro because they feared his dictatorial ways, they did not claim that he was a Communist. A Cuban exile whom U.S. officials highly respected, Felipe Pazos, assured them that no Communists held positions in the 26th of July hierarchy.[11] Cuban Rotarians and other business people in Santiago de Cuba assured Wollam that there was no connection between the rebels and the Communists. *The* Cuban question, they insisted, was the Batista dictatorship, not Communism.[12] British and Canadian diplomats posted in Havana did not report to their governments that Castro or the 26th of July Movement had submitted to Communism. Smith's "visceral conviction" that Castro had become a tool of the Communists lacked an essential test of a sound conclusion: evidence.[13]

Castroite insurgents in the mountains became so tired of journalists' and U.S. officials' unending questions about alleged Communist machinations in the rebellion that they poked fun at the ritual. "Have you seen any Communists here?" they joked and laughed with foreign visitors.[14]

The kidnapping episode and accentuated anti-Americanism gave more urgency to the repeated question. In August, for example, a unit of the State Department's Bureau of Intelligence and Research tackled the issue. The conclusion: "Although the Cuban Government has consistently charged that the 26th of July Movement is penetrated and influenced by communism, little evidence exists to prove these allegations, although there are continuing reports that some communists have entered the lower ranks of the rebel forces." M-26-7 lacks "any significant ideology." The PSP had joined the united front against Batista, but most rebels remained suspicious of the party.[15] Che Guevara himself said that although the PSP and M-26-7 cooperated in some activities, "mutual distrust" spoiled their relationship.[16]

At an embassy staff meeting in the fall, Smith asked: "Is Guevara a

Communist?" The assembled officers, many of them already quite at odds with their boss, mentioned the familiar rumors, and the new CIA station chief said he knew nothing about Che's alleged Communism. Smith bolted from his chair. "What the hell is this? Top career officers and not one of you can tell me whether Guevara is or is not a Communist?" The ambassador bellowed: "Well, I know that Guevara is a Communist even if you don't. And I'll tell you what I'm going to do. I'm going to burn up the wires to Washington and tell them that the Communists are marching through Cuba like Sherman's march to the sea. Now, how the hell do you like that?"[17]

Some Communists did join Raúl Castro's and Che Guevara's forces, and Communist propaganda sometimes did parallel that of other anti-Batista groups, but PSP relations with M-26-7 apparently remained cool and distant during most of the insurrection period. In September 1958, PSP leader Carlos Rafael Rodríguez joined the mountain forces, but the cautious PSP position vis-à-vis Castro's guerrilla movement was never clarified. Castro may very well have welcomed Rodríguez in order to coopt the PSP, to subordinate it to his wishes, or to use it to help him organize workers, because the Communists had some strength in labor unions and in the cities. But, contrary to Smith's charges, the PSP had not infiltrated M-26-7. Rather it seems that the 26th of July Movement had begun a pragmatic process of dominating the Communists, just as Fidel Castro had done so many times before to discipline, dominate, or eliminate other rival groups.[18]

That Castro later—after victory and repeated crises with the United States—declared himself a Communist cannot erase the *pre-1959 record* of minimal contact between the PSP and M-26-7. The reason that U.S. officials could not build a case that Castro or other rebel leaders had thrown in with the Communists was that the insurgents had in fact not done so. McCarthyite accusations erupted later that U.S. officials had failed to gather intelligence data and analyze the "Communist problem." Actually, U.S. officials repeatedly studied the question. Their true failure was that they dwelled on the wrong question, and thus they never fully understood why Cubans so detested Batista and why Castro was emerging triumphant. Castro became more radicalized and perhaps more enamored with Marxism as the rebellion intensified. But even on this point, many people who knew and observed Castro at the time found his ideology and his political programs ill-formed and in transition.

As late as October 31, Rubottom commented to a group of U.S. businessmen with interests in Cuba that "although there is no doubt the Communists are exploiting the opportunity provided by the rebellion, the Department had no conclusive evidence that the Movement is Communist-inspired or dominated." He added that "if we had had conclusive information to this effect, our attitude towards the Cuban situation would have been altered considerably."[19] Indeed, had serious links between Castro and the Communists in fact existed, and given the prevailing Cold War mood intensified by the Berlin crisis and Nikita Khrushchev's belligerence, the United States might very well have intervened directly to prevent a rebel triumph, to prevent a Communist victory, and to forestall another "who lost ——?" debate in the United States. One of the reasons why the Cuban rebellion had not pushed itself to the top of President Eisenhower's worry list was that U.S. officials had not discovered a Communist threat in Cuba.

On one occasion, even Ambassador Smith seemed to concede that his case that the 26th of July Movement had joined a Communist conspiracy lacked substantiation. In November he told the Department of State: "It is essential in my judgment, for sound US policy determination re Cuba, that our Government learn beyond any doubt whether and to what extent Castro movement is penetrated, supported, influenced or directed by international Communism. Our information of this subject is dangerously inconclusive."[20]

While U.S. leaders searched for traces of Communism in the insurrection, lectured the rebels on kidnappings, and protested mistreatment of North American property, they also found themselves once again in a spirited debate about U.S. military relations with the Batista government. Ambassador Smith, State Department officers, Cuban government officials, and anti-Batista groups debated the arms suspension (Should it be lifted?); the T-28 trainer planes (Could they be delivered without arousing protest that the United States was violating its pledge of neutrality?); the use of already shipped MAP matériel (Should the United States demand compliance with its request that the Cuban military sheathe such equipment?); shipments of weapons from other nations to Batista (Should the United States encourage or discourage?); and the military missions (Should they be withdrawn?).

Shortly after Raúl Castro released the kidnapped Americans, the Cuban government, with Smith's undisguised encouragement, pressed for new U.S. arms deliveries and the T-28s. The government could not maintain "public order" or defeat the "evil bands" in rural areas without more U.S. weapons and equipment.[21] Smith soon followed this Cuban appeal with his own: Washington must meet its "moral obligation" to the Batista government, which, Smith emphasized, had long been "friendly" to the United States, "benefited American investments," voted with the United States in international forums, outlawed the Communist party, and severed diplomatic relations with the Soviet Union. The United States must help the regime to prevent "international communism" from engulfing Cuba. The ambassador flew to Washington to argue his points.[22] As noted earlier, his efforts bore some fruit: The State Department approved the shipment of non-combat equipment.

The Cuban government continued to pursue the propeller-driven, single-engine T-28 trainers for its air force. Cuba owned the planes because it had paid for them, but U.S. authorities had suspended their shipment during the kidnapping crisis. Finally, in early September, Cuba gave up the contest and advertised the sale of the aircraft in the United States to the highest bidder.[23]

Ambassador Smith remained sharply at odds with the State Department on the T-28s and the armaments suspension, and, according to a purloined letter from the Cuban military attaché in Washington to his Havana superiors, a U.S. general said that "Ambassador Smith . . . is now a valuable co-operator with the American Armed Forces in his fight with the Department of State to defend the sale of arms to Cuba." Fidel Castro read a photostatic copy of the letter, and it apparently angered him.[24]

Unable to obtain weapons in the United States, Batista turned to Italy, the Dominican Republic, Nicaragua, and Britain. During August and September, with some hesitation because of volatile conditions on the island, the British agreed to sell Cuba 15 Comet tanks and 17 Sea Fury airplanes, propeller-driven fighters armed with cannon and machine guns.[25] The U.S. Military Mission officers' only regret over the Cuban-British deal was that the American efforts to subordinate the Cuban military through the standardization of equipment "will retrogress."[26] And that is exactly what the British hoped would happen: They eagerly sought to penetrate the military sales market so dominated by the United States by satisfying the Cuban "appetite" for arms.[27]

When the deal became public, M-27-6 urged Cubans to boycott British products, insurance companies, and motion pictures. Even more, Rebel Law No. 4 declared that British assets in rebel-held territory would be confiscated. Stung by public criticism in Britain and Cuba and fearful for its interests on the island, the British government eventually announced an arms suspension in late November—but not before 12 Furies and all of the medium tanks had been off-loaded at Havana harbor. Almost immediately some of the Sea Furies went into battle in eastern Cuba. Behind the scenes, and surely against State Department wishes, Smith quietly encouraged the British Ambassador in Havana to continue to sell arms to Batista.[28] But the British killed a Cuban order for $5 million worth of rifles and machine guns. Meanwhile, Batista placed an order for 10,000 Garand rifles with Italy.[29]

The economic sanctions took only a modest toll. Shell of Canada, owned by the Shell Group (60 percent Dutch and 40 percent British), became a target in large part because its general manager ardently backed Batista. The company suffered some damage, and Havana drivers boycotted Shell gas stations. British-made goods declined slightly in sales, while rumors spread that the British Ambassador would be kidnapped. Some Canadians, thought to be British, were harassed. Americans, British, and Canadians—the "ABC" colony in Havana—tended to cluster in similar businesses and clubs, so it is not surprising that Cubans sometimes mistook one for the other.[30]

The U.S. military missions on the island, meanwhile, lent themselves as always to pro-Batista purposes. On August 1, at Camp Columbia in Havana, U.S. Army Mission personnel led by Colonel Clark Lynn demonstrated field sanitary measures and the use of field kitchen equipment to Cuban Army units. During the two-hour demonstration, a battery of four 81 mm mortars fired several rounds. Exploiting the day for propaganda purposes, the Cuban Army issued a news release. Pictures that regime photographers took soon appeared in Batista's controlled press. That the event was staged to publicize U.S. backing of the regime, as Cuban exiles and Representative Porter charged, seems obvious. For what other reason would Cuba's military high command, including Generals Francisco Tabernilla and Roberto Fernández Miranda, attend a routine display of kitchen gear and instruction on sanitary methods?[31]

The affair reinvigorated pleas from Batista's opponents that the three U.S. military missions depart Cuba to end "a form of intervention in our internal affairs."[32] The controversial effects of such public shows of U.S. connections with the regime seem to have had little impact on the military attaché in the U.S. Embassy who, later in August, requested that the State Department approve a congratulatory message from U.S. Army Chief of Staff Maxwell Taylor to General Tabernilla on the occasion of Cuban Armed Forces Day. The State Department flatly denied the request.[33] High-ranking U.S. military officers such as Major General Ralph E. Truman nonetheless continued to entertain Cuban military leaders in public ceremonies, and MAAG officers went conspicuously to graduation exercises at Camp Columbia.[34] Embassy staff knew that Batista "has embarked upon a policy of using every opportunity to identify in the public mind continued U.S. military support of the regime."[35] The U.S. military itself seemed to endorse this policy.

At times the business-as-usual image that the U.S. military tried to sustain also characterized other Cuban-U.S. relationships and fed the popular Cuban view that the United States was in the trenches with the Batista regime. In August, for example, the two nations signed a convention for the conservation of shrimp; in September they initialed a comprehensive agreement for cooperation in developing atomic energy to generate power.[36] At the Waldorf-Astoria Hotel in New York City, the ambassador's wife, Florence Pritchett Smith, gave a gala ball to raise scholarship funds for Cuban students in the United States. Batista propagandists once again exploited an event to demonstrate U.S. support for the Cuban government.[37]

On October 13, the Cuban Senate held a special session to commemorate the centennial of the birth of Theodore Roosevelt. In conspicuous attendance was Batista, and sitting next to him as the president's guest of honor was none other than Ambassador Smith. Although the assembly heard a panegyric for the speak-softly-and-carry-a-big-stick Rough Rider and twenty-sixth U.S. president, surely some solons with nationalistic pulses quietly remembered that it was the bullying Roosevelt Administration that had imposed the Platt amendment on Cuba. At the same time, a brief ceremony at TR's bust in Santiago's Roosevelt Park drew criticism that the government did more to commemorate the U.S. president than it did to celebrate hallowed Cuban anniversaries.[38]

With the Cuban elections just a few days away, on October 30 Secretary of State John Foster Dulles went to the Cuban Embassy in Washington as Ambassador Arroyo's guest at a reception. The Cubans had contrived the evening for propaganda purposes. M-26-7's Betancourt fumed but thought it idle to lodge a protest with the State Department. Photographers snapped pictures which were then shipped by air to Cuba. The next day, controlled Havana newspapers printed photographs of the party, and their headlines emphasized that Dulles toasted Batista.[39] It is difficult not to believe that Dulles—and ceretainly his advisers—understood the import and message of the Secretary of State's presence at the affair: Cuba was a valued ally, and Cubans should vote for peaceful change.

These public demonstrations of Cuban-U.S. friendliness could not mask, however, Batista's seething frustration with the United States. His northern neighbor denied weapons to the regime, as when the United States withheld a destroyer escort from Cuba "pending stabilization of the political situation."[40] U.S. leaders also failed to block the arms that rebels managed to import from Florida. This despite U.S. assertions that authorities were diligently pursuing and intercepting illegal deliveries funded by M-26-7 collections in the United States.

The crash-landing of a small plane inside the territory of Guantánamo Naval Base on August 16 highlighted the issue. William Hormel (a U.S. citizen) and three Cuban passengers (members of DR, the Auténticos, and M-26-7) were flying from Miami to the Sierra de Escambray to drop 2000 pounds of arms and ammunition by parachute to insurgent forces. But foul weather set in and the gunrunner decided to abort the mission. Hormel then tried to land his Aero Commander at airstrips in both Raúl's and Fidel's Oriente fronts. But continued bad weather and an empty fuel tank forced Hormel to splash down in the bay of the naval base. Splitting up, the four men escaped from the Guantánamo reservation into Cuban territory. Hormel came across a peasant farmer sympathetic to M-26-7 who gave him dry clothes. Rebel contacts provided a car which got him to Santiago, where he boarded a bus for Havana.

Three days later Hormel, a M-26-7 sympathizer, entered the U.S. Embassy to tell his story. He claimed he was on his twenty-eighth trip, and all previous missions—to the Sierra Maestra and Sierra del Cristal—had proven successful. Receiving no help from embassy political officer William G. Bowdler and fearful that the Cuban police would kill him, Hormel sneaked out of Cuba through the rebel underground. Hormel was arrested upon his return to Miami, charged with smuggling arms to Cuban rebels, and then released after posting bond of $2500. The Cuban government sent a stinging rebuke to Washington for not halting continued rebel activities that blatantly violated U.S. law.[41]

Actually, U.S. authorities worked overtime to block illegal arms shipments from the United States. On September 9, at Fort Lauderdale, for example, U.S. customs agents and border patrolmen stopped the *Harpoon*, a 60-foot yacht loaded with medical supplies, arms, and ammunition. They arrested 30 insurgents allied with Prío and Castro. After their arrest, the rebels went on a hunger strike. The judge sentenced them to 60 days in jail and fined each of them $200. The U.S. case for vigilance, however, looked paler after a customs official in south Florida reported that although his agents had captured tens of thousands of dollars' worth of arms and ammunition in 1957–58, and gunrunning "is just about keeping us going night and day," probably 50 percent of all contraband shipments made it out of the country.[42] Andrew St. George himself discovered "a regular air supply shuttle from Florida to Southern Cuba."[43]

In early October, the *Miami News* reported that Prío-funded Cuban exiles were training with World War II-vintage P-51 fighter planes at the Broward County Airport, a point of departure for clandestine rebel flights to Cuba. U.S. agents who were spying on this activity told the newspaper that they were ready to nab the exiles the moment they used the planes against Cuba or for the transport of munitions. Following such newspaper reports, the Cuban government protested once again.[44]

U.S. officials did try to assuage the Batista government by sharing U.S. intelligence information. For example, through the State Department and U.S. Embassy, Cuban authorities were "discreetly" informed that a boat with 200 armed insurrectionists was about to depart the United States for the island. The Cuban Navy, which already knew about the voyage of the *Pimpernel* from Fort Lauderdale, captured the craft.[45] On another occasion, U.S. officials told Cuban authorities about a plot by a Prío group which intended to fly a jet plane over Havana to bomb the Presidential Palace when Batista was inside.[46]

For Cuban oppositionists who already believed that the United States was taking sides in the civil war, making a mockery of neutrality, this kind of cooperation with Batista's government only deepened the feelings of anti-Americanism that flourished in 1958. They very much appreciated the arms suspension, but Batista's legions were still using U.S. weapons to kill Cubans and the U.S. military missions still advised the officers of the tyranny. The MAP battalion soldiers remained in the field, despite U.S. protests. The Guantánamo Naval Base and its Marines, who had briefly entered and occupied Cuban territory, stood as a reminder to rebels that the United States and its many resources were quite near at hand and Washington might, just might, intervene to save the collapsing regime and U.S. interests. After all, U.S. leaders continued to toast Batista and his cohorts and Ambassador Smith constantly sounded like a partisan for the regime, the CIA still worked to build BRAC, and U.S. agents in Florida frequently intercepted rebel forays.

"The net result of our neutral position is to please no one," concluded Smith.[47] The Havana embassy country team wrote that "every act of the United States which affects the course of the struggle in Cuba can be, and is, interpreted as intervention. We are credited in Cuba with the very thing we seek to avoid, intervention, and we have lost for the present some of the friendliness Cubans normally feel toward the United States." More: "Our present posture of complete withdrawal from the struggle and attempted isolation from it ignores the realities of United States involvement in Cuba politically, economically and militarily."[48] The words were curious because they were contradictory. The United States had by no means withdrawn from the crisis—precisely because U.S. interests were so "involved" in Batista's Cuba.

IV

Dumping the Dictator, Blocking the Rebel

17

A Pox on Both Their Houses

THE U.S. EMBASSY did not anticipate "ideal democratic" elections, but they were "infinitely better than a violent overthrow of Batista and far better than no elections at all."[1] Few, now not even Ambassador Smith, expected the national elections of November 3 to end the revolutionary war, because the climate for the honest elections that President Batista had long promised could not have been more bleak. A full-scale civil upheaval was staggering Cuba. Constitutional guarantees remained suspended and the press muzzled. Batista's handpicked candidate, Andrés Rivero Agüero, running under the banner of the Progressive Action party, defended the regime. His two major opponents, the aging, semi-invalid, and discredited Auténtico Ramón Grau San Martín and the conservative lawyer-author Carlos Márquez Sterling of the Free Peoples party aroused little public enthusiasm. Both men may even have accepted financial aid from the government.[2] The rebels, moreover, urged voters to boycott a predicted fraud. Cynicism and polarization still ran rife through Cuban politics.

Election day passed quietly in large part because Cubans seemed resigned to Batista's prostitution of the outcome. Few voters went to the polls. Rebel threats scared off some. Consul Park Wollam estimated that perhaps only 5 percent of Santiago's electorate voted and other observers judged that the national figure reached no more than 30 percent. Batista's man Rivero Agüero won big, and government candidates also swept the congressional, gubernatorial, and municipal elections. The exact vote totals will never be known, first because fraud corrupted the process, and, second, because even before all of the ballots were counted, Congress in late November declared the electoral process closed and prohibited any challenges to the results.[3] "Fidel Castro has won the elections," concluded Márquez Sterling's disappointed running mate.[4]

Two days after the election, President Eisenhower uttered only his second public comment on the Cuban rebellion (his first had come in July during the kidnapping crisis). When asked at a news conference to explain U.S. policy toward the insurrection, he remarked that the United States was simply trying to protect U.S. citizens and "to keep out."[5]

As for Ambassador Smith, he felt that Batista had betrayed him, that he had broken his pledge to hold free and open elections, that he had killed whatever slim chances remained for a peaceful solution to the national crisis. By grabbing the election for his own candidate, Batista gave impetus to Castro's rallying cry that the dictatorship had to be banished from Cuba. This was, Smith recalled, Batista's "last big mistake."[6] Reluctantly and hesitantly the U.S. Ambassador soured on the Cuban leader he had long boosted. But Smith hoped that Batista could at least last until his term officially ended in early 1959.

Just a few days after the elections, Márquez Sterling, persuaded that he had actually won, sent word to Ambassador Smith that government fraud had stolen the elections. Now greater numbers of Cubans would be attracted to the 26th of July Movement. The United States had to act to block M-26-7's advance; it had to support the new government by renewing arms shipments. Smith obviously welcomed Márquez Sterling's advice, which he sent to the State Department with a barbed reminder that his superiors should stem the flow of rebel arms from Florida and cease Washington's "moral aid and comfort" to the rebels. Smith also intimated that he could nudge the new president-elect toward some political compromises. Embarrassed by the farcical elections, Smith now found a new hope, Andrés Rivero Agüero, and a new cause, U.S. backing for a Rivero Agüero government. Smith's goal was to head off a triumph by Fidel Castro. The State Department shared this objective, but asked to see some concrete plans.[7] After all, as Embassy Counselor Daniel Braddock reported, the "good-natured" president-elect counted Batista and the Cuban armed forces as his "best allies."[8] On the surface, it seemed doubtful that a solution to the Cuban crisis could come from such a loyal *batistiano.*

Smith and Rivero Agüero soon met over lunch at the ambassador's residence. The president-elect explained how he would end his nation's revolutionary strife. He intended to call a constituent assembly to restore public confidence in the electoral process and to return constitutional government to the island. Then he would have Fidel Castro, a "sick man" with a "syphilitic inheritance," killed or captured. To eliminate the rebel forces, of course, the Cuban government would need U.S. weapons.[9]

Meanwhile, U.S. authorities continued their efforts to thwart exile activities in the United States. In early November, U.S. Border Patrol agents arrested four men in Pennsylvania and West Virginia for smuggling arms to Cuba. A few days later, federal and local officers from Fort Lauderdale approached an arms-laden twin-engined Lockheed Lodestar at a remote airport. The surprised Cuban rebels opened fire. Some Cubans escaped, but the agents took 22 prisoners. In New York, a grand jury indicted Arnoldo Goenaga Barron, the head of Comité Ortodoxo de Nueva York, for failing to register as an agent for the Castro movement. A naturalized American citizen, he had led the expeditionaries whose boat U.S. authorities had seized off Brownsville the previous March.[10] Even Mexican officials, usually tolerant, confiscated aircraft and trucks loaded with contraband intended for Castro's forces.[11]

At this juncture Ambassador Smith returned to the United States. Before traveling to Washington, Smith stopped in New York City to attend a private testimonial dinner in his honor at the elite Brook Club, whose members included the financier Laurence S. Rockefeller, the columnists Joseph and Stewart Alsop, Ambassador Norman Armour, and other notables. His remarks, if he delivered them in the words he submitted earlier to State officials for clearance, were innocuous and superficial, stressing strong U.S.-Cuban ties. He did contrast Fulgencio Batista, "an intelligent man, a courageous man, with a firm chin and a ready smile," with Fidel Castro, the purveyor of insurrection and sabotage, whose 26th of July Movement had received Communist endorsement. At the end of his speech, the ambassador optimistically mentioned Rivero Agüero's determination to find a "peaceful solution" to the Cuban crisis.[12]

On November 22, Smith met in the Department of State with Rubottom, Snow, Wieland, and others.[13] The strong-willed ambassador judged these Latin American specialists as so pro-Castro that they had underestimated the Communist danger in the Caribbean. They, for their part, believed that Smith had become an extension of the Batista regime and had prolonged the Cuban crisis by failing to come down hard against government excesses. Smith began by summarizing the civil war as an "impasse," with neither Batista nor Castro able to defeat the other. The vast majority of Cubans—anti-Batista but not pro-Castro—longed for a peaceful order, he said. Economic distress for both Cubans and U.S. business interests threatened to undermine the newly elected administration. Smith then presented a plan quite similar to Rivero Agüero's: election of a constituent assembly, Batista's retirement from political life after his presidential term ended in February

1959, and "a token shipment of arms to his administration as a gesture of support for his efforts."

The deliberate, always cautious Rubottom spoke next. He wished the new government well, but "the initiative" for a solution to the Cuban crisis "must be theirs." He speculated: "If Batista should retire completely from the political scene, if Rivero [Agüero] makes approaches to various civic groups and wins their support, if the non-revolutionary opposition agrees to go along with the plan, if the [Roman Catholic] church will support it, if the military agrees—in short, if the country at large, and not necessarily including Castro and the 26th of July Movement, shows itself to be behind Rivero Agüero's efforts, the U.S. Government would be disposed to show its good will and support for him." The "ifs" were daunting. As for the resumption of arms shipments, "it could be a problem with regard to public opinion in the United States and in Latin America."

Before there could be a change in U.S. policy, Rubottom averred, Rivero Agüero must take steps toward a political solution, toward a government of reconciliation. Rivero Agüero's plan would fail unless Batista left the country for a few months. At one point, Wieland branded Rivero Agüero "purely a Batista stooge. This can't go down." Smith shot back that although the president-elect was "seemingly not a strong character," the alternative was a Communist government. The tired Communist issue rankled Wieland: "Well, why don't you give us the benefits of your intelligence? Maybe there's something we're not seeing in the reports." Reminding the ambassador that plenty of moderates had joined the insurrection, Wieland dared Smith to demonstrate that the Castro-movement was Communist-dominated. The choice, Smith insisted, was either Castro or Rivero Agüero. "It's not an either-or situation," snapped Wieland, who held hope that moderates would emerge. Smith shouted "nonsense," whereupon Rubottom tried to calm his colleagues by suggesting lunch.[14]

In the afternoon, Rubottom assured Smith that the ambassador and State Department officers agreed on a major point: they "would be very dismayed at the prospect of a 26th of July take over in Cuba." But how could they stop Castro?

Shortly after Smith's testy encounter in the State Department, the U.S. intelligence community prepared a "special national intelligence estimate" on Cuba. Several themes in the document reflected a growing official consensus. First, although Castro's 26th of July Movement enjoyed support, it had not yet gained a widespread following because many Cubans yearned for a peaceful transition to constitutional government. Nor had the guerrillas succeeded in defeating Batista's Army. In short, because the *fidelistas* suffered weaknesses and could not overthrow Batista before early 1959, when he was scheduled to step down from the presidency, time might permit development of a stop-Castro plan. Second, although the "anti-dictatorship line of the movement is a horse which the Communists know well how to ride," the PSP had not grabbed control of the M-26-7. In short, the Communist threat was neither evident nor imminent. Third, the authoritarian Batista regime was on the skids. Batista's Army, composed of poorly led and ill-equipped conscripts, lacked the will to fight. The Cuban military needed a steady flow of arms, more professionalism, and counter-guerrilla training. The elections were "little more than a sham." In short, Batista could hold out, but his successor faced substantial problems.

The November 24 document speculated about a military junta's taking power—one purged of Batista's top command and led by "younger and less venal officers." Some analysts who participated in the preparation of the report, however, could not "identify any prospective leadership in the Cuban Army competent to overthrow the Batista regime. . . ." If a successful junta persuaded the public of its good intentions to honor a democratic process, however, the Castro movement might lose momentum. Still, a junta, if it wished to end the civil strife, might have to negotiate with Castro. The policy option of a military junta offered no immediate peace, but it might get rid of Batista and hold off Castro.[15]

When Smith met again with Andrés Rivero Agüero, there was no talk of a military coup or the removal of Batista. The ambassador encouraged the president-elect to initiate political steps toward peace, but Rivero Agüero said he first intended to launch "successful military operations." Looking to the United States for help, he declared that the army needed more "fire power" to help blunt the Communist threat to Cuba.[16] Rivero Agüero's shopworn remarks hardly reassured U.S. officials who were searching for policy options. The president-elect offered no more than a wishful desire to finish off Castro. Any solution to the Cuban crisis from the likes of Rivero Agüero seemed dead.

In the last months of the year, although Batista's press censorship, inaccurate Cuban military press releases, repeated breakdowns of communications and transportation, and Radio Rebelde's suspicious—because unrelenting—victory claims made getting at the truth elusive, a pattern of rebel advance and Cuban Army dissolution emerged.[17] Cubans continued to sneer at the rigged elections, and Batista's enemies grew with each story of violence. Cities lost electricity and water, making it more difficult for the government to deliver basic services. The Cuban Hotel Association, whose members included the Havana Hilton and Riviera, appealed to the regime for a subsidy to help the leading hotels through the hard times of declining tourism. In cities such as Holgúin food supplies dwindled and unemployment rose. Each week more sugar plantations and mills became part of the "Free Territory of Cuba."

The insurgents spread across Oriente Province and laid siege to Santiago, home of the Moncada Barracks and a huge arsenal. Business barely limped along in that isolated yet still Army-controlled southeastern city. With rail and road traffic halted, food became scarce and expensive. Condensed milk shot up to one dollar a can. Rebels entered the Country Club golf course and detained some foreigners. Not far away, the Guantánamo Naval Base again found itself in the midst of the insurrection. Rebels cut the road to the base, forcing many Cuban workers to remain at the U.S. facility. After capturing the town of Imias on November 15, rebels brought 25 wounded government troops and a woman to the northeast gate of the military compound. U.S. doctors and medical corpsmen, worried about violating U.S. pledges of neutrality, nonetheless acted on humanitarian grounds and gave the prisoners-of-war emergency care before the Cuban Red Cross evacuated them via a Cuban Navy frigate. Throughout the province, one Army post after another fell to Raúl Castro's forces, as Fidel issued a stream of marching orders and public directives. "Paralyze" became his favorite order—paralyze the railway, the buses, fuel deliveries, telephones, the telegraph, the economy, the Army.[18]

Fidel himself, with Celia Sánchez loyally at his side, led unseasoned troops into victorious battle at Guisa, twelve miles from Bayamo. The Cuban Army sent several units to save the garrison, but each was repulsed. Guisa fell on November 30. The casualty numbers told how bloody the battle had been: The rebels lost 8 dead and 7 wounded; the Cuban military lost 116 killed and 80 wounded.

In the drive to the plains, the columns of Che Guevara and Camilo Cienfuegos had already pushed through Camagüey into the neighboring province of Las Villas, connecting with rebels of all sorts, including local Communists. By December 1, Che had negotiated an agreement with the DR for military collaboration and agrarian reform. He began hatching plans to seize the city of Santa Clara in central Las Villas. Everywhere, new recruits flocked to rebel columns while Batista's reluctant conscripts battled fitfully. Batista desperately juggled his commanders, even arresting some when SIM sniffed an officers' plot to oust him. By early December, the *fidelista* fighters in rural areas were adequately armed and swelling in numbers to some 7000.[19] Exiles in cities across the United States, in Caracas, Venezuela, and elsewhere continued to collect money for the cause.

Ever wary that victory could be denied or at least delayed at the cost of great casualties, Fidel Castro feared, first, a military coup to dethrone Batista and install an anti-Castro junta, and, second, U.S. intervention to save Batista or implant a counter-revolutionary regime. The 26th of July Movement *jefe* speculated early in the month that Batista gunmen might actually assassinate U.S. citizens and blame the rebels for the murders in order to prompt the delivery of U.S. weapons or spark a military intervention. If U.S. forces did enter the war, Castro vowed a bloody struggle. Cubans "would have no alternative but to die fighting."[20]

Castro had ample reason to worry about a military coup, for discussion about one to prevent a rebel victory flourished among Cubans and in the foreign community. In Havana, for example, several prominent U.S. businessmen, including executives of ESSO, Portland Cement, and the First National Bank of Boston, advised Smith that the United States should promote and support a military-civilian junta in order to weaken the "Communist-inspired" Castro movement. These executives had become convinced that Batista had to step aside to keep Castro outside. But Smith still believed that U.S. promotion of a military junta was too dangerous. He preferred instead another plan to persuade Batista to move on: Batista should turn over the presidency to Rivero Agüero as soon as possible; the Cuban military should back the new government; the new president would then form a "national union government" of "respected elements" (in other words, no Castroites), promise new elections within two years, and restore constitutional guarantees "as soon as feasible." The United States, Smith insisted, "should not be [a] silent spectator" in this process, because a failure to act would surely produce chaos from which only Communists would profit.[21]

As the diplomats searched for alternatives to Fidel Castro, some U.S. citizens and their properties remained "in the middle" of the Cuban civil war.[22] In the afternoon of November 1, Cubana Airlines flight 495 departed Miami for Varadero with 16 passengers and four crew members. Just before reaching the popular beach resort, four passengers pulled out guns and forced the pilot to fly to Oriente Province. The gunmen donned the olive uniforms of M-26-7. Unable to find a landing site, the plane crashed in shallow water in Bahía de Nipe near an American-owned sugar plantation at Preston. Only three people aboard survived. Six Americans perished. U.S. Vice Consuls Hugh Kessler and Wayne Smith hurried to the scene of the crash. The State Department denounced the hijacking as an "act of brigandage."[23]

Fidel Castro denied that the seizure of the Viscount aircraft had been authorized; he attributed the disaster to individuals who called themselves members of the 26th of July Movement but who had acted on their own initiative. Raúl Castro said that the two surviving rebels, if found, would be court-martialed for their heroic if stupid act.[24]

Rebel skyjackers, three men and three women, seized another Cubana flight on November 5. The DC-3, scheduled to fly from Manzanillo to Holguín, was diverted to the rebel airfield at Mayarí Arriba, where it was quickly pushed from the runway and camouflaged. Another DC-3 plane, seized in late October, was already tucked away there. The three crew members and 25 passengers received fair treatment. Like protective parents, the rebels particularly watched over Robert Montgomery, wanting no harm to come to the U.S. Navy machinist from the Guantánamo base. They would not even let him ride a horse for fear of an accident. He observed that the rebels smoked cigarettes from GTMO's Sea Store and he heard that although the DC-3 capture was ordered, the Viscount skyjacking was not. Within days the rebels turned the passengers over to the Red Cross. Among the released passengers was Amado Cantillo, the son of General Eulogio Cantillo, Batista's military chief in Oriente. An eight-month-old child with an obstructed windpipe died during the ordeal. Her parents said that rebel doctors did all they could to save the baby's life.[25] Still, such tragedies cost the rebels points in the ongoing propaganda war.

Consul Wollam reported that "the activities of Cuban Armed Forces are possibly more of a menace to American lives and property than the rebels have been."[26] If the rebels "appropriated" property, the army "requisitioned" it.[27] For weeks, B-26 bombers and P-47 fighters strafed and bombed suspected rebel positions, including those near U.S.-owned properties. At George A. Braga's Manatí Sugar Company, for example, B-26 aircraft severely damaged the mill during a fierce fight with a rebel unit. On at least one occasion Americans were forced to drive in front of an army unit sweeping an area for rebels. Whereas rebels gave receipts and promised compensation when they seized goods, Batista's soldiers looted. At Nicaro, Moa Bay, and Texaco facilities, where Cuban workers were overwhelmingly antigovernment, Army forces did not hesitate to shoot and destroy. The Cuban military also distributed leaflets warning rural folk to leave their farms for Santiago because the Air Force intended to bomb extensively to root out the "outlaws." To pay for the war, the Batista regime also raised taxes, including an extra levy of 10 cents on a bag of sugar.

U.S. businesses once again lay in the path of the rebel assault upon the island economy. At the Texaco refinery, rebels helped themselves to vehicles, tools, and gasoline, and the plant had to shut down for several days because its water system was sabotaged. Robert Thornton, American manager of Swift and Company, traveling in a Santiago suburb, barely escaped injury when eight shots pierced his car. After the United Fruit Company refused to pay the rebel "tax" on sugar, insurgents cut off water to the company plant at Preston for two weeks. Thirty-three American citizens there seemed endangered. In the southeast corner of Camagüey Province, rebels occupied the American-owned Francisco Sugar Company's property and closed the business down. "We're pretty much at the mercy of the rebels," complained the manager of another American sugar company.[28] Everybody wondered whether the *zafra,* the harvesting and processing of the 1959 sugar crop, would get under way without serious interruption.

Near the U.S. nickel plant at Nicaro in mid-November, during heavy fighting, government planes bombed and wrecked company offices, barracks, mines, and warehouses. Production stopped. Hundreds of refugees moved to company grounds, where food supplies proved inadequate. Cuban Electric's losses from sabotage mounted too. Increases in taxes and insurance premiums and higher costs arising from the operation of mobile equipment where transmission lines were down slowed the company's expansion program. ESSO's president, one of many American executives who held meetings with U.S. officials, complained that oil products could not be delivered because too many bridges had been destroyed. Compared with November 1957, sales of oil for November 1958 were down by 80 percent in Oriente and 35 percent in Camagüey. Moa Bay Mining also suffered insurrection-related obstacles as it tried to build its $75 million nickel plant in the midst of war.[29]

Heads of U.S. companies joined Ambassador Smith and Washington officials in believing that Batista had become a liability. The Batista government and military demonstrated day by day that they could not win the war, subdue the insurgency, or win popular favor after the electoral fraud. At the same time they used methods that destroyed the economy. "I've been pro-Batista all along," remarked one U.S. businessman. "But I now realize he's got to go. We can't support him now, [for] the Cuban people are unanimously against him."[30]

The Guantánamo Naval Base also became further ensnared in the Oriente conflict. On November 23, 25, and 27, rebels cut off GTMO's water supply from Yateras for short periods. On the ninth of the month, the Cuban Army soldiers guarding the water plant were withdrawn, because, said Batista, they were "bait for the rebels."[31] Ever the opportunist, Batista urged the United States to sell Cuba arms adequate for the defense of the aqueduct or provide radio communications equipment so that Army troops in the area could be summoned in an emergency. Washington turned down both alternatives but began to study the option of sending in Marines to protect the pumping station, as requested by the base commander. U.S. officials hesitated, for they feared the "bad propaganda" that would erupt if Marines once again entered Cuban territory.[32]

U.S. officials could not understand why the rebels would want to "irritate" the United States like this, and they released a stern statement about American "forbearance," the "gravity" of the Yateras question, and U.S. insistence that the "irresponsible acts" cease.[33] Radio Rebelde replied that battle conditions dictated the water shut-offs and that Batista and Smith were conspiring to provoke a fight between the rebels and U.S. forces."No, this time the events of 1898 and 1933 will not be repeated. We want to maintain the best of relations with the country of Abraham Lincoln, but they must be based upon mutual respect."[34]

The insurgents may have harassed the waterworks—they never sabotaged the facility but could have done so if they had wished—to lure Batista forces back into rebel-held territory or simply to demonstrate M-26-7 domination of the area. But the rebal commander in the region, Felix Lugerio Peña, cited "grievances" against the naval base as the primary motivation. The rebels were protesting the alleged U.S. call for Cuban Army troops to resume guard at Yateras, and they were also angry because U.S. officials had supposedly evicted a wounded comrade from the base hospital. Memories of the rocket heads also lingered. All that said, Wollam, on the basis of "reliable sources," reported that Fidel Castro on December 2 "was heard on the radio instructing his brother, Raúl Castro, in 'no uncertain terms' not to molest the Base or its water supply."[35] About the same time, Fidel let it be known through a published report by Andrew St. George that he welcomed the presence of a U.S. diplomatic representative in the Sierra Maestra so that rebel leaders, who were not yet insisting on recognition, could discuss a range of issues with U.S. authorities.[36]

American officials prepared to send a Marine contingent to Yateras should the waterworks suffer another stoppage. Smith was instructed to seek prior Cuban approval for such an operation to ensure rapid U.S. action. He received official permission. To Washington he also forwarded Batista's request that the U.S. base supply Cuban troops at Yateras with food and radio equipment. The State Department balked, because the base would appear to be giving logistical support to the Cuban Army. Washington's advice: Discourage the Cuban military from despatching reinforcements to Yateras and let the issue cool down. Batista, who probably could not have spared the soldiers in any case, did not send new troops, the rebels relented, and the problem passed.[37]

Washington officials in late November and early December were thinking well beyond the annoying yet manageable Yateras crisis. They were doing nothing less than plotting to remove President Fulgencio Batista from office. Fed up with Batista, Rivero Agüero, and Smith, and eager to block the *fidelistas,* Washington officials had come to a "pox on both their houses" attitude, reported the Canadian Ambassador to Cuba.[38] They maneuvered behind Ambassador Smith's back. They shed any remaining neutralist proclivities and decided to manipulate Cuban politics as never before during the insurrection. U.S. officials still thought they had time to blunt Castro's drive. All they needed was Batista's cooperation and some reliable anti-Castro

Cuban leaders who could be installed and sustained. The hunt for the elusive "third force" accelerated. The hegemonic power decided that the rebellion had become too anti-American, that it threatened extensive U.S. interests, and that Batista had become a liability to U.S. goals. Like other friendly tyrants before and after him who had gone too far in corrupting politics, practicing terror, losing popularity, and endangering economic development, Batista had to go.[39]

18

Batista Dismissed: Pawley's Plot and Smith's Blow

William D. Pawley had no doubts about how to stop Fidel Castro. After plumbing Cuban issues with CIA officials, he went directly to President Dwight D. Eisenhower to make his case. This Florida businessman with investments in Cuba had opposed the suspension of arms shipments to the Batista regime. Now he warned the president in private White House meetings that Castro would shackle Cuba with Communism. Eisenhower listened to this fervent anti-Communist millionaire, this outspoken "zealot," this sometimes "impetuous" analyst of Latin American affairs, because the president had talked with Pawley for years and respected his opinions.[1] Now Pawley, who thrived on salesmanship and adventure, pushed a bold plan designed to prevent the triumph of Fidel Castro.[2]

William Douglas Pawley, born in South Carolina in 1896, had spent some of his youth in Cuba, where his father sold supplies to the Guantánamo Naval Base.[3] In the 1920s the young, ambitious entrepreneur moved to Florida during the real estate boom and piled up riches. In Cuba and China he organized airlines and sold them to Pan American Airways. While in China he helped launch the famed Flying Tigers, the team of American pilots who served the Chinese government against Japan before U.S. entry into World War II. After the war, President Harry S. Truman appointed Pawley the U.S. Ambassador to Peru and then to Brazil. During the 1950s, Pawley invested in Havana's bus system. He also raised large contributions for Eisenhower's two presidential campaigns. This political work, probably as much as his knowledge of Latin America and experiences in Cuba, gained Pawley access to the White House.

The files the FBI compiled on Pawley, including records of interviews, are filled with his accusations against U.S. officials he considered Communists or Communist sympathizers. A favorite target was Spruille Braden, former U.S. Ambassador to Colombia, Cuba, and Argentina and former assistant secretary of state. Pawley and Braden had clashed in the 1940s as they competed for assignments in the foreign policy establishment. Although personal rivalry underlay their dislike for one another, Pawley claimed that Braden had fallen under the spell of known Communists. He hurled charges at others too, displaying the conspiratorial mentality that darkened the era of McCarthyism.[4]

On one occasion in fall 1958, after meeting with President Eisenhower, Pawley went over to the State Department to discuss Cuban issues with Rubottom and Wieland. He disputed their view—a view shared throughout the U.S. bureaucracy, including the CIA and FBI—that Castro was not a Communist.[5] Pawley distrusted Wieland in particular, judging him naïve, incompetent, pro-Castro, and blind to the Communist threat. Later, in the early 1960s, Pawley denounced this State Department officer to a Senate committee, blaming him and Herbert Matthews for Castro's triumph. Wieland "is either one of the most stupid men living" or he is "intentionally" helping America's "enemies," Pawley told the senators.[6] In Pawley's unpublished memoirs, he wrote that Wieland did everything he could to cover up Castro's "Communist stamp" while Rubottom offered "flabby acquiescence" in Wieland's anti-Batista, pro-Castro machinations.[7]

Although Wieland fed Charles Porter, a pro-Castro member of Congress, confidential State Department reports on conditions in Cuba, the record is clear that Wieland actually preferred a "third force" alternative to Castro.[8] Still, Pawley's charges—and those of the disgruntled Ambassador Smith—ultimately led a witch-hunting subcommittee of the Senate Judiciary Committee to question Wieland's "integrity and general suitability" and eventually to reprimand him as an "active apologist" for Castro.[9]

At Pawley's Miami home in late November 1958, several U.S. officials shared dinner after they had participated in a public forum which Pawley moderated. Into the morning hours, Deputy Assistant Secretary of State for Inter-American Affairs William P. Snow, former Assistant Secretary of State Henry F. Holland (Rubottom's predecessor), and Chief of the Western Hemisphere Division of the CIA's Directorate of Plans J. C. King ruminated about the Cuban crisis. Apparently no record of this meeting exists, and Pawley's recollection of the evening is sketchy, but by his own account he pressed for Batista's capitulation to "a caretaker government unfriendly to him, but satisfactory to us."[10] The United States would recognize and shore up this new regime with military assistance. The project's goal was sharply drawn: to prevent Castro's coming to power. The gray-haired Pawley, who spoke fluent Spanish and had met Batista on previous occasions in Cuba, volunteered to talk directly with the president in Havana.

Pawley's guests urged him to fly to Washington to persuade high officials to act. The former ambassador first sought Secretary of State John Foster Dulles's approval. Ill with cancer, Dulles nonetheless conversed with Pawley by telephone and gave his encouragement. At the White House, President Eisenhower liked what he heard. He apparently instructed Acting Secretary of State Christian Herter to devise a plan full of "inducements" so attractive that Batista would not hesitate to accept U.S. advice.[11]

For two days Pawley huddled with State Department officials and plotted a scheme. According to Pawley's account of this meeting—the only one available, for there seem to be no records of this and other meetings and the CIA will not declassify documents on its role in plots to oust the Cuban president—Batista would resign and retire with his family to their Daytona Beach, Florida, mansion. In Cuba *batistianos* would suffer no reprisals. U.S. arms deliveries to a caretaker regime made up primarily of military officers would begin immediately. The new short-term government would prepare for fair and free elections and leave office in 18 months.

"The keystone of the plan was that I be authorized to speak for the President," recalled Pawley, for Batista would hardly step aside at the request of someone who lacked authority.[12] Eisenhower gave the authorization, but State officers had second thoughts just before Pawley departed. Herter called him to the State Department, where Rubottom explained that the plan was being modified: Pawley could not disclose to Batista or anyone else that he spoke for the president. Pawley should say only that he spoke as a private citizen who would work to persuade his government to accept the plan. "I felt the point of a knife being thrust between my shoulder blades," Pawley remembered with the hyperbole common to his accounts.[13]

Pawley's mission was hardly kept a secret. Former Ambassador Arthur Gardner probably knew about the "Miami matter" to see "the Head Man," as he put it. It seems that Vice President Nixon, whose close friend was the Cuban expatriate Charles "Bebe" Rebozo of Miami, was also privy to the Pawley project.[14] CIA Director Allen Dulles might have informed some U.S. businessmen about the forthcoming mission. One of them apparently leaked the Pawley scheme to a prominent anti-Castro, anti-Batista Cuban lawyer, Mario Lazo, who then told Ambassador Smith about it on Thanksgiving Day at the Havana Country Club.[15]

A standard CIA joke had it that Cuba had three forms of communication:

telephone, telegraph, and tell-a-Cuban.[16] Given the Cuban reputation for spreading information with alacrity and given the existence of spy networks, most likely other Cubans, including Fidel Castro and Fulgencio Batista, got word of the Pawley venture.[17] Perhaps Castro was referring to the Pawley mission on December 12 when Radio Rebelde broadcast his declaration: "We Cubans are capable of finding solutions to our problems and will not tolerate intervention by a foreign power."[18] The rebel leader could also have been taking note of Ambassador Smith's return to Cuba, revealing Castro's fears that the North American diplomat had cooked up some new scheme to blunt the insurgency.

On December 7 the 62-year-old Pawley and his wife Edna flew to Havana and checked into the Country Club. The CIA Station Chief in Havana, James Noel, knew why Pawley was coming, but apparently most officers of the U.S. Embassy did not. The next day Pawley talked with Prime Minister Güell and outlined the plan for the caretaker government. Güell then informed Batista, who refused to see the U.S. emissary. Pawley insisted on a meeting, telling Güell that no Cuban president could snub a representative of the U.S. president. Batista relented, and in the evening of the ninth, according to Pawley's account, Pawley and Batista met at the Presidential Palace for three hours. Pawley repeated the details of the U.S. scheme to the Cuban leader.

Who did the United States recommend for the provisional government? Batista wondered. Prominent on Pawley's list was Colonel Ramón Barquín, the imprisoned military officer convicted in April 1956 for conspiracy to overthrow the Cuban government. Barquín had received some training in 1948 at the Strategic Intelligence School of the U.S. Army, had served in 1949 with the Inter-American Defense Board in Washington, and had been Military Attaché in the Cuban Embassy in Washington for 1951–56. U.S. officials thought him "friendly to the United States"—probably one of the moderates with whom they hoped to work.[19] Pawley also mentioned General Martín Díaz Tamayo, who had joined Batista in the 1952 coup but whom Batista had recently dismissed from the army and arrested; Major Enrique Borbonnet, imprisoned with Barquín on the Isle of Pines; and José Pepín Bosch, a former finance minister and then head of the Bacardi Rum Company. If Batista would accept them, Pawley remarked, "I'm confident that I can go back to Washington and get the plan approved as United States policy." The Cuban president, hardly welcoming the empowerment of his political foes, remarked that Castro would think him weak if he learned that Batista had accepted the U.S. plan, and "my family and I will never leave Cuba alive."[20] To Pawley's dismay, Batista balked.

Pawley's recollection of the meeting did not suggest any curiosity on

Batista's part about whether President Eisenhower endorsed the plan. Only later, when Batista was in exile in the Dominican Republic, did Pawley tell him directly that he had gone to Havana as President Eisenhower's representative. "If only you could have told me!" Batista reportedly replied.[21] This account is contrived and disingenuous, for, in early December, Pawley had violated his State Department instructions just enough to inform Güell that his mission had Eisenhower's imprimatur. Güell had told Batista that Eisenhower backed the plan, and Batista had even noticed that Pawley had not mentioned the American president's endorsement.[22]

Pawley said later that when he could not play his "ace card" (telling Batista that Eisenhower subscribed to the plan), he knew that his mission was doomed. He also charged that State officials deliberately modified the plan in order to destroy it.[23] The plan was surely doomed for many other reasons, not the least of which was Batista's determination to hang on. The Cuban president was not about to give up power yet. "I must carry on," he told Pawley. "I have a duty to remain at my post. And that I intend to do."[24] As he had done on a number of occasions, the U.S. client was disobeying the hegemon's wishes. Standing in the way of success, too, was Castro's resolve to resist outside interference and the supreme difficulty facing U.S. officials who sought to install a junta made up in part of prisoners who had no knowledge of the plan. As Smith and others had advised, a military junta was no panacea because it would still have to deal with the Castro movement. The Cuban's people's and the insurrectionists' unwillingness to tolerate a replay of the manipulated politics of the past could be expected to block U.S. efforts to control the outcome of the civil war. Cuban oppositionists of all kinds, sensitive to slights to their nationalism, would remain suspicious of a government brokered by the United States.

Although State Department officials in Washington may have given only lukewarm backing to Pawley's mission because they recognized the obstacles facing the creation of a military junta and they resented the obnoxious Pawley's intrusion into normal bureaucratic procedures, no evidence has come to light that they undermined his plot. In fact, they cooperated by helping to draft the plan and by ordering Ambassador Smith to leave Havana for Washington on December 4. In that way, if the plan went awry, the ambassador would not be implicated and U.S. officials could plausibly deny that the United States was intervening in Cuban affairs. Far out of Havana, of course, the prickly Smith, known to oppose a military coup, could not meddle in or torpedo the operation. During his stay in Washington, Smith found it difficult to pry details from government officials, but when he did he apparently endorsed Pawley's junta plan.[25]

Pawley's failure did not remove the acerbic Floridian from Caribbean or

Cuban affairs. In early 1960 Pawley journeyed with Senator George Smathers to the Dominican Republic to ask the dictator Rafael Trujillo to democratize his regime in order to prevent "another Cuba." In November of that year President Eisenhower summoned Pawley to the White House and asked him to tell Trujillo directly that he must step down. Pawley failed in this mission too, for Trujillo refused to leave.[26] In Florida, Pawley increasingly associated with the CIA-funded exiles who were seeking to overthrow Castro's government. "Pawley is obviously moving in top-level circles in this Cuban situation and apparently has some semiofficial backing," read a FBI report.[27] With Eisenhower's encouragement, Pawley helped organize the Bay of Pigs invasion. He typically blamed its failure on "leftist bureaucrats in the State Department" and President John F. Kennedy's "betrayal."[28] Many years later, on January 6, 1977, in the throes of an illness at age 80, Pawley committed suicide by shooting himself.[29]

By early December 1958, U.S. alternatives for blocking Fidel Castro short of armed intervention were dwindling fast. Smith's "plan" to induce Rivero Agüero to launch political reforms had fizzled. Pawley's mission to oust Batista in favor of a junta had foundered. Batista's arrest of Díaz Tamayo and other military officers placed a higher hurdle before any scheme to empower a junta. Nor would Batista budge, as Foreign Minister/Prime Minister Güell bluntly told Daniel Braddock on December 10. In fact, said Güell, Batista would probably remain active in military affairs for three to six months after Rivero Agüero took office. U.S. arms were still needed, too. The Embassy Counselor was aghast. Didn't Güell recognize that Batista was "the bone of contention," and that as long as he retained power, "responsible elements" in Cuba would not abandon Castro for another government?[30]

During this U.S. search for ways to stiff-arm Castro, Senator Allen J. Ellender, Democrat of Louisiana, blundered into Cuba.[31] It was the kind of episode Foreign Service Officers have always dreaded—having to pander to a politician on a junket, obligatory ceremonies and social functions, a hurried visit with photo opportunities, superficial study followed by uninformed statements on a delicate subject during an electric time. Senator Ellender had rankled Cubans in the past. An owner of sugar-cane land in Louisiana, Ellender had long advocated reducing Cuba's sugar quota to favor domestic U.S. producers.[32]

Ellender departed Miami for Havana the evening of December 9, the very time Pawley was meeting with Batista. For several days Ellender met with embassy officers, prominent Cubans, and U.S. businessmen, and he kept a diary of his observations. Ellender thought the embassy building "magnificent," but "over built . . . for a country the size of Cuba." Looking for budgetary savings, Ellender complained that embassy staffers wrote too many reports on the tourist trade and that the USIS was "not very effective in the light of all the trouble that is brewing in Cuba" and given "so much adverse United States feeling prevalent in Cuba." He concluded, too, that many members of the military missions were loafing.

"There does not seem to be much unrest in and about Havana," Ellender recorded in his diary. Members of the American Chamber of Commerce in the capital city told him that they feared Castro. The United States, they remarked, had stood "on the sidelines," letting pro-Castro newspaper reported like Matthews and Dubois turn the rebel into a hero. The Louisiana senator became persuaded that Batista could win if only he could obtain U.S. arms. Several other diary entries reveal the shallowness of Ellender's analysis: "It would be a tragedy for civil war to break out in our back yard"; "There is some unemployment and I cannot for the life of me see why there should be any"; and "Somehow that fellow [Fidel Castro] seems to wield some power." Apparently unaware of the Pawley mission or other conspiracies with which the United States was associated, Ellender thought that the United States should not "merely stand by and let things happen," because, he added, the Cubans "look upon us as big brothers."

One day Ellender flew on a company plane to Moa Bay. U.S. officials nervously awaited word of his journey into rebel territory, but he made the trip unmolested. From the air, the senator saw the crashed Cubana plane, its tail sticking out of the water. On his return flight he flew over the U.S. nickel plant at Nicaro. He was impressed by the extent of U.S. interests in Cuba. At a press interview in the USIS office on December 12, Ellender's unguarded and ill-informed utterances also worried U.S. officials.[33] For one, he called the rebels "bandit groups" and said that the U.S. press exaggerated their importance. He questioned the existence of a civil war ("I don't see any fighting, do you?") and suggested that arms shipments should be renewed to the "friendly" Cuban regime.

Cuban Ambassador Arroyo cheerfully told State Department officials in Washington that "the Cuban Government was very pleased to have such a powerful, independent, and influential man [Ellender] to represent these views to the American Government."[34] Batista press censors, of course, allowed Cuban newspapers to report the senator's remarks. *The Times of Havana* wrote critically, however, that "the trail of memories that he leaves behind will long outlive his brief visit," because most Cubans probably believed that he spoke for the U.S. government.[35] Radio Rebelde lambasted the senator's views. If rebel leaders read Ellender's statements as U.S. policy, then the senator's visit further fed rebel suspicions of the United States on the eve of victory.

Soon after Ellender departed and Smith returned to Havana, Washington instructed the U.S. Ambassador to disabuse Batista of any notion that the United States would endorse a government that Batista helped create. Although both Batista and Smith published recollections of their December 17 conversation at the president's well-guarded country estate at Finca Kuquine, scholars in the historical office of the Department of State have been unable to locate a contemporary record of this important meeting.

Prior to the meeting with Batista, Smith had informed Güell that the United States no longer supported the Cuban government. At the estate, the ambassador assured Batista that he had tried to persuade his Washington colleagues differently, but he had to report that the United States would not back the Rivero Agüero government under any conditions. Batista, however, could make it possible for moderate, anti-Castro oppositionists to rally behind a new, broadly based government by leaving Cuba. Could he go to Daytona Beach with his family? Batista asked. No, Smith replied, it would be better if he went to Spain, and soon. Breathing "like a man who was hurt," Batista remarked that the army would disintegrate if he left the island. He hinted that he could last until a new government broadened by Rivero Agúero was inaugurated on February 24 *if* he received U.S. weapons. When the Cuban president raised the question of a military junta, Smith refused to discuss it. The U.S. message stung: Batista had to abandon his country altogether. That Batista mentioned Daytona Beach at all tipped his hand—he was, in fact, already thinking about leaving. "I had dealt him a mortal blow," recalled Smith.[36]

At the same time, while publicly adhering to a non-intervention policy, the State Department was also trying to generate interest among Latin American nations for outside mediation of the Cuban crisis, perhaps under the auspices of the Organization of American States. U.S. Embassy officials throughout the Western Hemisphere probed for possibilities of multilateral cooperation. They found few. Latin American officials either sympathized with the Castro movement, recoiled from meddling in another nation's internal affairs, or thought the OAS incapable of acting effectively. Rubottom recognized that a mediation effort would fail so long as Castro opposed any role for outsiders. "At a fine hour," said Castro on Radio Rebelde, "these people with intentions of intervention or of calling in the Organization of American States are coming out of the woodwork." They did not worry when the "dictatorship" was committing violence against Cubans. "They do not have the right to worry now."[37] It became apparent, as one U.S. official reported, that the approach to other nations revealed "acceptance that the Cuban problem was an inter-American concern" but yielded "no concrete actions or agreements."[38]

Also abortive was a U.S. probe designed to prompt the Papal Nuncio in

Cuba to initiate a truce.[39] Apparent rebel overtures transmitted to U.S. officials also came to naught.[40] State Department officials could no longer put much stock in mediation or negotiations as vehicles to end the Cuban crisis. Politics had polarized too much, and the civil war had advanced too far. The only tools U.S. officials had to work with—and uncertain ones at that—were exiles and Cuban Army officers who agreed that Batista must be pushed out in order to prevent Castro's triumph.

The Cuban military sunk into greater and greater defeatism and demoralization as rebel units pinched off one garrison after another. Army soldier desertions and defections increased. Growing numbers of officers would no longer fight for the regime.[41] Indeed, Batista's army was in disarray at the top; some of his generals began to search for deals with the rebels and with U.S. officials. A few days before Christmas, General Tabernilla himself, worried that his soldiers would no longer fight, visited Ambassador Smith to recommend a military junta led by General Eulogio Cantillo. He also wanted to find out what the United States preferred for a post-Batista, anti-Castro government. Fearful of undermining Batista by dealing with Tabernilla, Smith said only that he would report the conversation to Washington.[42] At the same time, Batista began to identify airplanes that could carry him, his cohorts, and their families out of the country. His advisers readied three aircraft with 108 seats. Then he made up a list of the people he would permit to escape with him.[43]

In Oriente, Fidel Castro's legions continued to advance, seizing the Texaco refinery on their way to conquer Santiago, where Christmas approached without the usual festivities. It was just as well, wrote Park Wollam, because a "bearded Santa would probably have a difficult time with local authorities who suspect any hirsute individual."[44] The Guantánamo Naval Base, to whose gates flocked hundreds of refugees, stood as a U.S. island in a rebel ocean. Castro still worried that Batista would find a way to induce U.S. intervention, so the rebel leader warned his officers "not to fall into any provocation or trap." Don't "provoke the American enterprises," he ordered them.[45] In Camagüey Province, guerrillas gained control of the Central Highway while Prío's Second Front, led by Eloy Gutiérrez Menoyo and William Morgan, continued to squabble with M-26-7 forces.

In Las Villas Province, the forces of Che Guevara, Camilo Cienfuegos, and the DR encircled Santa Clara, a major city sitting astride the Central Highway. Batista's commander pulled troops from the countryside to defend the provincial capital, abandoning territory to the rebels. Some 6000 Army troops and several hundred police dug in. Just before the insurgents began moving toward Santa Clara on December 14, Batista abruptly fired his local commander. One small town after another fell to the advancing rebels. Inside

Santa Clara the DR and M-26-7 underground snipers harassed Army troops. Batista sent a train with 600 reinforcements and boxcars loaded with arms and ammunition. The Batista soldiers of the armored train at first decided to fight "in the style of the settlers of the North American West against the Indians," remembered Che Guevara, but they soon surrendered, supplying Guevara's unit with the weapons they needed for the pivotal battle of Santa Clara.[46]

The day after Christmas, as Che Guevara's forces prepared for the final assault against Santa Clara, Fidel Castro penned a letter to his Argentine cohort. Holding a flashlight, Fidel wrote: "The war is won, the enemy is collapsing with a resounding crash. . . ."[47] Still, he worried that a military junta, promoted and backed by the United States, would seize power and prolong the war. As a panoply of U.S.-backed conspiracies quickened, Fidel Castro's distrust of the United States rose.

19

U.S. Third-Force Conspiracies and Batista's Flight

CASTRO SMELLED JUNTA rats everywhere. Word spread about a host of schemes to block the 26th of July Movement's road to power. About mid-December, for example, senior military officers led by longtime Batista loyalist General Francisco Tabernilla approached one of Castro's underground leaders in Havana. They proposed to replace Batista with a civilian-military junta consisting of the commander of army forces in Santiago, General Eulogio Cantillo, an officer who had received training in the United States; Manuel Urrutia, Castro's president in waiting; an imprisoned anti-Batista military officer such as Barquín or Borbonnet; and two other civilians to be named by Castro. Tabernilla's emissary, Colonel Florentino Rosell, remarked that the United States would immediately recognize the new government. In fact, rebel intermediary Pepe Echemendía told Castro that he was informed that the proposal had been discussed with "the North American embassy." Echemendía also said he knew of other anti-Batista plots—including one by Montecristi's Justo Carrillo.[1]

Castro would not cut deals with people determined to deny him power. The rebels, after all, were winning. The major cities of Santiago and Santa Clara were under siege. He thundered against U.S. interference in the near-triumphant insurrection. This time, the United States would not steal Cuba's history. "North American-supported junta unacceptable," Castro crisply said on December 24.[2] Four days later Castro nonetheless met with General Cantillo at an Oriente sugar mill. Cantillo flew in on a Sikorsky helicopter, touching down near the white bed sheets the Castroites had spread out as a marker.

Raúl Chibás, who attended the meeting, recorded in his diary that General Cantillo mentioned that the U.S. Department of State supported his mission.[3] Cantillo would probably not have agreed to participate in the scheme except with U.S. knowledge or backing. Colonel Rosell had talked with Jack Stewart, the CIA agent, who had shown interest in the plot. U.S. government records either remain classified or they do not exist, making it difficult for the historian to know for sure to what extent U.S. officials signaled a green light for this conspiracy.[4]

In his talk with Cantillo, Castro staunchly ruled out a junta and lectured Cantillo about Batista Army failures and the inevitability of rebel victory. After listening to three hours of Castro's passionate defense of the rebellion, Cantillo promised to surrender his troops and join the insurrection.[5] Castro wondered if he had really gained a major concession, and for good reason, for it is not at all clear that Cantillo intended to fulfill his pledge. Actually few people deemed General Cantillo reliable. U.S. military advisers in Havana, for example, questioned Cantillo's "sincerity and trustworthiness."[6] The United States, however tenuously, had thrown in with a "third force" of dubious reputation and efficacy.

Upon hearing of Cantillo's journey to Castro, Batista upbraided Tabernilla but did not force him to resign. Cantillo, exhausted and weary of the confusion in the disintegrating government, returned to Havana. Batista ordered him to report posthaste. The president told him that he was thinking of leaving Cuba: Cantillo would have to assume the responsibilities of national leadership. Cantillo now sent word to Castro that Cantillo's planned surrender would have to be postponed. "That son of a bitch," exploded Castro.[7]

Justo Carrillo was also conspiring, and also with U.S. connivance. An economist and former head of a government agricultural credit bank, he led the Montecristi group. His name frequently appeared on lists of potential presidents. Had Colonel Ramón Barquín's 1956 plot to overthrow Batista succeeded, Carrillo would likely have become the Cuban president.[8] As early as July 1958, during the kidnapping crisis, the anti-Castro and anti-Batista Carrillo initiated what he called "la operación Isla de Pinos"—Operation Isle of Pines.[9] In exile in Miami, he collected men, arms, and airplanes for a mission to free Barquín from prison and install him in power. Because he feared that U.S. authorities would catch and jail his conspirators for violating U.S. law, Carrillo sought help from the CIA.

Montecristi's Andrés Suárez talked with CIA agent Basil Beardsley in Havana. Sometime in November Carrillo and the CIA linked up.[10] CIA contact Robert Rogers huddled with Carrillo in Miami, and on December 21 another contact, Willard Carr, discussed the Isle of Pines expedition with

Carrillo at the Hotel Commodore in New York City. Head of an aluminum company and former college classmate of CIA Director Allen Dulles, Carr explained how an airfield in Mérida, Mexico, would be used so that the Carrillo forces could take off without fear of interception by U.S. customs agents.

A third U.S.-backed scheme was also a CIA operation. The CIA was plotting with Miami representatives of Prío's Segundo Frente de Escambray to fly in weapons to Gutiérrez Menoyo's guerrilla band in Las Villas. The CIA's Jack Stewart, with State Department knowledge, approved the mission for December 31. That day an American-piloted B-26 crash-landed in the Escambray Mountains and discharged weapons, ammunition, and more insurgents who intended to take Havana before M-26-7 did.[11]

A fourth December plot for a "third force" sprang from the exiled Manuel Antonio Varona, Prío's former prime minister.[12] This project received both Prío and U.S. encouragement and the support of José Miró Cardona, the exiled president of the Havana Bar Association. By this time, it seems, some U.S. officials had elevated Prío from an enemy of the United States to an anti-Castro ally and possible instrument of U.S. policy. While still prosecuting Prío in the courts, the United States had now come to the point of teaming up with projects which had the potential, however slight, of bringing the wealthy corruptionist and Auténtico leader to an influential position in Cuba.

"Tony" Varona conspired to depart Miami by air with men and weapons and land near Camagüey City, where his contingent hoped to collaborate with a disaffected Army garrison to rebel against Batista. If these forces managed to gain control of Camagüey Province, Varona reasoned, they could compel Castro to deal with the Auténticos. After Prío offered Varona a plane, Varona sent an emissary to William Wieland. Would U.S. authorities look the other way so that Varona's aircraft could fly without incident Wieland promised to "pass this up the line."[13]

On December 26, Wieland, who had grown skeptical that any anti-Batista, anti-Castro venture could work at such a late stage, told the Varona people that their project had in fact earned CIA approval. A few days later, failing to hear from the CIA agent who was supposed to help him fulfill his plans, Varona left Florida in his twin-engine Cessna and landed near the King Ranch in Camagüey. The aircraft had too little fuel to return and pick up the rest of Varona's group in the United States. Varona rested at the main house of the sprawling ranch while he futilely tried to make contact with the Army captain who was supposed to spearhead the rebellion. Local forces of the July 26th Movement soon showed up, and they demanded weapons. At first Varona refused, but, fearing a confrontation, finally surrendered ten rifles.

While these conspiracies unfolded, confusion, uncertainty, disarray, and dismay gripped the highest echelons of the U.S. government. Secretary of State Dulles was extremely ill with cancer. Taking Dulles's place, Acting Secretary Herter trailed rather than led discussions on Cuba. President Eisenhower seldom pressed the Cuban issue. His leadership style of delegating authority and relying on underlings, moreover, surely encouraged the chief executive's distance from the Cuban question.

At National Security Council meetings, CIA Director Allen Dulles made periodic reports about the accelerating Cuba crisis, but not until the December 18 meeting did Eisenhower seem particularly alert to the subject. The president "found it difficult to understand how the rebel forces gained strength so rapidly." He remarked that "perhaps Batista should be induced to turn power over to his successor."[14] Had Eisenhower forgotten about the Pawley plot, and had Allen Dulles not informed the president about CIA contacts with Cuban exiles and potential junta leaders? Had Eisenhower not read and understood the December 16 report which saw little hope for Batista and which wondered about "where the requisite [junta] leadership would come from"?[15]

Although President Eisenhower revealed little grasp of the Cuba question at the December 18 meeting, perhaps he knew the details yet thought that knowledge of the plots should be confined to only a few top U.S. officials and not mentioned at a large NSC meeting. A few days later, for example, Eisenhower instructed Allen Dulles not to provide the specifics of covert operations at meetings of the National Security Council.[16] The documentary evidence shows that CIA, State, and his own White House staffers kept him informed about events in Cuba. Yet they did not convey to him the severity and immediacy of the Cuban crisis—perhaps because they believed that one of the anti-Castro plots would work.

At the December 23 meeting of the National Security Council, the president heard Allen Dulles explain that Batista was on his way out and, if Castro won, some Communists would participate in the new government. Eisenhower wondered whether the Defense Department had been contemplating possible U.S. military action in Cuba. No, Herter answered. The discussion wandered through various aspects of the Cuban crisis until Director Dulles stated: "We ought to prevent a Castro victory." The president perked up again, remarking with some irritation: "This was the first time that statement had been made in the National Security Council." Dulles explained that Castro was too radical. Vice President Richard Nixon then allowed that the

United States could not "take a chance on Communist domination of Cuba." The president seemed attracted to a "third force" option to Batista and Castro. Herter suggested that a "contingency paper" was needed.[17]

Soon after the NSC meeting, Rubottom and Snow drafted and Herter sent to the president a lengthy memorandum on Cuba. It reviewed events, explained why the United States had withdrawn support from Batista, outlined U.S. efforts to move Batista out of Cuba, and mentioned CIA-State cooperation in trying to find "means short of outright intervention" to solve the Cuban political crisis. The document hardly counted as a contingency paper of plans. It read more like a white paper justifying a failed policy.[18]

The next day, Special Assistant for National Security Affairs Gordon Gray was puzzled about the "indeterminate discussion" at the NSC meeting. "I pointed out to the President that I had not been informed as to what was going on but that I had not pressed for any kind of directive in the meeting because it was not clear to me whether there were not some programs which he had approved." The president remarked that "the situation had been allowed to slip somewhat."[19] Soon Eisenhower lamented to Allen Dulles, among others, that "for one reason or another the main elements of the Cuban situation had not been presented to him." Eisenhower recommended "better coordination."[20]

When Santa Clara fell on December 30 and many high-ranking Cuban officers began packing their bags, U.S. options dwindled even more. Wealthy individuals converted pesos into dollars. The U.S. Embassy began processing a greater number of visa applications from Cubans associated with the regime. The number of tourists in Havana dropped further, but those who remained seemed to ignore the war and prepared for New Year's Eve celebrations at their hotels. One American journalist, however, thought Havana looked like a "dying city, scarcely breathing, bruised and sullen behind a facade of wholly synthetic gaiety."[21] In beseiged Santiago, Consul Wollam initiated the first phase of an evacuation plan to return American dependents to the United States. Observers sensed the end when Batista's two young sons arrived on the thirtieth at New York's Idlewild Airport, where demonstrators harangued them as "war criminals."[22]

In Washington, the U.S. Joint Chiefs of Staff desperately tried to resuscitate arms shipments to the regime, while a Defense Department official called Herter to find out "just what planning is going on." Couldn't the State Department at least make its "wishes" known?[23] Herter worried that "Defense is getting nervous about Cuba," so he called a meeting of high-level officials.[24] Near the end of the year the administration still lacked "coordination" on Cuba.

Much of the task for inter-departmental activity fell on Assistant Secretary Rubottom, who seemed overburdened. He read and cleared telegrams on the Cuban crisis and tried to keep himself informed on the island's fast-paced events. He also struggled to learn what the CIA was doing. While handling Cuba, he still had to attend to other issues—measures for hemispheric economic cooperation and an inter-American development institution, Argentina's economic malaise, Brazil's social unrest, and relations with Mexico's new president. Venezuela sat atop Rubottom's agenda too. The South American nation was adopting nationalistic policies toward foreign oil interests and had just elected a leftist president, Rómulo Betancourt.

In the morning of December 31, Rubottom appeared before a closed-door, executive session of the Senate Foreign Relations Committee. The senators urged Batista's ouster at the same time that they worried about both U.S. military intervention and a Communist takeover. Rubottom did not tell them about the U.S.-assisted plots to move Batista out, but he mentioned "rumors" that some military officers might seize power.[25]

At 4:00 p.m. Rubottom accompanied Christian Herter to a meeting with the White House's Gordon Gray, the JCS's Admiral Burke, and Defense, CIA, and Navy representatives. The discussion rambled, tempers flared, and nobody went away thinking that the United States had strong options. When Rubottom said that he did "not think that it will be possible to pin a Communist label on the Castro movement," Burke tried to get him to say otherwise. The leather-lunged admiral also complained that by not helping Batista, the United States had boosted Castro. It was "not quite so simple," Rubottom retorted, for Batista "put on a disgraceful performance with the arms given him." Gordon Gray asked for a clarification of U.S. policy. Rubottom, for whom concise statements always seemed elusive, answered: "The U.S. has been trying to get Batista to recognize that he, Batista, cannot defeat Castro as such, but that a third force is needed to defeat Castro politically." Again, the mysterious "third force."

What happens after Batista? That was the key question. Speculation abounded on whether Prío, Varona, and Miró Cardona could become a united force to oppose Castro. There was no "strong man" behind whom the United States could stand, although Burke assured his colleagues that in two or three days he could get arms to such a leader. The United States desperately needed to identify a viable third force, "because once identified it will serve as a rallying point." Herter closed the meeting with the advice that all of the departments should "keep in close touch."[26] In the last days of 1958, then, the United States was still haplessly attempting to locate a credible anti-Castro third force.

Castro's rebel forces swept forward, dashing hopes that he could be blocked. Batista's press censorship had masked Cuban army defeats, and some

U.S. media, relying upon official Cuban accounts, actually had the government winning. "Forces loyal to President Batista of Cuba have defeated a large rebel force at Santa Clara," said one CBS news report.[27] The Associated Press's Larry Allen reported that Batista's Army had "hammered retreating rebel forces"—this just as the president was about to flee his country.[28]

On the last day of 1958, Prime Minister Güell met with Ambassador Smith to inform her that Batista would resign in favor of a junta—upon the condition that the United States would support the new government, especially with weapons.[29] Smith promised to press Washington to help, but he was not optimistic. Shortly before 10:00 p.m., top-ranking State Department officials lectured Smith in a telegram that the United States had done all that it could. Batista had refused to take "constructive political measures" to save himself. The regime, moreover, had had enough military weapons to crush the "minor" Castro movement at the start. Batista proved militarily and politically ineffective. The United States did not "set out to weaken Batista. The unfortunate sapping of his remaining bases of strength [was] due [to] his own decisions" that limited the "latitude of means available to US to cooperate with him." It was too late to send military aid to any government Batista anointed as his successor, concluded the telegram. In short, Smith had backed the wrong man.[30]

"Tonight we go," Batista told one of the Tabernilla officers in the early evening of December 31.[31] Batista invited government leaders to a New Year's Eve party at Camp Columbia in Havana. Nobody refused such an invitation. On the guest list were those loyalists chosen to fly out of the country with him. Three aircraft were fueled and ready. General Cantillo was summoned and told that he must stay behind to lead the armed forces and a junta. SIM officers began loading their Oldsmobiles with suitcases and families to escape from Havana.[32]

At Camp Columbia the prominent guests joylessly moved around the buffet table. Once cocky and at the pinnacle of power, they now faced perilous futures. Shortly before midnight Batista made a grand entrance. He extended New Year's greetings, ate a dish of *arroz con pollo* (chicken and rice), and then slipped into his office to meet with the military brass. By 2:00 a.m., he had abdicated, turning the government over to General Cantillo. Rivero Agüero, the president-elect, had not even been informed, but a seat was reserved for him and his wife on one of the airplanes. "It is the end of the world," he moaned.[33] Members of the select group scurried home to fetch belongings and family and then frantically raced to the airfield. They fled so fast that they left sheaves of documents behind—evidence of corruption and crimes that the victorious rebels would soon use to demonstrate why the insurrection had been necessary. The records also exposed U.S. cooperation with the dictatorship.

Shortly after Ruby Phillips of the *New York Times* dropped off Herbert and Nancie Matthews at their hotel, she heard planes overhead. Strange for that late night hour, she remarked.[34] David Phillips (no relation to Ruby), working for the CIA under cover as an executive in a Havana public relations firm, was lounging on a lawn chair sipping a glass of champagne when he heard the rumble of the aircraft. He immediately telephoned his CIA case officer. "Batista just flew into exile," he said. "Are you drunk?" came the reply. "If Batista's not aboard I'll eat your sombrero," wagered Phillips.[35]

In the early morning hours of the first day of the new year, Batista's plane roared down the runway, his "road of sorrow" now taking him to the Dominican Republic.[36] Rivero Agüero likened the DC-4 to "a huge casket carrying a cargo of live corpses." When the plane flew over the Sierra Maestra, he looked down, wondering aloud how Castro would treat them if they happened to crash in the rebel stronghold.[37] Other planes headed for the United States, avoiding Miami because the many exiled anti-regime Cubans there might stage a dangerous demonstration. Some *batistianos* jumped into boats and sped toward Florida, while others sought asylum in foreign embassies. Ambassador Smith knew that the United States had given Batista and his cronies the final shove. At least, he thought, "this shortened [the] war and saved many lives."[38]

The U.S.-supported "third force" plots, meanwhile, self-destructed one by one. Justo Carrillo's plan stalled when, on December 30, CIA contact Albert Graham showed up at Carrillo's Coconut Grove, Florida, home to report that there would be no military operation. Instead, the CIA preferred to pay the warden at the Isle of Pines prison $100,000 to release Colonel Barquín. Aware that Che Guevara's forces were closing in on Santa Clara, Carrillo groaned that this new plan came too late. But he promised to have one of his people in Cuba extend the bribe.[39] On the island, however, the Montecristi agent Felo Fernández discovered that the warden at the Isle of Pines institution had been transferred. Judging his mission now too risky, he jettisoned the bribery scheme. In Miami, Carrillo waited anxiously. Soon after he heard that Batista had fled, Carrillo rushed to the Miami airport to board a Cuban flight to Havana. He dreamed of installing Barquín, but his plans went awry once again. The Havana airport was closed, and his plane had to return to Miami.[40]

Years later, after Castro claimed power, the highly nationalistic Carrillo remembered U.S. meddling and ineptitude during the December crisis. Carrillo wrote that Pawley's mission constituted yet another instance of U.S.

"Plattism." Once again, "the United States thought it had the right and the responsibility to decide when a Cuban president would abandon power and what type of government would follow." Indeed, "the U.S. still believed that it could not only remove presidents but also appoint new ones."[41] True, but Carrillo, like so many other anti-Batista Cubans, had invited U.S. intervention. They themselves had asked North Americans to help. They themselves had sought a *North American* solution to the *Cuban* problem.

As Carrillo's plan disintegrated, so did Varona's. At the King Ranch in Camagüey Province, "Tony" Varona failed to locate the Army officer who was supposed to consummate the power grab with the former prime minister. Desperate, Varona decided to go to the enemy—the provincial 26th of July *commandante*—in hopes of making some deal. Armed with machine guns, Varona and a few compatriots set off for the rebel base. They arrived on the morning of January 1 just in time to learn that Batista had fled. Varona jumped in his car and headed for an army camp in search of someone who could help him gain power in post-Batista Cuba. On the way the engine overheated and the block cracked. He abandoned his disabled car, commandeered another, and then rushed to military garrisons, only to discover that they had already surrendered to 26th of July forces. Without an army or a mobilized political constituency of his own, he abandoned his quest for the Presidential Palace. He drove to his mother's house instead. Varona stayed on in Cuba after Castro's victory, demanding free elections and opposing the agrarian reform program, but later went into exile in the United States and became one of the political leaders behind the CIA's Bay of Pigs operation.[42]

With Varona's failure in the last days of 1958, Carlos Prío's campaign to topple Batista and stop Castro collapsed, ending Prío's already slim chances of returning to power. After Castro's triumph, Prío returned to the island from the United States. In February 1959 his U.S. trial for conspiracy was transferred from New York to Miami, but not until December did U.S. authorities suspend Prío's sentence after he pleaded guilty to charges that he had violated U.S. laws by organizing anti-Batista activities on American soil. In post-Batista Cuba, the former president tried to make amends with Castro but wielded little influence. Denouncing Castro as a Communist in early 1961, Prío departed once again for the United States and began to aid anti-Castro saboteurs for what Prío called a "holy war" to overthrow the regime. Inveterately anti-Castro to the end, Prío in April 1977 killed himself with a gun—just three months after William Pawley had committed suicide.[43]

On New Year's Eve in Palma Soriano in Oriente, Fidel Castro, Celia Sánchez, and other rebels gathered for a party. Over morning coffee at the sugar mill they learned that their arch-enemy Batista had fled, and that Cantillo, their would-be ally, had assumed power. "This is a cowardly betrayal," boomed Castro.[44] The Communist Carlos Rafael Rodríguez was in Palma Soriano too. So was Errol Flynn.[45]

In Washington, Assistant Secretary of State for Economic Affairs Thomas Mann poked his head into the room of Robert Stevenson, the new Officer in Charge of Cuban Affairs. "You fellows better batten down the hatches," Mann said, "because there's going to be some real stormy weather."[46] The staunchly anti-Castro Mann later replaced Rubottom as Assistant Secretary of State for Inter-American Affairs.

20

Madhouse: Castro's Victory, Smith's Defeat

AN EERIE QUIET enveloped Havana in the early morning hours of January 1.[1] People knew that Batista had fled. Telephones had been buzzing with excited conversation for hours. Looking out their windows, *habaneros* could see SIM and police cars, once the terror of the streets, race away to unknown destinations. Cubans, remembered Ruby Phillips, seemed "afraid to trust their liberation. They were afraid of the empty streets and the silence."[2] Suddenly, before noon, they burst out of their homes. M-26-7 rebels who had lived in the underground now hugged and cheered openly, not yet resenting the many demonstrators around them who had become instant revolutionaries. Cars decorated with 26th of July symbols cruised the streets honking salutes to the good news. Looters appeared, too. Some broke into the homes of known *batistianos.* Others smashed shop windows, carrying away clothes and furnishings meant for sale to rich Cubans and the ABC colony.

The surging crowd also attacked parking meters, slot machines, and casinos, symbols of Batista's ill-gotten wealth and sweetheart deals with unsavory foreigners. This destruction counted as more than looting and more than a symbolic beheading of Batista: It represented protest against North American influence in Cuba. Herbert and Nancie Matthews were staying in the Sevilla Biltmore, two blocks from the *New York Times* office. Rioters barged into their hotel and trashed the casino. At the Capri Hotel, however, George Raft faced the mob at the main entrance of the fancy establishment. In his Hollywood, gangster-like style, Raft snarled, "Yer not comin' in my casino."[3] The crowd turned around and left. Not all casinos were destroyed; at some, hotel employees—some 10,000 Cuban workers depended upon the tourist trade—defended their jobs by confronting the invaders.

Sporadic gunfire—from police officers, happy celebrants, and insurrectionists—reminded everybody in Havana that a civil war still raged. A nervous Nancie Matthews saw Red Cross workers tending the killed and wounded. Havana, she said, had become "like a madhouse."[4] Stern-faced M-26-7 militia and members of the Civic Resistance soon restored order. "Fierce, bearded men . . . who looked like pictures of Australian bushrangers or characters in a film of the Klondyke [*sic*] gold rush" filled the streets and warned people to go into their homes or be killed.[5] Triumphant dissidents also freed political prisoners from Batista's overstuffed jails. CIA agents began to make contact with the rebels, trying to determine who had authority. Everybody and nobody did. As Fidel Castro said a few days later about the pell-mell collapse of the Batista regime, "everything is completely mixed up."[6]

Soon after the *batistianos'* aircraft lifted from Cuban soil, General Cantillo called Ambassador Smith and asked for permission for the planes carrying former Cuban leaders to land in the United States. Smith promised cooperation. Later in the morning Smith led a delegation of ambassadors to Camp Columbia to ask Cantillo to respect the rights of asylees—those Cubans who escaped to foreign embassies to save themselves from the swelling anti-Batista anger. The general also telegraphed Army commanders in Santiago to contact Castro and appeal for a ceasefire. All across Cuba, 26th of July Movement rebels, DR insurrectionists, Communists, and other insurgents claimed power in government offices, labor union headquarters, radio stations, police precincts, and Army posts—and not always cooperatively. At the CMQ Network in Havana, for example, PSPers asked the M-26-7 rebels who had reached there first to allow them some air time. The *fidelistas* denied the request.[7]

All over the world exiled Cubans cheered Batista's fall. In Miami, automobiles loaded with Cuban exiles waving the red and black banners of M-26-7 paraded on Flagler Street. In Jacksonville, where one of the planes carrying *batistianos* landed, anti-Batista Cubans tangled with the new arrivals. In Washington, D.C., Ernesto Betancourt, the M-26-7 representative who had so often argued the rebel case at the State Department, walked into the Cuban Embassy on Sixteenth Street to oust Batista's ambassador, Nicolás Arroyo, and seize incriminating files. Former U.S. Ambassador Arthur Gardner stood beside Arroyo, demanding that the documents remain in Arroyo's possession. But the takeover actually proved easy, for many of the embassy staff had already thrown in with the rebellion. In New York City, the Cuban Consul General took down the pictures of Batista, emptied his desk, and turned his office over to leaders of the 26th of July Club, who promptly taped likenesses of Fidel Castro on the walls.

In Mexico City, the rebel supporter Teresa Casuso put on her best dress and walked to the United Press International office to confirm reports of Batista's departure. Then she coolly approached the Cuban Embassy and claimed it for Castro. When vengeful anti-Batista exiles arrived to sack the embassy, she opened the gates, cautioning the excited crowd to respect the property of their now liberated nation. She calmed them too by promising to torch the Batista banners, trophies, and pictures she had piled conspicuously in the garden.[8]

In Havana, Ambassador Smith rallied his embassy staffers the morning of January 1. Many had partied all night and felt none too fit. They nevertheless began the task of protecting and evacuating U.S. citizens caught in the Cuban whirlwind. "I wasn't to get home again for three days," recalled Vice Consul Wayne Smith.[9] He and others handed out U.S. flag identification decals to be pasted to cars and dwellings. They trooped from hotel to hotel to count heads and warn U.S. citizens to stay indoors. Some 2500 tourists were in Cuba at the time, most of them in Havana. At the Riviera Hotel, when the general strike called by the 26th of July had deprived the management of its workers, Meyer Lansky himself had to pitch in to make sure that his anxious guests were fed. Although rumors had it that the gambling bosses had fled with Batista, Lansky stayed behind to cooperate with the U.S. Embassy in evacuating Americans. The mobster himself did not fly to Miami until January 7.[10]

Americans caught in the midst of the turmoil in early 1959 pressed embassy personnel for help.[11] U.S. tourists ran out of cash because the general strike had shut down the banks. Restaurants also halted service, and food—at least the kind American travelers expected—became scarce. Americans ventured onto the streets to mingle with the curious *barbudos* who filtered into the capital city, yet the inability of most Americans to speak Spanish left them feeling isolated. Cruise ships kept coming into port, while members of the U.S. Congress called the State Department to request protection for constituents. One Missouri congressman appealed to the State Department to help a touring circus owner who was eager to get himself and all of his animals out. Canadians also asked U.S. officials to arrange their exit from the island.

But Cuba's transportation system was collapsing, and the airport was closed. People rushed to the docks hoping to climb or drive aboard the Havana-Key West ferry, *City of Havana.* Those Americans wanting to evacuate were "getting nervous, petulant, and impatient," especially when M-26-7 authorities would not at first let the ferry enter Havana harbor.[12] Might not *batistianos* flee by this route? Might not its docking violate the general strike?

In Washington, State Department, JCS, and White House officials, during the evening of January 1, decided to despatch from Key West two submarine tenders and three destroyers to waters off Havana, out of sight of land. A Marine battalion was also ready to be airlifted to Cuba if necessary. U.S. officials agreed that no publicity would surround the expedition, but Press Secretary James Hagerty nonetheless mentioned it to reporters. Might not jittery Cubans now think the United States was intervening? As it turned out, the U.S. warships were not needed. The airport reopened and embassy staff gathered evacuees at the Hotel Nacional before convoying them to chartered planes. By early evening on January 4 commercial aircraft flying to Miami and New York and the ferry to Key West had evacuated 1,772 Americans. The U.S. warships began returning home on that day, allaying fears that they had been sent to meddle in Cuban affairs.[13]

From Oriente Province on January 1, Fidel Castro commanded Guevara and Cienfuegos to march on Havana. He also broadcast appeals for order on Radio Rebelde, declared a general strike, and bitterly denounced Cantillo's assumption of power, which he believed the U.S. Embassy had helped arrange.[14] "Revolution, *yes!* Military coup, *no!*"[15] He also prepared to seize Santiago, home of Moncada Barracks and symbolic capital of the insurrection.

In Havana, meanwhile, General Cantillo pondered his options. The Directorio Revolucionario controlled the University of Havana. Rebel forces were marching toward the city, wildly applauded by onlookers. Much of the capital city and the rest of the island had already fallen to insurrectionists and the general strike was paralyzing transportation. Many Army troops were shedding their uniforms and disappearing into the populace, while exiles from abroad were beginning to return. General Cantillo recognized that he had no political constituency and that he faced certain defeat. He recoiled from fighting the battle-tested units of Guevara and Cienfuegos. But he wanted to save the armed forces for another day. He conferred with the head of the U.S. Army Mission.[16]

Cantillo decided to invite another military man—an anti-Batista officer—to take over the armed forces. During mid-afternoon on January 1, he ordered the release of Colonel Ramón Barquín from the Isle of Pines prison and sent a plane to bring him to Havana. U.S. officials could not have been more pleased. Barquín became their immediate hope for an effective third force. They briefed President Eisenhower that Barquín might very well "strengthen the military's position, vis-à-vis Castro, and add a certain amount of stability to the situation."[17] The suspicious *fidelistas* sensed that Barquín, although anti-Batista, was an anti-Castro ally of the United States. "His background was a problem: a career officer, military attaché in Washington, educated in the United States," Carlos Franqui remembered. "His arrival helped finish off Cantillo, but ensconced at the Columbia base, he [Barquín] himself became a threat."[18]

Barquín, dressed in prison garb stamped with a large "P" and accompanied by other prisoners, including Major Borbonnet and M-26-7's Armando Hart, landed at Camp Columbia.[19] During the brief flight to Havana, Barquín set his immediate agenda: establish contact with Manuel Urrutia and Fidel Castro. Once on the ground, Barquín asked for information. "We are faced with chaos," answered one officer.[20] Needing political advice, he inquired about his ally Justo Carrillo. Andrés Suárez, now at Barquín's side, volunteered to reach the Montecristi leader in Miami. Before long Barquín despatched an Air Force plane to Miami to pick up the exile. Barquín also sent conciliatory word to Castro: Urrutia should come to Havana to install a new government. Barquín then ordered the arrest of General Cantillo. Like SIM, police, and other military officers, Cantillo faced charges of "war crimes."

During the evening of January 1 in Oriente, Castro met with Colonel José Rego Rubido, the district commander in Santiago, who shared with Castro the desire to avoid needless bloodshed now that Batista was gone. Shortly before 10:00 that night, a much relieved Park Wollam wired Washington: "Santiago completely controlled by 26th. The *cuartel* [barracks] has formally surrendered. 26th organizing provisional government of local control. Considerable celebrations but everything peaceful to this moment."[21] Castro's forces strode into Santiago without firing a shot. A few hours later, at a turkey dinner prepared by Connie Wollam, the consul's wife, Castro remarked that Americans would face no difficulties in the new revolutionary Cuba.

That night and into the morning Castro gave a victory address in Santiago. This time, Castro insisted, would not be like the 1890s, "when the North Americans came and made themselves masters of our country." Nor would it be like the revolution of 1933, when Batista replaced the Machado dictatorship with his own, nor like 1944, when Prío's "thieves" came to power. No traitors and meddlers would thwart revolution this time. "The people," he kept saying, would not be denied.[22] As Castro's words emphasized, the insurrectionists still worried that victory would be snatched from them.

When Carrillo arrived in Havana and parleyed with Barquín, it became suddenly clear to the Montecristi politician that he and Barquín had no power. Rebels controlled most of the country, the armed forces were falling apart, and Castro would not accept a military junta.[23] Barquín had inherited a "dead army," he complained to Ambassador Smith.[24] "All they left me with is shit," the colonel groaned to a CIA agent.[25] He recalled the joke about the surgeon who, when asked to assess the condition of a patient and recommend action, replied: "Operar no, autopsiar, sí." (Operate no, conduct an autopsy, yes.")[26] Soon Commandante Cienfuegos's column was at the gates.

In the evening of January 2, Barquín let the rebel soldiers into Camp Columbia, where the bearded, long-haired Camilo Cienfuegos, a veteran of *Granma,* claimed power for the 26th of July Movement. Cienfuegos encountered a MAAG member who was observing the transfer of power. The new Cuba, snarled the triumphant *barbudo,* "has no need for a mission that was not able properly to train an army which disintegrated in the face of the enemy."[27]

Barquín, Washington's last "third force" hope to block Castro, resigned January 3.

Guevara's unit soon seized Havana's La Cabaña fortress. The Second Front of the Escambray entered Havana too. Its troops took quarters in the luxury hotels—the Nacional, Sevilla Biltmore, and Capri—and wondered about their place in the new Cuba. There could be no mistaking one fact: the 26th of July rebels possessed the most power—they ruled Oriente Province; they controlled the 19 police stations and the major military installations in Havana; they held the principal radio and television facilities and the airport; they enjoyed the tumultuous acclaim of Cubans across the island. In Pinar del Río Province, where the former U.S. Army man and youthful adventurer (and future historian of Latin America at the University of Florida) Neill Macaulay marched with rebel forces, peasant families lined the country roads to applaud the victorious *fidelistas.* When the rifle-touting band stopped in La Coloma, townsfolk set up tables in the plaza and offered a much welcomed feast.[28]

The naming of Judge Urrutia's cabinet reassured observers who had warned against Castroite radicalism. Many of the appointees were professionals, moderates from the middle class. Included was the new prime minister, José Miró Cardona, president of the Havana Bar Association, whom U.S. officials thought a "strong leader and moderating influence."[29] The foreign minister was Roberto Agramonte, Ortodoxo presidential candidate in 1952—"considered friendly to the U.S."[30] The new interior minister was Luis Orlando Rodríguez Rodríguez, an Ortodoxo and publisher whose newspaper Batista had shut down after it had editorialized against Ambassador Gardner. U.S. officials were particularly pleased with the finance minister, Rufo López-Fresquet. Luis Buch and Raúl Chibás became presidential advisers. Justo Carrillo was named the president of the Agricultural and Industrial Development Bank. Montecristis and Ortodoxos were represented in the new government, but not Prío Auténticos or DR, Second Front of the Escambray, or PSP members. The M-26-7 representatives in the new regime were drawn primarily from the urban underground. The U.S. Embassy's Braddock rated it a "good" cabinet—"basically friendly toward the United States and oriented against communism."[31]

U.S. officials may have liked the composition of the new government, but members of the Directorio Revolucionario did not. They held the Presidential Palace and they were armed. Word soon circulated that they would not turn the palace over to Urrutia until the government was broadened to include all revolutionary organizations, as M-26-7 had promised in the Caracas Pact. A showdown seemed possible, with anti-Batista groups slashing at one another. Camilo Cienfuegos joked that a couple of cannon balls ought to be shot into the palace to scare DR off. Carlos Franqui, now editor of the new 26th of July newspaper, *Revolución,* thought that a good idea, since the palace was a hated symbol of Batista's power. But Che Guevara and Faure Chomón negotiated to defuse the crisis. DR, like everybody else in January, bowed to the reality of *fidelista* power.

Urrutia flew to Havana on January 5 and took possession of the ransacked palace. He announced that elections would be held in 18 months, that gambling was forbidden, and that "only decent Americans will live in Cuba."[32] Ambassador Smith suddenly appeared, uninvited. The atmosphere turned frigid. Worried about the *batistianos* who were seeking asylum, the U.S. envoy sought assurances that they would enjoy safe passage out of the country. Urrutia promised that traditional international law would be honored. Smith left as suddenly as he had arrived.

Executions across Cuba soon reduced the numbers of former Batista officers. When Earl E. T. Smith heard that Cantillo might be executed, the ambassador entered Cienfuegos's Camp Columbia office to plead for the general's life. Smith, dressed in a white suit, felt uncomfortable in the company of the unkempt, unshaven rebels. "The group reminded me of pictures I had seen of the Dillinger gang," he recalled.[33] Spared the firing squad, Cantillo was sentenced to 15 years at the Isle of Pines prison.

Meanwhile, Communists broke into BRAC offices and grabbed truckloads of documents that detailed the anti-Communist bureau's rough treatment of dissidents, its connections with the CIA, and, of course, Communist activities. Eventually the files were burned, thereby feeding a conspiracy thesis that the *fidelistas* and Communists were covering up a devilish plot to bring Marxism-Leninism to Cuba.[34]

If such a conspiracy existed, few in Havana or Washington knew about it. Indeed, U.S. officials moved quickly toward the recognition of the new Cuban government. The moderate nature of Urrutia's cabinet persuaded even Ambassador Smith's business advisory group to recommend speedy diplomatic recognition. The government, which enjoyed strong popular support, "was much better than they had dared hope for."[35] ESSO and Texaco actually placed product ads in the new publication *Revolución.*[36] The businessmen backed recognition despite—or perhaps because of—an incident at the headquarters of the Cuban Electric Company, where an employee committee made up of M-26-7, PSP, DR, and Auténtico representatives staged a "takeover of the management" and demanded a 20 percent wage raise.[37] FBI officers, meanwhile, informed J. Edgar Hoover that Urrutia was both anti-Communist and anti-gambler.[38] An American banker praised "the way their troops have behaved," which "certainly throws dust on the fear that they are a bunch of Communists."[39]

Secretary Dulles urged President Eisenhower to extend recognition: "The Provisional Government appears free from Communist taint and there are indications that it intends to pursue friendly relations with the United States."[40] William Wieland remembered that U.S. officials hoped that recognition would enhance the competitive power of the moderates, for at the start Castro did not appear to have firm control of the government.[41] As well, the major Latin American nations—Mexico, Venezuela, Brazil—had already acted to welcome the new government. At 5:00 p.m. on January 7 the United States recognized the new Cuban government.

Five days earlier, leaving his brother Raúl in charge of Oriente and surrounded by 26th of July warriors, regular Army defectors, and the machines of war, Fidel Castro began a slow-moving march of victory from Santiago to Havana.[42] A captured helicopter flew overhead, carrying messages, finding detours around broken bridges and other obstacles, showing off. All along the roads people embraced the victors and cried for speeches. Exhausted yet exhilarated, and savoring the genuine popular fervor for his remarkable accomplishment, Fidel Castro obliged the crowds in one town after another. In Cienfuegos he spoke from a flatbed truck from 11:30 p.m. until 3:00 a.m. "I think we've learned not to sleep," he remarked to a comrade. "We have lived for some time under great emotional stress."[43]

American journalists rushed to interview him. The *Chicago Tribune's* Jules Dubois reached him first. Using a chartered Piper Apache aircraft, Dubois caught up with Castro at Holguín on January 3. Asked about his attitude toward the United States, Castro replied that "if I have had to be very cautious about my statements in the past, from now on I am going to have to be even more careful."[44] Herbert Matthews found Castro at Camagüey on January 5, but they had little chance to talk. In Matanzas, Ed Sullivan interviewed the 32-year-old leader for his Sunday-night television show.[45]

In the twilight of January 8, Fidel Castro's victory caravan, slowed by massive, worshipful crowds, finally reached Havana. Everyone seemed awestruck. "It is all like the second coming of Christ—beard and all," thought Nancie Matthews.[46] The British Ambassador went even further: Castro "represents a mixture of José Martí, Robin Hood, Garibaldi, and Jesus Christ."[47] In the capital city, Fidel Castro climbed down from his tank to exchange greetings with President Manuel Urrutia. Trusted rifle slung over his shoulder, the M-

26-7 guerrilla leader spoke briefly to the throng from the terrace of the Presidential Palace. When he was ready to lead his green fatigued *barbudos* to Camp Columbia, the mass of joyous Cubans parted to clear a path.

At the camp, a bleary-eyed Castro rambled through an incoherent speech, extolling "the people" and calling on all insurgents to lay down their arms (an undisguised warning to the DR). "FIDEL! FIDEL! FIDEL!" the crowd chanted. That evening the young rebel leader slept in the Havana Hilton, soon to become his headquarters under a new name—the Havana Libre.

"Dr. Castro, you are a good neighbor," television talk-show host Jack Paar told Fidel Castro. "My suite here [at the Hilton] is just below you, and you haven't been down once to borrow sugar."[45] Tired and hoarse, Castro did not warm to the humor. He had barked earlier to Celia Sánchez that he was tired of North Americans and "their way of twisting everything he says."[49] Once, in earshot of an Associated Press reporter at the Hilton, Castro reacted fiercely to the comment that U.S. intervention in Cuba still remained a possibility. If the North Americans invaded, he flashed, "two hundred thousand gringos will be killed."[50] Well schooled in his nation's history and still angry about the "third force" machinations of December, Castro fully expected the United States to plot once again to deny him his hard-earned power and the Cubans their long-awaited revolution. He vowed not to flinch.

Nobody, Cuban or North American, seemed to know Castro's plans for the future, other than that he intended to break the old order that had been dominated by a domestic and foreign elite succored by the military and the U.S. government. *El jefe* spoke in specifics about the past, in generalities about the future. Everybody knew he had dreams; few knew if he had any precise programs. Preoccupied with the immediate, running without sleep, enthralled by the public euphoria, consumed by the sheer joy of having won against such steep odds—Fidel Castro himself may not have known exactly where he was going to lead the nation. U.S. Embassy personnel sensibly observed that his promises of justice, good government, land reform, universal education, and democracy seemed "utopian."[51] The day after U.S. recognition, Rubottom remarked that the new government officials "have not asked for anything and probably do not know yet themselves what they need."[52] "I kept wondering what Fidel was thinking," recalled Carlos Franqui. "He improvises and never shares power."[53]

On a few issues Castro had made up his mind. En route to Havana he had said he would permit a free press, even the Communist organ *Hoy.* To fulfill rebel decrees issued during the insurrection, the new government would execute "war criminals." Castro insisted that the U.S. military missions leave. They did, explained one U.S. official, because a superpower like the United

States could not appear to be "begging" to save them.[54] Castro had apparently decided, too, that his rival Carlos Prío Socarrás must be banned from politics.[55] And Castro pledged to subsidize the debt-ridden Havana Sugar Kings. He wanted the International League to remain in Cuba, "even if I have to pitch."[56]

Castro also intended to ban all games of chance. "We are not only disposed to deport the gangsters," he remarked, "but to shoot them."[57] Under pressure from tourist industry laborers, however, Castro reopened the casinos in February. The terms were stiff, however: the casinos had to pay heavy taxes to the government, one-armed bandits, or slot machines, were outlawed, and casino operators and all North American personnel had to obtain a letter from the U.S. Embassy attesting to their "good reputation."[58] According to an FBI report, Fidel Castro even asked Errol Flynn to help him find a manager for the Sans Souci casino. Flynn brought somebody to Havana, but the casinos did not long survive.[59] In October 1960, Meyer Lansky learned that his Riviera Hotel had been nationalized. Disgruntled U.S. racketeers left the island, some later to plot with CIA officers to assassinate Fidel Castro.

As Castro marched toward Havana in early January 1959, Washington decided to replace Ambassador Smith, who was criticizing U.S. recognition as hasty.[60] The victorious rebels and their followers, in both Cuba and the United States, urged his removal. They remembered his 1957 denunciation of police brutality in Santiago, but they had watched him become an ardent ally of the regime. A lawyer who handled M-26-7 business in the United States explained that it "would be inconceivable" that Smith remain in Havana.[61] The snobbish British Ambassador once described Smith as "a good shot and golfer. But he is not particularly intelligent." Cubans "tend to regard him as aloof and even rude."[62]

Acting Secretary of State Herter ranked Smith a "playboy extraordinary" who ought to be replaced before the Cubans declare him persona non grata.[63] Yet when Herter asked for President Eisenhower's approval to request Smith's resignation, the president, once again revealing his shallow understanding of the Cuban crisis, asked "why it wouldn't be better to ask the new Government whether Ambassador Smith would be acceptable." Herter explained the obvious: the United States would be embarrassed because the answer would surely be "No."[64]

Herter informed Smith that he would have to step down. But Eisenhower continued to worry that Smith, whom the president insisted had done a good job, might be offended because it looked like he was being "kicked out."[65] Learning about the president's concern and averse to leaving under fire, Smith appealed to Herter and Eisenhower. Like Arthur Gardner before him,

Smith tried to save his job. He pointed out that the Urrutia government was not requesting his recall, that "precipitate resignation could be interpreted as acknowledgment that US policy in Cuba has not been correct and that US had been favoring one side," that he could establish "favorable relations" with the new Cuban government, and that he was being credited by Cubans with having facilitated Batista's sudden departure. Smith requested a full "review" of the question of his resignation.[66] Annoyed State Department officials reconsidered the issue but did not change their minds. As Dulles told Eisenhower, "the fact is it will be extremely difficult to carry on relations with a new gov[ernmen]t without a new amb[assador]."[67] Bowing to Washington's orders, Smith submitted his resignation on January 10.

Two days later, Eisenhower approved the appointment of Philip W. Bonsal as the new ambassador. A career Foreign Service Officer who had served earlier as Cuban desk officer in the State Department (1939–40), Ambassador to Colombia (1955–57), and Ambassador to Bolivia (1957–59), Bonsal spoke Spanish, knew Cuba and Latin America well, and seemed quite acceptable to the Cubans. But Republican politics, which had produced the unfortunate appointments of Arthur Gardner and Earl E. T. Smith, almost derailed Bonsal. Just after Eisenhower picked Bonsal, Henry Luce, publisher of *Time* and *Life,* telephoned Secretary of State Dulles to tell him that his wife, Clare Booth Luce, "would be inclined to take on Cuba." The next day, hearing from his advisers that she was not the person for Havana, Dulles firmly informed the president that the Cuban assignment required a "trained FSO." The secretary complained that "the pressure is mounting. There are a number of people who are almost trying to purchase it. . . ."[68] In the end President Eisenhower took Dulles's advice, and Clare Booth Luce was offered the ambassadorship to Brazil instead. The Senate confirmed Bonsal on Feburary 16 and he arrived in Havana three days later.

Before Ambassador Earl E. T. Smith departed Cuba, *Bohemia* magazine excoriated him as a Batista henchman. Smith angrily cabled the State Department: "There are two things I will not be called—one is a crook, the other is a fairy."[69] If his memoirs provide a clue, soon after he left his post Smith set himself the task of discrediting his detractors—especially State Department and embassy officers with whom he had disagreed. He settled back in Florida, and though a Republican, Smith gave Democratic presidential candidate and Palm Beach friend John F. Kennedy advice in 1960 on how to deal with the issue of Catholicism in the campaign.[70] When Kennedy became president in early 1961, he offered Smith the ambassadorship to Switzerland. But the Swiss balked, fearful that their interests in Cuba would come under attack if they received him. President Kennedy then "did something very petty," recalled Under Secretary of State Chester Bowles. Kennedy issued an order "that nobody was to go to the Swiss embassy—stay away from the Swiss, make them suffer, make them sweat."[71] A year later Smith published his angry, distorting memoir, *The Fourth Floor.* In the 1970s the former ambassador served as mayor of Palm Beach, and later President Ronald Reagan appointed him to the Presidential Commission on Broadcasting to Cuba, which helped launch the Voice of America's Radio Martí, which was aimed against the Castro regime. Earl E. T. Smith died at age 87 in 1991.[72]

V

Losing a Client

21

A Complete Break: How Did the United States Let This One Get Away?

"FIDEL CASTRO CLEARLY established himself as the dominant military and political figure in the revolution," the U.S. Embassy reported in early January 1959.[1] As U.S. officials wondered what Castro would do next, traditional hegemonic assumptions guided their wary observations. The new Cuban leaders "had to be treated more or less like children," CIA Director Allen Dulles told the National Security Council. "They had to be led rather than rebuffed. If they were rebuffed, like children, they were capable of almost anything."[2] U.S. diplomats found Castro restless, headstrong, opportunistic, and driven by an "undeviating urge for fame and political power." He was prone to violence and independent action, but he was not a Communist.[3] "Castro has taken Cuba by storm," reported embassy diplomat Daniel Braddock.[4]

Although some North Americans marveled at the young revolutionaries' "austere Puritanism that ruled out any license," and some predicted that Castro would not harm U.S. investments and trade because "Cuba is accustomed to U.S.-type goods and U.S. operations," embassy officials noted that "some danger flags are already up and need to be closely watched."[5] The legalization of all political parties, including the Partido Socialista Popular, counted among the alarms. In that tense Cold War time of the Berlin crisis, Mideast unrest, and nuclear arms race, President Eisenhower's aides dutifully and regularly alerted him to this open door to Communist activity.[6]

But much more than the re-emergence of Communists in Cuba rattled North Americans. With the support of the vast majority of Cubans, the

Castro government began to rid the nation of a *batistiano* system of corruption that infected almost every institution. Executions of *batistianos* eventually reached more than several hundred.[7] The new regime also began quickly to cut the legs out from under both its Cuban rivals and U.S. interests, likely partners in any anti-Castro campaign. "No mere palace coup or military coup, the Castro revolution has brought a complete break," Braddock astutely observed within the first two weeks of Castro's rule.[8]

Castro would not cooperate with a United States determined to direct the new Cuba. He intended to bury Plattism once and for all, and he had long challenged Washington's presumption that its word was fiat in the Western Hemisphere. "The Platt Amendment is finished," and the new Cuban leaders would "neither sell themselves, nor falter nor become intimidated by any threat," Castro insisted.[9] "We no longer live in times when one had to worry when the American Ambassador visited the [Cuban] Prime Minister," Castro once proudly informed Bonsal.[10] "What happened in Guatemala will not happen here," claimed Che Guevara.[11]

This stunning result—a triumphant Fidel Castro idolized by the Cuban people, facing little countervailing power from island rivals, insisting on tossing out the old guard, tolerating Communists, and articulating a vitriolic anti-Americanism—is exactly what U.S. leaders did not want and had worked hard to prevent in the 1950s. For a long time after the *Granma* voyage the United States had backed Fulgencio Batista's strong-armed response to the insurgency. When Batista faltered and became ever more repressive, Washington cut back on arms shipments in a vain attempt to coerce him to reform. When he proved incorrigible, the Eisenhower Administration schemed to push the unpopular dictator out before Castro could come in. The plots designed to block Castro came too late, and they lacked leaders who could rally the Cuban people behind a viable third force.

How did the United States, the hegemonic giant in the hemisphere, let this one, just 90 miles off the Florida coast, get away? The world's mightiest power had not intervened militarily to save a longtime ally or deny power to a young insurgent known for his hostility toward the United States and its influential institutions on the island. The Sixth Fleet patrolled the Mediterranean and the Seventh Fleet plied Asian waters. Some one million U.S. military personnel, part of the three-million-strong armed forces, were stationed abroad to protect U.S. interests. As Castro struggled toward victory in 1958, the United States dispatched troops to Lebanon to settle political questions in that faraway Mideast land. That year, too, the Eisenhower Administration tenaciously defended the U.S. position in Berlin against Soviet threats. By the time of Castro's victory, the United States had intervened in Southeast Asia, on the way to major commitments in Vietnam. Then why not Cuba? The answers are many, springing from the Cuban, North American, regional, and international contexts of the late 1950s.

The methods traditionally available to a hegemonic power that seeks to influence the outcome of a civil war in a small neighboring state seem exhaustive: military intervention, arms deliveries, economic coercion, bribery, covert operations, political subversion, threats, diplomacy. A hegemon such as the United States, eager to apply the containment doctrine against a perceived enemy like Fidel Castro, uses these means to check, coopt, or roll back. In the Cuban case, however, U.S. officials found that they could achieve none of these results. Castro's military advance could not be stopped; he would not be wheedled into compromises; he rejected juntas contrived to deny him power; he outmaneuvered more moderate rivals upon whom U.S. officials counted. The United States failed to erect a breakwater to blunt and disperse the waves of Cuban nationalism and rebellion that carried Castro to victory. The U.S. experience with Cuba, of course, was not atypical. Hegemonic powers, as the United States later learned in Vietnam, Iran, and Nicaragua, always face limits imposed by their own ideas and politics and by local, regional, and international conditions.

Timing, for one, determines outcomes that a hegemon may not want. When Castro sparked the insurrection in late 1956, his tiny guerrilla army seemed no threat to the U.S.-backed structure of power on the island. The United States felt no urgency to act against the upstarts, against a "back-door harassment."[12] When the Castro movement seemed to have shot its bolt after the feeble general strike of spring 1958, Batista ordered an offensive that appeared to have a good chance of pinning down and overwhelming the *barbudos* in the mountains. The United States sensed advantage, not imminent danger. "By the middle of '58 Castro seemed very low," remembered the State Department officer Robert Stevenson. "His star was waning and it looked as though he was never going to make it."[13] Batista seemed to have regained control. But then, in a matter of a few months, he fell very very fast, much faster than U.S. officials expected. When the summer offensive sputtered and the autumn national elections proved fraudulent, the United States was not prepared for Batista's rapid descent. By the time U.S. officials managed to concoct several third-force conspiracies, it was too late.

Terrain also places limits on the projection of hegemonic power. Castro's insurgents were holed up in a rugged, isolated, mountainous area that few regular forces could penetrate and high-flying airplanes could not subdue. CIA Director Allen Dulles recognized the problem: "In those mountain areas what you need against guerrillas are guerrillas. . . . It is rough country, and there is no use sending tanks and heavy artillery up there."[14] More, the poor local farmers welcomed the insurrectionists and nurtured them, providing food, shelter, intelligence, and labor. Despite their meager resources and much to the astonishment of Batista and U.S. observers, then, M-26-7 warriors survived and flourished. In discussing the "difficulties of dealing with this guerrilla-type activity," one U.S. intradepartmental committee recalled "the days of Pancho Villa," when U.S. troops invaded Mexico earlier in the century and failed to capture the legendary anti-American Mexican.[15] As the U.S. military knew full well during the July 1958 hostage crisis, sending an invasion force after a guerrilla army not only endangered U.S. citizens and interests (the rebels could always retaliate against U.S.-owned properties) but also risked a protracted war in an inhospitable environment. One U.S. military man doubted that even a "division of American parachute troops could eliminate Castro unless aided by luck and treachery."[16]

Any hegemon also needs effective local instruments through which to project power. But the United States had aligned itself with a flawed, weak, corrupt, ineffective, and unpopular regime it ultimately could not control. A dictator who could not be dictated to and who ultimately could not dictate, Batista could never overcome the illegitimacy and corruption of his government. "However corrupt Grau and Prío were," remarked a Cuban student, "we elected them and therefore allowed them to steal from us. Batista robs us without our permission."[17] The notorious corruption that afflicted so many institutions, Batista's unsavory alliance with business and deals with foreign investors, and the scab of casinos and mobsterism that encrusted Havana—all undermined the Batista regime. Families which had suffered the brutality of Batista's SIM, police, and army could neither forgive nor forget.

Batista did not fall because he lacked weapons. He fell because he never marshaled popular support, and once his power began to erode, few would fight for such a debased system. The United States went down with him because U.S. interests were too intertwined with practices and institutions the Cuban people detested. Like a beached whale—losing bearings and balance, too stranded to be revived by nourishing waters, vulnerable to the elements, and watched by curious, confounded onlookers who wondered how such a large, powerful beast could end up so helpless—Batista self-destructed.

Had Batista's inflated rhetoric about sterling economic progress rung true, he might have had a better chance of survival against the disgruntled Cuban people. But Cuba was mired in a dependent sugar economy; unemployment ran high; and peasant farmers struggled at the margin of subsistence far from the bright lights and casinos of Havana. Middle-class professionals, believing themselves closed out of an export-oriented economy dominated by foreigners and by a Cuban elite, grew frustrated in the political system of naked corruption and illegitimacy. If ever there was an authoritarian government ripe for toppling, Batista's was it.

The United States suffered the problem of trying to persuade an unreliable, disreputable local government to reform, to make concessions to critics in order to obviate blatant U.S. intervention. This was not the first or the last time the United States was bound to the fortunes of an unpopular and collapsing ally whom American officials ultimately decided to jettison. In 1949 Washington abandoned Chiang Kai-shek (Jiang Jieshi) as Mao Zedong's Communist revolutionaries neared triumph. In 1963, U.S. officials gave the green light to military coup leaders in Vietnam, who then overthrew and murdered Ngo Dinh Diem, the president the United States had installed years before and backed with substantial aid. No longer finding the Somoza regime in Nicaragua a useful instrument, the United States helped push it out as the Sandinista insurgency claimed victory in 1979. Seven years later, Ferdinand Marcos in the Philippines also learned that the United States could shift loyalties and sacrifice a besieged, dependent ally who was no longer seen as capable of serving U.S. interests.

The other side of the Cuban coin, of course, was the remarkable ability of Castro's hunted band of ill-equipped invaders to overcome numerous setbacks, to move from a rugged mountain hideout to the capital city, to defeat an entrenched army long supplied by a superpower, to attract a fervent political following. Fidel Castro and the 26th of July Movement knit together a national coalition across classes. Although the insurrectionist leadership was primarily middle-class, the movement against Batista constituted an amalgam of *campesinos,* workers, intellectuals, students, and professionals (doctors, lawyers, and teachers among them) both in the cities and in the countryside. Relations among these groups were often rocky, but their common cause against Batista provided enough unity to win.

Castro repeatedly outmaneuvered his political and military competitors, often coopting them through manifestos and pacts. Sometimes Batista's henchmen helped Castro by murdering or jailing his rivals. Determined, improvising, boundless in energy, and hard-nosed, Castro demanded and usually received loyalty from his disciplined network. He sometimes said what others wanted to hear; he sometimes said what he deeply believed. In most cases, he said it with an intensity that many found hypnotic. Fidel Castro proved a remarkable, captivating leader with steely purpose. Perhaps revolution would have swept over Cuba without him, but that is very difficult to imagine, given the mark of his personality on Cuba's history beginning at least as early as 1953. The United States lost to him because it could not control or crush this strong leader or stem the cascading popular support that his movement generated.

Hegemons face limits not only when they encounter an unusually popular, determined, and skilled opponent such as Castro, but also when that opponent can employ the force of nationalism against outside intruders. In mobilizing a popular following, Castro was astute in exploiting traditional resentments against the United States.[18] The Platt amendment, U.S. military occupations, and Washington's support for dictators had fueled nationalistic attitudes long before Castro attacked the Moncada Barracks. Distaste for U.S. cultural forms that challenged a Cuban identity and a dependent's frustration over foreign economic domination also galvanized an anti-American component in Cuban nationalism. Not all Cuban dissidents held anti-Yankee views, of course; in fact, the moderates that U.S. officials so admired often applauded features of the United States—especially its economic success and democratic tradition. Although the moderates believed that the new Cuba should work with, not against, the United States, they too wanted to reduce Cuban dependency.[19] Casinos, Coca-Cola, Cuban Electric, Batista's gold telephone, Guantánamo Naval Base, rocket heads, U.S. tanks and planes, military decorations for *batistianos*—the symbols of outside influence were everywhere. Castro did not invent anti-Yankeeism and not all Cubans harbored anti-Americanism, but for many Cubans, Batista and Uncle Sam were such conspicuous partners that Castroite rebels could make a persuasive case against both.

Cuban anti-Americanism gained momentum because the United States appeared to violate its own pledges of neutrality and non-intervention. From the perspective of many anti-Batista groups, the United States was two-faced. The record, in fact, speaks against what some scholars have called an "evenhanded" or "passive" U.S. policy.[20] The United States was neither neglectful nor complacent toward Cuba, and the list of unneutral U.S. acts the Cuban rebels compiled became lengthy: Export-Import Bank loans, openly pro-Batista businessmen Gardner and Smith as ambassadors, the arrest and prosecution of former President Prío and other anti-Batista dissidents, the interdiction of rebel arms shipped from U.S. ports, cooperation with Cuban authorities in identifying and capturing Cuban exiles active in supporting the rebellion, denials of visas to insurgent Cubans, military missions and weapons deliveries, shipments of non-combat equipment to Cuban forces after the arms suspension, the MAP battalion, sending the Guantánamo Marines on to Cuban territory to guard the Yateras waterworks, conspicuous displays of U.S. support for the regime, and awards for Cuban military officers. The Castroite rebels particularly resented the "third force" conspiracies. "You Americans keep complaining that Cuba is only ninety miles from your shore," Castro once remarked to Herbert Matthews. "I say that the United States is ninety miles from Cuba, and for us that is worse."[21]

But what if U.S. leaders and agencies had been more intelligent, more efficient, better informed, more carefully organized, less divided? Could they have forced the different sides of the civil war to compromise? Could they have saved Batista or pushed him out to lock Castro out? Would a "real understanding" of the sources of the revolution and Castro's popular appeal have changed the outcome?[22] In fact, the U.S. "system" did not work smoothly. Policy evolved in a tug-of-war environment of several groups—the ambassadors, the embassy staff, the military, CIA, State Department, White House, members of the House and Senate, and journalists—who squabbled as they maneuvered to influence decisions.

Hurried, awkward, and uncoordinated decisionmaking high in the administration in the last days of the Batista era undermined U.S. hegemony. The operational style of the Eisenhower Administration, which delegated authority to lower echelons, and the cancer that wore down Secretary of State John Foster Dulles's health helped muddle U.S. policymaking. "The man in the White House when it started should have stopped it at the beginning," snapped former President Harry S. Truman. Remember, said the tough-talking Missourian, that President Grover Cleveland had intervened in Venezuela in the 1890s and Truman himself had acted in Greece, Turkey, and Berlin in the 1940s.[23] Truman typically overlooked complexities to embrace the simple answer, but the record shows that not the dying Dulles, nor a hesitant Acting Secretary Herter, nor an inattentive Eisenhower grappled closely with the Cuban problem until just before Batista's flight.

U.S. agencies did not function well during the insurrection. The CIA was quite active on the island, but it bungled anti-Castro plots and probably aided both sides. The USIS failed to create a Cuban public opinion favorable to the United States. Customs officials interrupted numerous arms shipments destined for the rebels, but they could not patrol every mile of shoreline; they could not choke off gunrunning from U.S. territory. Although U.S. officials in both Havana and Washington ultimately decided to dismiss Batista when it became urgent to block Castro, they failed to identify or cultivate leaders who could replace the dictator. U.S. officials, moreover, frequently underestimated Castro's strength and appeal and often overestimated Batista's ability to quell the rebellion. On the very eve of Batista's collapse, the CIA cited the failed general strike to report that Castro had been unable to garner widespread support among Cubans.[24]

U.S. officials also overestimated the ability of the Cuban armed forces. Although well armed, the Cuban military was corrupt, ill-trained, and poorly commanded. The U.S. military missions became cozy with Cuba's military brass and failed to shape an effective fighting force, especially one trained to counter guerrillas. When the rebel forces pressed hard, the Cuban armed forces bent and became demoralized. The oft-mentioned "will to fight" was lacking in Batista's legions, especially when compared with the dedicated rebel groups.[25] Because of Cuban government censorship (in place for 18 of the 25 months of the insurrection), poor communications, and official falsification of data, it was difficult for observers—even Castro and Batista—to get a clear picture of the performance and condition of the Cuban armed forces.[26] Batista himself seemed stunned in fall 1958 by the "surprising frequency" of defections and desertions from his military.[27]

Ambassadors Gardner and Smith took the Batista point of view and never fully plumbed the depths of Cuban discontents. These Republican loyalists were sent to Cuba as political appointees because the island seemed orderly, subservient, and predictable under Batista rule. Their conservative businessman's perspective, hobnobbing with upper-class Cubans and U.S. company executives, and suspicion of career Foreign Service Officers blinded them to the structural flaws of the Cuban economy and the steady decay of the Batista regime. Their fixation on the Communist issue inhibited all but the most simplistic analysis. When these two ambassadors, neither of whom spoke Spanish, failed, and Castro won, their malicious, exculpating explanation was that subversives in the Department of State had sold out Batista in favor of Castro.

Earl E. T. Smith in particular, at the critical moments, was at war not only with State officers in Washington but also with many of his own embassy advisers, including the CIA mission chief. The atmosphere in the embassy became "poisonous," remembered one diplomat.[28] Like the heads of the U.S. military missions, Ambassador Smith misread Cuba and sometimes misled Washington. "A real pro of an Ambassador," USIS officer Richard Cushing said years later, "could have convinced Batista to step aside when things looked bad—and there might not have been a reason for a Fidel Castro to take over."[29] Although Smith's closeness to Batista probably helped delay the U.S. decision to push the Cuban president out, Cushing saw the "bad" before many others did and his conclusion overlooks the host of obstacles to a U.S. resolution of the crisis, including Batista's own stubbornness and refusal to heed U.S. advice.

Not all officials in Washington, of course, bought the Smith case in 1957–58. Rubottom, Wieland, and others grasped much of the essence of the Cuban question—especially Batista's highly unpopular rule—and once they had failed to nudge Batista toward reforms through lectures, advice, and arms cutbacks, they tried to distance the United States from the vulnerable regime and ease him out before Castro could come in. Yet nothing moved Batista to compromise with his adversaries. Here too is a rub: many Washington officials naïvely believed that compromise was possible in Cuba or that a middle way could be found. But Batista and Castro would not move to the middle, and by the time U.S. officials had fully soured on Batista, no viable third force existed. Politics had become too polarized. The United States had seldom protested when Batista imprisoned or killed Castro's rivals, driving others into the arms of the 26th of July Movement. U.S. officials seemed unable to fathom that neither Batista nor Castro would abide a brokered deal. Nor did they grasp the extent to which the spirit of compromise had been extinguished in years of urban and guerrilla warfare, forced exiles, assassinations, B-26 raids on rural folk, and SIM torture sessions.

Public opinion at home and abroad also conditioned the U.S. decisionmaking environment on Cuba policy, making the United States hesitant to intervene openly to save Batista or stop Castro. The United States was so sensitive to foreign opinion because it was beginning to decline as a power in the Cold War international system.[30] In the 1950s, allies like Charles de Gaulle's France challenged the United States, the Third World emerged as a formidable adversary, the non-aligned movement gained momentum, and the Soviet Union, slowly climbing out of the rubble of World War II, boldly contested the United States by offering foreign aid to developing nations. In Latin America itself, Guatemalans tried but failed to reduce U.S. influence, nationalistic Panamanians demanded more control of the Canal Zone, and the Organization of American States increasingly balked at U.S. direction.

Throughout Latin America and the United States, reasoned voices rose to criticize U.S. alliances with military dictators, whose regimes increasingly faced the hostility of their own people. In 1956–60, all told, ten military dictatorships fell, including Batista's. The stone-throwing rioters who interrupted Vice President Richard Nixon's visit to South America; the U.S. senators—Mike Mansfield, Hubert Humphrey, Wayne Morse, among others—who bombarded U.S. officials with complaints against the shortsighted, failed policy of endorsing dictators primarily because they were anti-Communist or because they protected U.S. economic interests; the U.S. journalists who exposed the ugly consequences of military rule; and the articulate Cuban exiles who helped shape U.S. opinion critical of Batista—all influenced State Department officers.[31] So did conservative Republicans such as Nixon and the president's brother Milton S. Eisenhower, president of the Johns Hopkins University. They advised U.S. policymakers to change at least their style in dealing with notorious dictators—give a cool handshake to dictators, a warm *abrazo* to democratic leaders.[32] U.S. decisions on Cuba, argued some analysts, "were spotlighted and subject to more than usual Congressional and public scrutiny."[33] American public opinion was hardly intense or uniform on Cuba; many members of Congress ignored the insurrection; and U.S. officials seemed willing to endure criticism if necessary. Still, they endured anti-Batista brickbats that conditioned their decision to abandon the dictator.

Why not treat officially with Fidel Castro before 1959? Why not an *abrazo,* or why not use diplomacy to tame him, to coopt him? Although U.S. officials maintained contact with the rebels and Consul Wollam negotiated with them to gain the release of kidnap victims, they studiously avoided formal links with the insurgents. Should the United States have established formal diplomatic ties with Castro during the insurrection in order to have positioned itself to compete with the Communists?[34] Several problems dog this general question. First, U.S. officials did not rank Castro as a democratic leader with whom they could easily work. Second, Castro did not want North Americans in his camp, unless he could exploit *them.* Third, there was no evidence that M-26-7 was embracing Communism or that Communists influenced the movement. And last, hegemonic powers do not truck with insurrectionists seeking to overthrow a client government. Had the United States opened formal links with the insurgency or recognized the Castro movement and the news had leaked, as it surely would have, U.S. officials would have undermined Batista long before Washington was ready to dispense with him.

U.S. leaders shared the distorting hegemonic presumption of "backyardism," which itself ironically limited the effective exercise of power. It clouded their vision, promoting superficial analysis. High-level U.S. officials lacked a keen interest in the Cuban insurrection so long as Cuba seemed manageable—especially at a time when Berlin, Quemoy (Jinmen) and Matsu (Mazu), the Arab-Israeli conflict, and other Cold War and Third World flashpoints and the nuclear arms race demanded their attention.

One State Department official recalled that Latin America did not rank high on the U.S. crisis agenda "as long as things were rocking along, and while many of the Latin American issues seemed to be between the 'ins' and 'outs' rather than ideological."[35] The widely circulated magazine *Foreign Affairs,* for example, did not carry one article specifically on Latin America in 1957, and only two in 1958 (both on Mexico). The prestigious Council on Foreign Relations in New York, publisher of the journal, heard few speakers address Western Hemispheric issues. Few newspapers covered Latin American affairs, and then only when an earthquake or some other catastrophe struck.[36] Most National Security Council meetings heard not a word about Cuba in 1957–58.[37]

North American "indifference" to Cuba, wrote two government advisers who later reviewed U.S. policy, stemmed not so much from a preoccupation with urgent business elsewhere in the world but from a "less obvious and more basic" reason: "We mistakenly looked at Cuba through traditional eyeglasses and regarded it as *controllable.*" Gordon Chase and John Plank added that "we were probably so used to thinking in the standard stereotyped terms of 'immense reservoirs of good will' toward the United States, so used to thinking that the Cubans regarded us with boundless affection and esteem and would soon 'come to their senses,' so fully persuaded of Cuba's total dependence upon the U.S., that we could not recognize the force of Cuban nationalist pride and apparently found it difficult to take Cuba or Castro seriously."[38] A combination of American arrogance—what the scholar Piero Gleijeses has aptly called "imperial hubris"—and ignorance doomed to failure Washington's stop-Castro strategy.[39]

Self-imposed restraint, poor leadership, sensitivity to public opinion, shallow analysis, miscalculation, incompetence, untoward timing, inhospitable terrain, a formidable, heroic opponent, and alliance with a corrupt, vulnerable ally—all derailed the U.S. effort to block Castro. More generally, the United States hesitated to rearrange Cuban politics until it was too late because, like most hegemonic powers through history, U.S. officials tended to keep hands on, but fingers out, of their dependent. That is, dominant powers do not intervene so long as they believe that local elites are keeping order and protecting interests—economic, military, political, cultural. Batista seemed to be doing just that until the summer of 1958, until just before he speedily collapsed. U.S. leaders hoped that after the fall elections an orderly transfer of political power could take place.

Batista's rapid self-destruction forced the United States to intervene. But by late fall 1958, by the time Washington had decided to push Batista from his perch, the United States, although it did not know it, was left with few local tools with which to work. Castro was alienated from North Americans, Batista was repudiated and tumbling down, Barquín was in jail, Prío and the Auténticos were discredited, the Ortodoxos were subordinated to the 26th of July Movement, and exile third-force alternatives such as Carrillo were unable to build strong constituencies and armies. Nor could the Organization of American States, less and less subservient to U.S. dictates, be induced to intervene, if only to mediate.

Then why not a military intervention by either U.S.-favored Cuban exiles or U.S. forces? Why not a Bay of Pigs in 1958? Besides the impediments imposed by terrain, such an extreme measure never seemed necessary to U.S. leaders until after Castro was several months in power. After all, at first, they did not tag him a Communist, they judged the new government minimally acceptable, and they anticipated that the so-called moderates would box in the young rebel. Try as they did when Ambassador Smith harped on the issue, they had never discovered a threatening Communist or Soviet role in the insurrection that would justify such intervention.[40] The judgment on Fidel Castro was almost uniform: He was not a Communist. Had he been, U.S. warning flags might have shot up earlier and perhaps prompted some vigorous blocking operation, whatever the difficulties. In 1958, moreover, Castro had moderated his public statements about the nationalization of foreign-owned property and had promised democratic elections. Castro's radicalism—whatever it was—was muted.

U.S. officials, including Ambassador Smith himself, even after Castro marched into Havana, did not believe that Cuba had moved so far from the U.S. orbit that the United States should undertake the drastic measure of military action. The victors did not violently punish North Americans for prolonging the civil war—as they said they might. Ending the general strike of early January 1959, moreover, the rebels soon imposed order. There was nothing like the violence and destruction that followed the collapse of the Machado regime in 1933. The ABC colony and Washington applauded the rebels' discipline in forcing looters and rioters off the street. In early January, U.S. officials in both Havana and Washington, as well as members of the North American business community in Cuba, although apprehensive about the M-26-7 triumph, nonetheless uttered optimistic predictions about the new government. Some companies even pre-paid their taxes as a friendly gesture. The Urrutia cabinet, conciliatory and anti-Communist, seemed in fact to be a government with which the United States could deal to sustain U.S. interests. The once intense American disappointment over the failure of the third-force conspiracies eased.

Many North Americans, in fact, expected that the so-called moderates would eventually control the new Cuba and discipline Fidel Castro. Castro's radical tendencies would be tempered because he would have to share power. Assistant Secretary Rubottom, less than 24 hours before Batista quit Cuba, told the Senate Foreign Relations Committee: "It has been hard to believe that the Castros alone, that the 26th of July Movement alone could take over, because they have not had enough broad support in Cuba to do this job by themselves. . . . I would not be happy with Castro solely in command. I cannot quite visualize that at this stage."[41]

Rebel leaders themselves seemed surprised by their sudden, almost uncontested Olympian standing. Celia Sánchez remarked that they never anticipated that they "would be so strong and so popular. We thought we would have to form a government with Auténticos, Ortodoxos, and so forth. Instead we found that we could be the masters of Cuba."[42] Another oppositionist, José Pardo Llada, recalled that he fully expected that the United States would intervene to prevent the revolution. "Watch out for the Americans" was a phrase he had learned in his childhood days. "Everything up to the Chiclets that we chewed came from the North," so it was difficult to think that the United States—the "pro-consul"—would not resolve the Cuban crisis on North American terms.[43]

During the insurrectionary period, U.S. officials had interviewed many oppositionists, hearing a repeated message from Cubans they respected: Moderates would get the upper hand in any new government and keep change within acceptable bounds. "Trust us, John, trust us," the future Minister of the Treasury and "avowed friend of the United States" Rufo López-Fresquet kept saying to the U.S. Embassy's John Topping. "If Fidel turns out to be a problem, we politicians and moderates will be able to handle him once we get into office."[44] President-designate Urrutia once told López-Fresquet that "mature men like ourselves" would form Cuba's new government. "Castro is now doing the muscle work; later we will do the intellectual work."[45]

The exiled former president of the Cuban National Bank, Felipe Pazos, remarked to a State Department officer in fall 1958 that because the 26th of July Movement was "middle class," it "could not stray too far to the left without losing a considerable amount of this support."[46] The Bank of America officer A. F. Montero reported from Havana three days before Batista fled that "the middle class is growing stronger, and will emerge with much more economic and political power out of this conflict. Once the political issue is settled, the country will rapidly return to full production again."[47] When Castro would not bow to U.S. wishes, then, U.S. leaders expected the "moderating influence" of pro-U.S. middle-class Cubans to bring him into line.[48] Trusting in moderates, Rubottom and his colleagues never appreciated the profound Cuban fascination with Fidel Castro, the captivating, heroic nature of his triumph, and the popularity of his message that Cuba must chart a sharply new course.

Another self-described moderate, the anti-Batista dissident Mario Llerena, titled his memoir *The Unsuspected Revolution.* He recalled that to many middle-class people like himself Castro was a "crude force that could be put to good use if it were properly harnessed and guided."[49] Many of these anti-Batista moderates inflated their capacity to influence policy in the new government. Perhaps they did so because they believed that the United States would come to their aid.[50] In turn, U.S. officials, after listening to the self-confident claims of these admired Cuban patriots, became convinced that they wielded bargaining power. After Philip Bonsal assumed his ambassadorship, he was amazed to witness just how little defense middle- and upper-class Cubans actually put up when Castro began to move against them. Years later, when Bonsal asked an anti-Castro exile why Cubans who recoiled from the radicalism of the revolution had not resisted it more strenuously, the exile answered that "we did not believe you, the United States, would let Castro get away with it."[51] As Llerena came to regret, a "dangerous delusion" misled moderate Cubans.[52]

A New York banker once commented that "anybody who gets in there [the Cuban government] rides a gravy train. No matter how radical he might be, the way things are done will make him turn conservative."[53] Castro was well aware of this prevailing attitude. The "ruling classes," he remarked, "believed . . . that all revolutionaries change and become conservatives with the passage of time."[54] As North Americans were wont to say, every Cuban leader had his price and any Cuban government that wished to survive had to respect the United States and its interests.[55] The reasons seemed compelling: sugar, tourism, investments, cultural and family ties, military training, geography, the need for foreign aid. No Cuban government would jilt the United States for fear of losing its North American economic lifeline. The Cubans "will give way in time to a general desire, based largely on self-interest, for good relations with the United States."[56] A mixture of wooing, coaxing, and pushing from the stern, traditional father in Washington would supposedly keep the Cuban "children" in the family and the new Cuba within the North American definition of a tolerable revolution.

Steeped in the anti-Americanism that had developed on the island throughout the twentieth century, Fidel Castro considered such thinking anathema. As he had repeatedly said, tossing out Batista but leaving the United States and its substantial interests in place would once again rob Cuba of an historic moment to affirm its independence. So, one down, one to go.

22

Failing the Tests: The United States and Cuba in the Castro Era

Not long after Castro came to power and set a revolutionary course to roll back U.S. interests, the United States asserted its hegemonic role and applied "a series of tests" to his regime, for "if we gave in to Castro all along the line," remarked a State Department officer, "we would get kicked around in the hemisphere."[1] Washington became persuaded that Fidel Castro so threatened U.S. security and economic interests and core values that he had to be removed from the Cuban government. The regime's execution of *batistianos* after showy trials, agrarian reform program that undercut North American economic interests, control of utilities industries, postponement of elections, tolerance of Communists, calls for revolution throughout Latin America, hints that Cuba would follow a neutralist course in the Cold War, and vituperative anti-American rhetoric widened the chasm between Havana and Washington.

To rising U.S. criticisms of the executions in early 1959, Castro snapped: "What do Americans know about . . . a tyrant's atrocities, except in the novels and movies?"[2] The North Americans had never shown moral disgust toward Batista's crimes. Why now against legitimate punishments? At a huge rally, Castro blistered his critics with a lesson in U.S. and world history: "What was done at Hiroshima and Nagasaki? In the name of peace two cities were bombed and more than three hundred thousand human beings killed. We have shot no child, we have shot no woman, we have shot no old people. . . . We are shooting the assassins so that they will not kill our children tomorrow, and when all is said and done the total of assassins we shoot will not be more than four hundred, which is about one assassin for every thousand men, women, and children assassinated in Hiroshima and Nagasaki."[3] The fervent anti-Americanism incubated during the rebellion flared again and again. Wollam reported that the United States was continuing to "take the rap" for Batista's record.[4]

U.S. intelligence officials speculated that the nationalist attacks on the United States would dissipate because Cubans knew that a "serious disruption" in relations would hurt them. "Cuba's economic dependence on the US, especially as a market for its sugar," dictated Cuban caution. So did "the realities of administering a country closely linked commercially, politically, and historically, with the US."[5] Cuba's "self-interest," in other words, gave the United States leverage.[6] Anyway, the "moderates" could be expected to "stabilize the situation" because they "realize that it is to Cuba's advantage to work with the United States."[7] It was important for Castro to understand, too, Ambassador Bonsal warned him, that in that "critical time" of the Cold War, the United States was quite sensitive to Cuban hints of "neutralism."[8] Bonsal advised Washington that the United States did not yet have to begin "unlimbering our artillery against Castro" because the Cubans might likely "straighten themselves and Castro out."[9]

When Castro visited the United States in April 1959 under the auspices of the American Society of Newspaper Editors, Eisenhower deliberately snubbed him by leaving town to play golf in Georgia. Before the visit, when Secretary Herter had complained at a National Security Council meeting that Castro had displayed "singularly bad behavior" by accepting the society's invitation without consulting the State Department, the president suggested that Castro be denied a visa.[10] But a counterargument prevailed within the administration: Cuba needed loans to restore a broken economy, relieve unemployment, and shore up dwindling cash reserves, as the new president of the Cuban National Bank, Felipe Pazos, was making clear.

Although Pazos argued later that the United States was quite eager to help Cuba out of its economic doldrums through loans, Rubottom and other U.S. officials actually expressed restraint, had no assistance plans to offer, and seemed content to wait until Castro changed his behavior.[11] The Cubans might receive a balance of payments loan, a U.S. briefing paper advised, but they would first have to negotiate a stabilization agreement with the International Monetary Fund (which the United States largely controlled). As for economic development loans, they would be treated on a case-by-case basis only after the availability of private capital had been determined. Cuba might first have to alter some of its policies that "reduce the inflow of private capital for investment."[12] More, CIA Director Allen Dulles observed, if

Castro did not become more cooperative, Congress might well reduce the sugar quota, and then the Cuban leader "would indeed be in the soup."[13]

Castro resolved not to enter the United States as a supplicant. Pazos recalled that Castro did not want to appear "as one more Latin American leader 'sold out' to imperialism."[14] So the large Cuban contingent that traveled with Castro to Washington received stern instructions not to ask for foreign aid. "The Americans will be surprised," Castro told Rufo López-Fresquet.[15] During an earlier trip to Caracas, according to President-elect Rómulo Betancourt, Castro had asked Venezuela for a $300 million loan so that Cuba could free itself from its dependence upon the North American sugar market, U.S. banks, and international lending agencies.[16] As Castro remarked during his U.S. visit, Cuba did not wish to rely upon North American private capital to restore the island's economic health. "Relying on private investment in building the economic future for Cuba," Castro told the Council on Foreign Relations in New York, "is like curing cancer with mercurochrome."[17]

Vice President Nixon met with Castro and judged him "either incredibly naïve about communism or under communist discipline—my guess is the former." In paternalistic fashion, Nixon hoped the United States could lead the young leader "in the right direction."[18] "I talked to him like a Dutch uncle," Nixon crowed.[19] Herter informed Eisenhower that Castro was "very much like a child" who became "voluble" and "wild" when he spoke Spanish. Eisenhower said this sounded like a story he had heard about the neutralist leader Jawaharlal Nehru of India, who "accumulated emotional frenzy" as he spoke.[20]

U.S. officials worried that Castro's foreign policy was moving not toward alignment with the Soviet Union but toward the Third World neutralist posture that Nehru and Egypt's Gamal Abdul Nasser, so admired by Castro and so detested by Washington, had helped formalize in the Non-Aligned Movement. Would Castro play East against West, too?[21] As well, almost immediately, Castro appealed for revolution across Latin America. The victorious Cuban rebels wanted to get even with adversaries such as the Dominican Republic's Rafael Trujillo and Nicaragua's Luis Somoza Debayle, dictators who had supplied the Cuban dictatorship with weapons. But this call for exporting the Cuban example sprang also from the perceived need to develop allies who could stand with the new Cuba against the United States when the anticipated Cuban-American tussles arose. Castro personally aspired, moreover, to become a hemispheric if not global anti-imperialist leader.[22] Washington could only read this thrust as a danger to U.S. hegemony in Latin America—as a "Nasser-like ambition" toward the Caribbean.[23] And might not Cuba become a symbol for the rest of Latin America—evidence, as the Communists were saying, that even a small country in the shadow of the United States can defeat "the powers of reaction and imperialism"?[24]

During the rest of 1959 and through 1960, with Ambassador Philip Bonsal's thinking that Castro suffered "mental unbalance at times" and Eisenhower's concluding that the Cuban leader "begins to look like a madman," Havana and Washington traded punch for punch.[25] One government report saturated with the hegemonic presumption later concluded that "no sane man undertaking to govern and reform Cuba would have chosen to pick a fight with the US."[26] In November 1959 the Eisenhower Administration decided to work with anti-Castro groups within Cuba to "check" or "replace" the revolutionary regime, and thus end an anti-Americanism that was "having serious adverse effects on the United States position in Latin America and corresponding advantages for international Communism."[27] The United States could not "do business" with the Castro government, Rubottom told Bonsal.[28] The Castro government had failed the U.S. tests. It had to go.

In March 1960, Eisenhower ordered the CIA to train Cuban exiles for an invasion of their homeland—this shortly after Cuba signed a trade treaty with the Soviet Union. The CIA, as well, hatched assassination plots against Castro and staged hit-and-run attacks along the Cuban coast. Cuba pushed its land reform while the United States protested that inadequate compensation was being offered for confiscated properties. As American-owned industries were nationalized, the United States in July suspended Cuba's sugar quota and then forbade American exports to the island, drastically cutting commerce. On January 3, 1961, certain that the American embassy was a "nest of spies" aligned with counter-revolutionaries who were burning cane fields and sabotaging buildings, Castro heatedly demanded that the embassy staff be reduced to the small size of the Cuban delegation in Washington.[29] The United States promptly broke diplomatic relations with Cuba. The rupture in relations elevated covert action—especially an invasion by Cuban exiles—as one of the few means left to topple Castro.

The consequences of this escalating contest are now all too familiar. While Cuba accelerated a bitterly anti-American revolution, the United States stepped up U.S. covert actions, including the disastrous Bay of Pigs expedition of April 1961.[30] Castro moved Cuba steadily toward Communism and military alliance with the Soviet Union, and President Kennedy launched a multitrack program of covert, economic, diplomatic, and propagandistic elements designed to bring down Castro.[31] Ambassador Bonsal explained in his memoirs that "Russia came to Castro's rescue only after the United States had taken steps designed to overthrow him."[32] President John F. Kennedy's Secretary of Defense Robert McNamara later remarked: "If I had been in

Moscow or Havana at that time [1961–62], I would have believed the Americans were preparing for an invasion."[33]

Critical to understanding the frightening missile crisis of fall 1962 is the relationship between U.S. activities and Soviet/Cuban decisions to place on the island missiles that could strike the United States. Soviet and Cuban officials first discussed in May 1962 the idea of placing nuclear-tipped missiles on the island; in July, during a trip by Raúl Castro to Moscow, a draft agreement was probably initialed; in late August to early September, during a trip by Che Guevara to Moscow, an accord was probably put into final form.

By 1962 more than two hundred anti-Castro, Cuban exile organizations operated in the United States. Many of them banded together under the leadership of José Miró Cardona. He met with President Kennedy in Washington on April 10, 1962, and the Cuban exile left the meeting persuaded that Kennedy intended to use U.S. armed forces against Cuba. Indeed, after Miró Cardona returned to Miami, he and the Revolutionary Council—the government-in-exile—began to identify possible recruits for a Cuban unit in the U.S. military.

If Havana worried about meetings like these, it grew apprehensive too about the alliance between the exile groups and the CIA, whose commitment to the destruction of the Havana government knew few bounds. Hit-and-run saboteurs burned cane fields and blew up oil storage tanks. One group, Agrupación Montecristi, attacked a Cuban patrol boat off the northern coast of the island in May. Directorio Revolucionario Estudiantil, another exile organization, used two boats to attack Cuba in August. Alpha 66 attacked Cuba on numerous occasions. CIA officers and "assets" were at the same time plotting to assassinate Fidel Castro. Many of these activities came under the wing of "Operation Mongoose," the covert effort engineered by Attorney General Robert Kennedy to disrupt the Cuban economy and stir unrest on the island.

Intensified economic coercion joined these covert activities. The Kennedy Administration, in February 1962, banned most imports of Cuban products. Washington also pressed its North Atlantic Treaty Organization allies to support the "economic isolation" of Cuba.[34] In early 1962, too, Kennedy officials engineered the eviction of Cuba from the Organization of American States.

At about the same time, American military planning and activities, some public, some secret, demonstrated a determination to cripple the Castro government. The director of "Operation Mongoose," Brigadier General Edward Lansdale, noted in a top-secret memorandum to the president that he designed his schemes to "help the people of Cuba overthrow the Communist regime from within Cuba and institute a new government." But, he asked: "If conditions and assets permitting a revolt [timed for October 1962] are achieved in Cuba, and if U.S. help is required to sustain this condition, will the U.S. respond promptly with military force to aid the Cuban revolt?" Lansdale gave the answer he preferred: "The basic plan requires complete and efficient support of the military."[35] Another contemporary document, this one from the chairman of the Joint Chiefs of Staff, General Maxwell Taylor, noted in the spring of 1962 that the "Operation Mongoose" plan to overthrow the Cuban government would be undertaken largely by "indigenous resources," but "recognizes that final success will require decisive U.S. military intervention."[36] Because the plan also required close cooperation with Cuban exiles, it is likely that Castro's spies picked up from the leaky Cuban community in Miami at least vague suggestions that the U.S. military was contemplating military action against Cuba.

American military maneuvers heightened Cuban fears. One well-publicized U.S. exercise, staged during April 1962, included 40,000 troops and an amphibious landing on a small island near Puerto Rico. Some noisy American politicians, throughout the year, were calling for the real thing: an invasion of Cuba. In the summer of 1962, finally, the U.S. Army began a program to create Spanish-speaking units; the Cuban exiles who signed up had as their "primary" goal a "return to Cuba to battle against the Fidel Castro regime."[37]

By summer 1962, then, when Havana and Moscow were contemplating defensive measures that included medium-range missiles, Cuba felt besieged from several quarters. The Soviet Union had become its trading partner, and the Soviets, after the Bay of Pigs, had begun military shipments that ultimately included small arms, howitzers, armored personnel carriers, patrol boats, tanks, MIG jet fighters, and surface-to-air missiles. Yet all of this weaponry, it seemed, had not deterred the United States. And given the failure of Kennedy's multitrack program to unseat Castro, "were we right or wrong to fear direct invasion" next? asked the Cuban leader.[38] As he said in July 1962, shortly after striking a missile-deployment agreement with the Soviets: "We must prepare ourselves for that direct invasion."[39]

Had there been no exile expedition at the Bay of Pigs, no destructive covert activities, no assassination plots, no military maneuvers and plans, and no economic and diplomatic steps to harass, isolate, and destroy the Castro government in Havana, there would not have been a Cuban missile crisis. The origins of the October 1962 crisis derived largely from the concerted U.S. campaign to quash the Cuban Revolution. To stress only the global dimension (Soviet-American competition), as is commonly done, is to slight the local or regional sources of the conflict. To slight these sources is to miss the central point that Premier Nikita Khrushchev would never have had the opportunity to install dangerous missiles in the Caribbean if the United States had not been attempting to overthrow the Cuban government.

After the nuclear brinkmanship of the missile crisis and the withdrawal of the Soviet missiles from Cuba, with Fidel Castro still in power, Washington created a new State Department office, the Coordinator of Cuban Affairs, and sought to "tighten the noose" around Cuba.[40] Meanwhile, angry at the Soviet Union for backing down and for barely consulting with the Cubans during the crisis, Castro made gestures toward détente with the United States. He sent home thousands of Soviet military personnel and released some political prisoners, including a few Americans, as steps toward rapprochement.[41]

Then, in late April 1963, the *jefe máximo* departed for a four-week trip to the Soviet Union, where he patched up relations with Khrushchev and won promises of more foreign aid.[42] Washington stirred against Moscow's "grandiose" reception of Castro, the latter's "vehemence" in denouncing the United States, his "tone of defiance rather than conciliation," and the refurbished Soviet-Cuban alliance.[43] Soon Robert Kennedy asked the CIA to "develop a list of possible actions which might be undertaken against Cuba."[44] In mid-June 1963 the NSC approved a new sabotage program. The CIA quickly cranked up new dirty tricks and revitalized its assassination option by making contact with a traitorous Cuban official, Rolando Cubela Secades. Code-named AM/LASH, he plotted with the CIA to kill Fidel Castro. In Florida, American officials intercepted and arrested saboteurs heading for Cuba, but they usually released them and seldom prosecuted. Alpha 66 and Commando L raiders hit oil facilities, sugar mills, and industrial plants.[45]

In fall 1963, Cuba continued to seek an accommodation. Through contact with a member of Ambassador Adlai Stevenson's United Nations staff, William Attwood, the Cuban government signaled once again its interest in improving relations. The President authorized an eager Attwood to work up an agenda with the Cubans.[46] In late October, when Kennedy met with the French journalist Jean Daniel, the president spoke in both hard-line and conciliatory tones about Cuba. Daniel journeyed next to Havana to interview Castro, who later claimed that Daniel carried a "private message" from Kennedy asking about the prospects for a Cuban-American dialogue.[47] On November 18, however, Kennedy sounded less the conciliator and more the warrior. In a tough-minded speech, he reiterated the familiar charges against Castro's "small band of conspirators."[48] The president, reported his aide McGeorge Bundy, sought to "encourage anti-Castro elements within Cuba to revolt" and to "indicate that we would not permit another Cuba in the hemisphere."[49] In Havana, while Daniel and Castro were discussing chances for Cuban-American détente on November 22, the news of the assassination of President Kennedy reached the two men. "*Es una mala noticia*" ("This is bad news"), Castro mumbled repeatedly.[50]

On the very day that Kennedy died, AM/LASH rendezvoused with CIA agents in Paris, where he received a ball-point pen rigged with a poisonous hyperdermic needle intended to produce Castro's instant death.[51] Like all other assassination plots against Castro, this one failed.

At the time of Kennedy's death, U.S. Cuba policy was moving in opposite directions—probing for talks but sustaining multitrack pressures. "How can you figure him out?" Castro had wondered in late October 1963.[52] "What can I do about Cuba that won't get me in trouble?" the new President Lyndon B. Johnson began to ask. "The answer is little," McGeorge Bundy told a colleague, "but he [Johnson] needs to be taken up and down the hills we've all been on so many times."[53] The Johnson Administration soon decided to put the "tenuous" and "marginal" Cuban-American contacts through Attwood "on ice."[54] From that time onward, except for a thaw in the mid-1970s that was cut short by Cuba's sending of troops to Africa to help political allies, U.S.-Cuba relations remained frozen.

From the 1960s through the 1980s the United States consistently demanded two Cuban concessions: an end to support for revolutions abroad, especially in the hemisphere, and removal of the Soviet military from the island. Castro just as consistently demanded two U.S. concessions: abolition of the economic embargo and American respect for Cuban sovereignty. Neither Havana nor Washington would compromise.[55]

In the United States, debate intermittently erupted about who "lost" Cuba, with Republicans and Democrats exchanging barbs. The exile community in Florida multiplied as tens of thousands fled Castro's dictatorial regime and settled in the United States to become a vocal political force against normalization of relations. Exiles continued to organize paramilitary groups and, sometimes in collusion with the CIA, harassed, raided, and destroyed. U.S.-funded Radio Swan and Radio Martí beamed propaganda at Cuba. The naval base at Guantánamo remained in U.S. hands despite Cuban demands for the territory. U.S. officials also closely monitored Latin American affairs, sometimes invading nations (Dominican Republic and Grenada), subverting governments (Chile and Nicaragua), and endorsing military regimes (Brazil and Guatemala) to prevent "another Cuba." And drawing a lesson that they applied elsewhere, including Vietnam, U.S. leaders sought to avoid conditions that might repeat the past of substituting a Castro for a Batista.[56]

As for Cuba, in the face of the U.S. economic embargo and unrelenting hostility, Castro came to rely on Soviet assistance and trade to implant socialism and sustain a fragile economy, which, despite efforts to diversify, remained dependent on exports of sugar.[57] Castro also turned to the Soviets for a "shield against U.S. power."[58] Soviet-Cuban relations ran hot and cold, but Castro ultimately bowed to the reality of dependency and supported the

Soviet invasion of Czechoslovakia in 1968 and sent troops to buttress Soviet-backed political factions and regimes in Africa. Also defining itself as an independent Third World leader, Castro's Cuba aided liberation movements and anti-Yankee leftists, especially in Central America and the Caribbean. The Reagan and Bush administrations vowed to bring down Castro by tightening the economic embargo, installing the propagandistic TV Martí, and courting the vehemently anti-Castro Cuban American National Foundation. Havana defiantly declared that "they will never forgive . . . that we have made a socialist revolution under their very noses."[59]

Cuba's troubles mounted when the Communist governments in Eastern Europe and the Soviet Union collapsed, bringing the Cold War to an end. The Cuban economy, already beset by corruption, natural disasters, hard currency shortages, and inefficiency, declined further when former allies toughened trade terms and Russia drastically cut aid. The Cuban Revolution's successes in medicine and education faced reversal. Fuel and food supplies dwindled, and popular unrest heightened. Predictions of Castro's rapid downfall flourished. Miami Cubans covered their car bumpers with stickers: "Next Christmas in Havana."[60] Cartoonists sketched "Robinson Castro" very alone on a desolate isle.[61] "Today we hear the breaking and crumbling of Castro's dictatorship," crowed President George Bush.[62] "Castro's final hour" had come.[63]

Fidel Castro insisted that he would stay the course. Cubans would blunt the "imperialists' " attempts to "crush us," leaving the attackers no more than "the blood-soaked dust of our earth."[64] A Cuban official wondered why "it is only with Cuba that America continues the Cold War."[65] A Canadian diplomat explained the pull of a hegemonic past: "The Americans want Castro's head and a government of America's choice."[66] The new Clinton administration, backing market democracies and wary of the political clout of the Cuban community, hesitated to initiate a turnaround. Still, Cuba and the United States inched toward less hostile relations as the 35th anniversary of Castro's triumphant march into Havana neared. Breaking ranks with hardliners, some Cuban-Americans urged lifting the embargo and opening talks with Castro. Quietly Washington and Havana cooperated to return to Cuba exiles held in U.S. prisons for felony crimes, to improve telephone service, and to join forces to curb cocaine trafficking. U.S. officials eased restrictions on travel to Cuba, while Castro legalized the dollar and encouraged foreign investment.[67]

In the post-Cold War age, once fierce enemies discovered shared interests. When Israelis and Palestinians could begin to settle their profound differences at the negotiating table and Vietnam and the United States could take steps toward normalizing relations, was it too farfetched, asked some, to think that Cuba and the United States could also reach an accommodation?

NOTES

Abbreviations Used in the Notes

ACSI	Assistant Chief of Staff for Intelligence
AWF	Ann Whitman File
BW	*Business Week* (magazine)
CIA	Central Intelligence Agency
CR	*Congressional Record*
DA	Department of the Army
DCR	Department of Commerce Records
DDE	Dwight D. Eisenhower
DDEL	Dwight D. Eisenhower Library, Abilene, Kansas
DDEP	Dwight D. Eisenhower Papers
DDRS	Declassified Documents Reference Service, Carrollton Press
Desp	Despatch
DS	Department of State
DSR	Department of State Records
DSB	*Department of State Bulletin*
FBIR	Federal Bureau of Investigation Records
FBIS	Foreign Broadcast Information Service (CIA)
Fdr	Folder
FO	Foreign Office (Great Britain)
FOIA	Freedom of Information Act (documents declassified under)
FRUS	*Foreign Relations of the United States* (Department of State)
HaEmb	American Embassy, Havana, Cuba
HAR	*Hispanic-American Report*
IHALBSP	Institute of Hispanic American and Luso-Brazilian Studies Papers, Hoover Institution Archives, Stanford, CA
JCS	Joint Chiefs of Staff

JFKL	John F. Kennedy Library, Boston, Massachusetts
JFKP	John F. Kennedy Papers
JW	Joint Weeka (weekly, numbered report from American Embassy, Havana, to Department of State, filed as 737.00 (W) with date in Department of State Records)
LBJL	Lyndon B. Johnson Library, Austin, Texas
LC	Library of Congress, Washington, DC
Memo	Memorandum
MemoConv	Memorandum of Conversation
MemoPhone	Memorandum of Telephone Conversation
NA	National Archives (Washington, DC, unless otherwise indicated)
NAC	National Archives of Canada, Ottawa, Ontario, Canada
nd	No date indicated on document
NSC	National Security Council
NSF	National Security File
NYT	*New York Times*
OH	Oral history interview
OSANSA	Office of the Special Assistant for National Security Affairs
PPP	*Public Papers of the Presidents* (with President and year)
PRO	Public Record Office, Kew, London, England
RG	Record Group
SS	Secretary of State
Tel	Telegram
Vol	Volume
WHCF	White House Central Files
WHO	White House Office
WP	*Washington Post*
WSJ	*Wall Street Journal*

Preface

1. A.S. Fordham to Henry Hankey, March 19, 1958, AK1015/16, Fdr 132164, FO 371, PRO.

Introduction

1. Oct. 28, *PPP, DDE, 1959,* 271.

2. National Security Council paper, April 1982, quoted in Walter LaFeber, *Inevitable Revolutions: The United States in Central America* (New York, 1993; 2nd ed. rev.), 271.

3. April 11, 1898, *FRUS, 1898,* 757.

4. 1902 statement quoted in David Healy, *Drive to Hegemony: The United States in the Caribbean, 1898–1917* (Madison, WI, 1988), 133.

5. Quoted in David H. Burton, *Theodore Roosevelt: Confident Imperialist* (Philadelphia, 1968), 106.

6. Quoted in Healy, *Drive,* 132; Burton, *Roosevelt,* 106.

7. Quoted in Louis A. Pérez, Jr., *Cuba Under the Platt Amendment, 1902–1934* (Pittsburgh, 1986), 323–24.

8. U.S. Congress, Senate, Committee on the Judiciary, *Communist Threat to the United States Through the Caribbean,* Part 9 (Aug. 30, 1960), 700.

9. CIA, "The Caribbean Republics," National Intelligence Estimate 80-54, Aug. 24, 1954, CIA Records, FOIA.

10. Lieutenant General W. K. Harrison, "Analysis of Situation in the Caribbean Command," Nov. 14, 1955, Box 2, Caribbean Command Records, Records of Joint Commands, NA.

11. For analyses of racist views toward Latin Americans, including Cubans, see Michael H. Hunt, *Ideology and U.S. Foreign Policy* (New Haven, 1987), 58–68; Rubin F. Weston, *Racism in U.S. Imperialism: The Influence of Racial Assumptions on American Foreign Policy, 1893–1946* (Columbia, SC, 1972), 137–82; Frederick B. Pike, *The United States and Latin America: Myths and Stereotypes of Civilization and Nature* (Austin, TX, 1992), 187–91. For examples of cartoons, see John J. Johnson, *Latin America in Caricature* (Austin, 1980).

12. Robert A. Stevenson to Ambassador Hugh S. Cumming, April 7, 1958, 737.00/4-758, Box 3078, DSR, NA; Mann to Draper, "Cuba: Briefing Book for the Use of Inspectors," April 24, 1953, Box 2, Lot File 57 D 59, *ibid.;* Stephen G. Rabe, *Eisenhower and Latin America: The Foreign Policy of Anticommunism* (Chapel Hill, 1988), 20.

13. "Notes from Oral Briefing at Room P-53, on January 31, 1955," Box 1, Central American Trip, 1955, Series 361, Vice President, Pre-Presidential Papers, Richard M. Nixon Papers, Federal Archives and Records Center, Laguna Niguel, CA.

14. DS, Public Studies Division, "American Opinion Series Report on U.S. Relations with Latin America," No. 32, Oct.–Dec. 1956, Box 40, DSR, NA.

15. Ali A. Mazrui, "Uncle Sam's Hearing Aid," in Sanford J. Ungar, ed., *Estrangement: America and the World* (New York, 1985), 181.

16. Quoted in Warren Miller, *90 Miles from Home: The Face of Cuba Today* (Boston, 1961), 101. For the nationalistic response of pre-Castro Cuban scholars and intellectuals to U.S. interference on the island and to the hegemonic presumption, see Robert Freeman Smith, "Twentieth-Century Cuban Historiography," *Hispanic American Historical Review* XLIV (Feb. 1964), 44–73; Duvon C. Corbitt, "Cuban Revisionist Interpretations of Cuba's Struggle for Independence," *ibid.* XLIII (Aug. 1963), 395–404; James H. Hitchman, "The Platt Amendment Revisited: A Bibliographic Survey," *The Americas* XXIII (April 1967), 343–69; Louis A. Pérez, Jr., "The Meaning of the *Maine:* Causation and the Historiography of the Spanish-American War," *Pacific Historical Review* LVIII (Aug. 1989), 293–322; Louis A. Pérez, Jr., *Cuba and the United States: Ties of Singular Intimacy* (Athens, GA, 1990), 291–303.

17. Harry A. Scott to SS for External Affairs, April 13, 1953, File 513-40, Vol 2750, RG 25, NAC.

18. The few public opinion polls taken suggested that a majority of Cubans did not consider the United States an evil force on the island, but in the repressive atmosphere of the Batista era, public opinion polls are suspect sources. Alfred A. Padula, Jr., "The Fall of the Bourgeoisie: Cuba, 1959–1961" (Ph.D. dissertation, University of New Mexico, 1974), 542n.

19. Dec. 2, 1958, *FRUS, 1958–1960,* V, 69–70.

20. Louis A. Pérez, Jr., *Cuba: Between Reform and Revolution* (New York, 1988), 183.

21. A "hegemonial system" is one in which "one state is powerful enough to maintain the essential rules governing interstate relations." Robert O. Keohane and Joseph Nye, *Power and Interdependence* (Glenview, IL, 1989; 2nd ed.), 44.

22. For the anti-Yankee elements in Cuban nationalism, see Ramón E. Ruíz, *Cuba: The Making of a Revolution* (New York, c. 1968; 1970); Edward González, *Cuba Under Castro: The Limits of Charisma* (Boston, 1974); C. A. M. Hennessy, "The Roots of Cuban Nationalism," *International Affairs* XXXIX (July 1963), 346–58; Carlos Alberto Montaner, "The Roots of Anti-Americanism in Cuba: Sovereignty in an Age of World Cultural Homogeneity," *Caribbean Review* XIII (Spring 1984), 13–16, 42–46; and Pérez, *Cuba and the United States.* Anti-Americanism, of course, has been a worldwide phenomenon. See "Anti-Americanism: Origins and Context," *The Annals* CDXCVII (May 1988), 9–171.

23. Montaner, "Roots of Anti-Americanism," 13.

24. For suggestive discussions of the borderlands phenomenon, see David Thelen, "Of Audiences, Borderlands, and Comparisons: Toward the Internationalization of American History," *Journal of American History* LXXIX (Sept. 1992), 436–44; Jorge A. Bustamante, "Demystifying the United States-Mexico Border," *ibid.,* 485.

25. Padula, "Fall," 505, 507.

26. March 18, 1953, *FRUS, 1952–1954,* IV, 6.

27. June 19, 1958, *FRUS, 1958–1960,* V, 29.

28. See Thomas G. Paterson, *On Every Front: The Making and Unmaking of the Cold War* (New York, 1992; rev. ed.).

29. Hunt, *Ideology,* 116–17, 123–24; Richard J. Barnet, *Intervention & Revolution: The United States in the Third World* (New York, 1972; rev. ed.).

30. Jorge I. Domínguez, *To Make a World Safe for Revolution: Cuba's Foreign Policy* (Cambridge, MA, 1989), 8–9; Jorge I. Domínguez, *Cuba: Order and Revolution* (Cambridge, MA, 1978), 54–55, 133.

31. Robert Taber, *M-26: Biography of a Revolution* (New York, 1961), 301.

1. Dependencies

1. Legat Havana to Director, Jan. 30, 1956, FBIR, FOIA.

2. Letter to SAC, nd, but probably Jan. 1956, *ibid.*

3. Director to Assistant Attorney General William F. Tompkins, July 26, 1956, *ibid.*

4. Tad Szulc, *Fidel: A Critical Portrait* (New York, 1986), 290. See also Chapter 2 below.

5. Quoted in Pérez, *Cuba Under the Platt Amendment,* 331.

6. For the events of 1933–34, see *ibid.* and Luis E. Aguilar, *Cuba 1933: Prologue to Revolution* (New York, c. 1972, 1974).

7. John Gunther, *Inside Latin America* (New York, 1941), 454.

8. Quoted in Padula, "Fall," 512.

9. MemoConv, "Batista's Address to Cuban Nation," March 10, 1952, Box 3368, 737.00/3-1052, DSR, NA.

10. Willard L. Beaulac to DS, Tel 599, March 11, 1952, Box 3368, 737.00/3-1152, *ibid.*

11. Adrian Holman to FO, No. 11, March 11, 1952, AK1015/8, Fdr 97516, FO 371, PRO.

12. March 22, 1952, *FRUS, 1952–1954,* IV, 868–70.

13. HaEmb to DS, Tel 525, Oct. 13, 1953, Box 3371, 737.00/10-1353, DSR, NA.

14. Szulc, *Fidel,* 323–24.

15. Rolando E. Bonachea and Nelson P. Valdés, eds., *Revolutionary Struggle, 1947–1958,* Vol. I of *The Selected Works of Fidel Castro* (Cambridge, MA, 1972), 257.

16. Victor Franco, *The Morning After: A French Journalist's Impression of Cuba Under Castro,* trans. by Ivan Kats and Philip Pendered (New York, 1963), 95–97. The manual is reprinted in Jay Mallin, ed., *Strategy for Conquest: Communist Documents on Guerrilla Warfare* (Coral Gables, FL, 1970), 319–62.

17. Quoted in Szulc, *Fidel,* 326.

18. Carlos Franqui, *Diary of the Cuban Revolution* (New York, 1980), 94–95.

19. Herbert L. Matthews, *Fidel Castro* (New York, 1969), 86; Director to SAC, New York, March 12, 1956, FBIR, FOIA.

20. Vicente Cubillas, Jr., "Mitin Oposicionista en Nueva York," *Bohemia* XLVII (Nov. 6, 1955), 60.

21. A copy of the photograph is in Box 2, Herbert L. Matthews Papers, Columbia Univ. Library, New York, NY.

22. Bonachea and Valdés, *Revolutionary,* 281–84.

23. Alfred Padula makes this point in his "Financing Castro's Revolution, 1956–1958," *Revista/Review Interamericana* VIII (Summer 1978), 235.

24. Edward S. Little OH, Foreign Affairs Oral History Program, Lauinger Library, Georgetown University, Washington, DC, 6.

25. Bonachea and Valdés, *Revolutionary,* 164–221.

26. Franqui, *Diary,* 96.

27. Memo on July 26 Club of New York, May 4, 1956, New York Office, FBIR, FOIA.

28. Castro in Franqui, *Diary,* 95.

29. Bonachea and Valdés, *Revolutionary,* 285.

30. Quoted in Jules Dubois, *Fidel Castro, Rebel—Liberator or Dictator?* (Indianapolis, 1959), 138.

31. Castro quoted in Szulc, *Fidel,* 343.

32. *Ibid.,* 343.

33. Dan Wakefield, "Puerto Rico: Rebels Find a Welcome," *The Nation* CLXXXV (Nov. 23, 1957), 384–86.

34. Padula, "Financing," 234–46; Rufo López-Fresquet, *My 14 Months with Castro* (Cleveland, 1966), 11.

35. Mario Llerena, *The Unsuspected Revolution: The Birth and Rise of Castroism* (Ithaca, NY, 1978), 219.

36. Memo, Assistant Attorney General William F. Tompkins to Director, Oct. 15, 1956, FBIR, FOIA; Miami Office Memo, Aug. 26, 1957, *ibid.;* Memo, Tompkins to Director, Jan. 9, 1958, *ibid.*

37. Ernesto Che Guevara, "A Decisive Meeting," in Jane McManus, ed., *From the Palm Tree: Voices of the Cuban Revolution* (Secaucus, NJ, 1983), 73.

38. Franqui, *Diary,* 96, 98.

39. Quoted in Szulc, *Fidel,* 336.

40. Author's interview with Francisco López Segrera, April 9, 1985, Instituto de Relaciones Internacionales, Ministerio de Relationes Exteriores, La Habana, Cuba; Author's interview with Hugo Pons, April 10, 1985, Departmento de Investigaciones sobre Estados Unidos, Universidad de La Habana, Cuba.

41. Quoted in John M. Kirk, *José Martí: Mentor of the Cuban Nation* (Gainesville, FL, 1983), 56.

42. Quoted in Pérez, *Cuba and the United States,* 80.

43. Franqui, *Diary,* 9.

44. Late 1957 statement in Bonachea and Valdés, *Revolutionary,* 355.

45. MemoConv, Nov. 25, 1957, Box 2, Lot File 59 D 573, DSR, NA.

46. Richard Cushing to Herbert Matthews, nd, but probably 1957, Box 3, Matthews Papers.

47. Quoted in Dubois, *Fidel,* 99.

2. Confusionist Cubanism: The Political Mess

1. *FRUS, 1952–1954,* IV, 900.

2. "The Caribbean Republics," Aug. 24, 1954, National Intelligence Estimate No. 80-54, CIA Records, FOIA.

3. "Notes from Oral Briefing . . . 1955," Nixon Papers.

4. James C. Hagerty Diary, March 11, 1955, Box 1, James C. Hagerty Papers, DDEL.

5. Arthur Gardner to Henry Holland, April 21, 1955, Box 2, Lot File 57 D 295, DSR, NA.

6. Edward G. Miller, Jr., to Willard Beaulac, June 10, 1952, *FRUS, 1952–1954,* IV, 875.

7. Cuban politics is discussed in Pérez, *Cuba,* 281–95; Domínguez, *Cuba,* 110–33; Charles D. Ameringer, *The Democratic Left in Exile: The Antidictatorial Struggle in the Caribbean, 1945–1959* (Coral Gables, FL, 1974), 189–94; Charles D. Ameringer,

"The Auténtico Party and the Political Opposition in Cuba, 1952–57," *Hispanic American Historical Review* LXV (May 1985), 327–52; Jules R. Benjamin, *The United States and the Origins of the Cuban Revolution: An Empire for Liberty in an Age of National Liberation* (Princeton, NJ, 1990), 119–32; Harold D. Sims, "Cuban Labor and the Communist Party, 1937–1958: An Interpretation," *Cuban Studies* XV (Winter 1985), 43–58; Samuel Farber, *Revolution and Reaction in Cuba, 1933–1960: A Political Sociology from Machado to Castro* (Middletown, CT, 1976), 145–201; Samuel Farber, "The Cuban Communists in the Early Stages of the Cuban Revolution: Revolutionaries or Reformists?," *Latin American Research Review* XVIII (No. 1, 1983), 59–84; Ramón Bonachea and Marta San Martín, *The Cuban Insurrection, 1952–1959* (New Brunswick, NJ, 1974); "Introduction" in Bonachea and Valdés, *Revolutionary,* 1–119; Padula, "Fall."

8. "Report of Sir Adrian Holman on the Conclusion of His Tour of Duty as Ambassador at Havana," April 14, 1954, AK 1016/1, File 10463-AH-40, Vol 2494, Department of External Affairs Records (RG 25), NAC.

9. *NYT,* May 10, 17, 22, 23, 26, and June 5, 1956; Ameringer, *Democratic Left,* 178.

10. Jorge García Montes quoted in Padula, "Fall," 70. From the start of U.S. tutelage in the early twentieth century, U.S. officials tolerated corruption. See José M. Hernández, *Cuba and the United States: Intervention and Militarism, 1868–1933* (Austin, 1993), 148–49.

11. A. S. Fordham to Selwyn Lloyd, "Leading Personalities in Cuba," July 4, 1958, AK1012/1, Fdr 132163, FO 371, PRO.

12. Statement of 1940 quoted in Jesse H. Stiller, *George S. Messersmith: Diplomat of Democracy* (Chapel Hill, 1987), 149.

13. Pérez, *Cuba,* 289.

14. For the history of the PSP, its rocky relations with the 26th of July Movement, and its weak ties with the Soviet Union in the 1950s, see Farber, "Cuban Communists"; Farber, *Revolution and Reaction,* 161–68; Andrés Suárez, *Cuba: Castroism and Communism, 1959–1966* (Cambridge, MA, 1967), 26–29; K. S. Karol, *Guerrillas in Power: The Course of the Cuban Revolution* (New York, 1970), 55–156; Jacques Lévesque, *The USSR and the Cuban Revolution: Soviet Ideological Perspectives, 1959–77* (New York, 1978), xii–xv, xiv–xx, 5, 8–10; Nicola Miller, *Soviet Relations with Latin America, 1959–1987* (Cambridge, UK, 1989), 5–10; Cole Blasier, *The Giant's Rival: The USSR and Latin America* (Pittsburgh, 1983), 78, 99–102; George J. Boughton, "Soviet-Cuban Relations, 1956–1960," *Journal of Inter-American Studies* XVI (Nov. 1974), 436–53; D. Bruce Jackson, *Castro, the Kremlin, and Communism in Latin America* (Baltimore, 1969), 11–14; Stephen Clissold, ed., *Soviet Relations with Latin America, 1918–1968: A Documentary Survey* (London, 1970), 20–21, 36–37; Edward González, "Castro's Revolution, Cuban Communist Appeals, and the Soviet Response," *World Politics* XXI (Oct. 1968), 44; Sauripada Bhattacharya, "Cuban-Soviet Relations Under Castro, 1959–1964," *Studies on the Soviet Union* IV, No. 3 (1965), 27–36; Szulc, *Fidel,* 418, 430, 440–41, 444, 451–55; Theodore Draper, *Castroism: Theory and Practice* (New York, 1965), 26–34.

15. Juan M. del Aguila, *Cuba: Dilemmas of a Revolution,* rev. ed. (Boulder, CO, 1988), 39.

16. PSP statement of Aug. 1953 quoted in Karol, *Guerrillas,* 139.

17. Feb. 1957 statement quoted in Suárez, *Cuba,* 26.

18. Quoted in Miller, *Soviet Relations,* 5–6, 10.

19. Quoted in Boughton, "Soviet-Cuban Relations," 438.

20. Richard H. Immerman, *The CIA in Guatemala: The Foreign Policy of Intervention* (Austin, TX, 1982), 185; Piero Gleijeses, *Shattered Hope: The Guatemalan Revolution and the United States, 1944–1954* (Princeton, 1991), 186–89, 280–81, 391; Herbert S. Dinerstein, *The Making of a Missile Crisis: October 1962* (Baltimore, MD, 1976), 1, 6, 10–14, 17–19.

21. O. Kuusinen at the 21st Congress of the Communist Party of the Soviet Union on Feb. 3, 1959, quoted in Lévesque, *USSR and Cuban Revolution,* 8.

22. Llerena, *Unsuspected,* 40.

23. Enrique Carrillo in MemoConv, "Cuban Situation and Activities of Revolutionary Groups," May 8, 1959, 737.00/S-859, Box 3082, DSR, NA.

24. CIA, "The Situation in Cuba," Special National Intelligence Estimate No. 85-58, Nov. 24, 1958, CIA Records, FOIA.

25. Quoted in Lee Lockwood, *Castro's Cuba, Cuba's Fidel* (New York, 1967), 162. See also Domínguez, *To Make,* 29–32.

26. Vilma Espín, "Deborah," in McManus, *From the Palm Tree,* 51; DS, "Draft White Paper on Cuba," May 12, 1959, p. 30, 611.37/1-159, DSR, FOIA.

27. For a detailed account of Castro in Bogotá, see Szulc, *Fidel,* 173–81.

28. "Report of Sir Adrian Holman," April 14, 1954.

29. May 8, 1956, *FRUS, 1955–1957,* VI, 831–32.

30. Franqui, *Diary,* 100–101.

31. A. H. Belmont to L. V. Boardman, Dec. 5, 1956, FBIR, FOIA.

32. American Embassy, Mexico, to DS, Desp 28, July 10, 1956, 712.00/7-1056, Box 2950, NA, DSR. See a similar conclusion in American Embassy, Mexico, to DS, Desp 152, Aug. 10, 1956, 712.00/8-1056, *ibid.*

33. Bonachea and Valdés, *Revolutionary,* 323; Szulc, *Fidel,* 354–64; *NYT,* June 26, 1956.

34. Bonachea and Valdés, *Revolutionary,* 84–85.

35. Teresa Casuso, *Castro and Cuba* (New York, 1961), 131–32.

36. *Ibid.,* 112–15; Peter G. Bourne, *Fidel: A Biography of Fidel Castro* (New York, 1986), 14–18, 130–31; Franco, *Morning,* 76–77.

37. Christopher Andrew and Oleg Gordievsky, *KGB: The Inside Story of Its Foreign Operations from Lenin to Gorbachev* (New York, 1990), 466.

38. In a 1974 interview, Prío said that at McAllen he had promised $100,000. Warren Hinckle and William W. Turner, *The Fish Is Red: The Story of the Secret War Against Castro* (New York, 1981), 3–10, describe the meeting based upon the interview. See also Dubois, *Fidel,* 133–34; Casuso, *Castro,* 111–12; Bonachea and San Martín, *Cuban Insurrection,* 66.

39. Quoted in "Trip to Cuba—Oct. 6–20, 1967," Notes, Box 27, Matthews Papers.

40. Quoted in Franco, *Morning,* 97.

41. Szulc, *Fidel,* 366–71.

3. Sugar and North American Business

1. For the Cuban economy—especially its dependency upon sugar, see the Cuban Economic Research Project, *A Study of Cuba* (Coral Gables, FL, 1965); Ismael Zuaznábar, *La economía cubana en la década del 50* (La Habana, 1986); Norman A. LaCharité, *Case Studies in Insurgency and Revolutionary Warfare: Cuba, 1953–1959* (Washington, DC, 1963); Carmelo Mesa-Lago, *The Economy of Socialist Cuba: A Two-Decade Appraisal* (Albuquerque, NM, 1981); Heinrich Brunner, *Cuban Sugar Policy from 1963 to 1970,* trans. by Marguerite Borchardt and H. F. Broch de Rothermann (Pittsburgh, 1977), 4–28; Pérez, *Cuba,* 295–307; Pérez, *Cuba and the United States,* 226–33; Leland L. Johnson, "U.S. Business Interests in Cuba and the Rise of Castro," *World Politics* XVII (April 1965), 440–59; Morris H. Morley, *Imperial State and Revolution: The United States and Cuba, 1952–1986* (New York, 1987), 40–71; Farber, *Revolution,* 153–56; Domínguez, *Cuba,* 54–109; Hugh Thomas, *The Cuban Revolution* (New York, 1977), 311–409; Jorge F. Pérez-López, *The Economics of Cuban Sugar* (Pittsburgh, 1991).

2. Quoted in Irving P. Pflaum, "Aspects of the Cuban Economy, Part I: The Financing of Castro's Reforms," in *AUFS Reports* (August 1960), vol. V (New York, 1966), 1.

3. U.S. Department of Commerce, *Investment in Cuba: Basic Information for the United States Businessman* (Washington, DC, 1956), 3.

4. SS to the President, "Proposed Modification of Sugar Act Unwise and Unfair," June 7, 1954, Box 2, Dulles-Herter Series, AWF, DDEP, DDEL.

5. For U.S. investment in Cuba and examples of company operations, see Philip C. Newman, *Cuba Before Castro: An Economic Appraisal* (India, 1965), 6–7, 58; Padula, "Fall," 593–97; "Joint Corporate Committee on Cuban Claims, Membership (As of September 30, 1983)," Bangor Punta Corporation Files, Greenwich, CT; Eugene Desvernine to author, April 29, 1985; Sears, Roebuck and Company, *Annual Report 1960* (1961), 26; Memo, "[Ford] Operations in Cuba," Aug. 11, 1959, Fdr 14, AR-67-14:1, Ford Industrial Archives, Redford, MI; "Facts About Compañía Ganadera Becerra, S.A.," Series 120, Braga Brothers Collection, University of Florida Library, Gainesville; Foreign Claims Settlement Commission of the United States, Claim of King Ranch, Inc., Decision No. CU 5751, "Proposed Decision," Aug. 19, 1970, King Ranch, Inc. Files, Kingsville, Texas.

6. *BW,* No. 1459 (Aug. 17, 1957), 104.

7. *HAR,* IX (Jan. 1957), 582; X (Feb. 1957), 18; X (April 1957), 128; X (Aug. 1957), 353; X (Nov. 1957), 533.

8. *BW,* No. 1440 (April 6, 1957), 144; *HAR,* X (Oct. 1957), 470; "Cuba—Notes Ending August 30, 1957," Box 0143920, Boise Cascade Records, Boise Cascade Company, Boise, ID.

9. U.S. Department of State official quoted in Morley, *Imperial,* 49. See also *NYT,* Feb. 5, Nov. 30, 1957; Commerce, *Investment; HAR,* X (March 1957), 72; X (Oct. 1957), 469.

10. Newman, *Cuba,* 15.

11. Summary of meeting of Discussion Group on Political Unrest in Latin America, Nov. 18, 1952, vol. XLV, Records of Groups, Council on Foreign Relations Archives, New York, NY.

12. Quoted in Morley, *Imperial,* 53.

13. Quoted in *NYT,* March 2, 1958.

14. Padula, "Fall," 28–29.

15. *Ibid.,* 67–69; *BW,* No. 1491 (March 29, 1958), 48; Irving P. Pflaum, "Aspects of the Cuban Economy, Part V: Fidelista Finances," in *AUFS Reports* (Sept. 1960), V, 8–9.

16. For technical assistance, see *FRUS, 1952–1954,* IV, 881, 893, 919; *FRUS, 1955–1957,* VI, 337; "Mutual Security Program, Fiscal Year 1958 Estimates, Latin America," p. 64, Box 42, Confidential File, WHCF, DDEL; U.S. Senate, *A Review of United States Government Operations in Latin America* (Washington, DC, 1959), 528–34.

17. U.S. Export-Import Bank of Washington, *Report to Congress for the Period July–December 1958* (Washington, DC, 1959).

18. For Nicaro, see later chapters and U.S. General Services Administration, "Nicaro Nickel Plant, Nicaro, Cuba," March 22, 1957, Box 20, Gerald D. Morgan Records, DDEL; U.S. General Services Administration Staff Paper, "Policy Regarding the Continued Operation and the Disposal of the Nicaro Nickel Plant," Feb. 28, 1958, *ibid.;* Franklin Floete (GSA Administrator), "Memorandum to Governor Adams," July 3, 1958, *ibid.; FRUS, 1952–1954,* IV, 885–89; *FRUS, 1955–1957,* VI, 848–50; Newman, *Cuba,* 81–88; Padula, "Fall," 595; JW 1, Jan. 3, 1957, DSR, FOIA; "Rebel Raid at Nicaro," July 23, 1958, Box 25, Lot File 61 D 411, DSR, NA.

19. Arthur Gardner to Henry F. Holland, June 7, 1955, Box 2, Lot File 57 D 295, DSR, NA.

20. International Bank for Reconstruction and Development, *Report on Cuba* (Baltimore, 1951), 359. See, for the same point, Maurice Zeitlin, *Revolutionary Politics and the Cuban Working Class* (Princeton, 1967), 45–65.

21. Ruíz, *Cuba,* 149–53.

22. Padula, "Fall," 59–60.

23. LaCharité, *Case,* 9; Thomas, *Cuban Revolution,* 330; Pérez, *Cuba,* 297; Pérez, *Cuba and the United States,* 227–31.

24. "The Cuban Revolution and Latin America," Oct. 8, 1962, meeting, vol. LXXXIX, Records of Groups, Council on Foreign Relations Archives.

25. Pérez, *Cuba and the United States,* 230.

26. *CR,* CII (Feb. 7, 1956), 2202.

27. March 15, 1955, *FRUS, 1955–1957,* VI, 797.

28. Joaquín Meyer, July 13, 1955, *ibid.,* 821.

29. Arthur Gardner to Senator George A. Smathers, Jan. 6, 1956, Box 5, George A. Smathers Papers, University of Florida Library.

30. James A. McConnell, Feb. 2, 1955, *FRUS, 1955–1957,* VI, 786.

31. Henry A. Holt to Arthur Gardner, Nov. 30, 1955, Box 2, Lot File 57 D 59, DSR, NA.

32. Arthur Gardner to Henry A. Holt, Dec. 5, 1955, *ibid.*

33. The act provided that if U.S. consumption exceeded 8,350,000 tons, any increase over that figure would be divided so that 55 percent went to domestic producers and 45 percent to foreign suppliers. Cuba's share of the 45 percent was set at 29.59 percent. See Public Law 545, 70 Stat., Ch. 342, May 29, 1956, pp. 217–21, and Cuban Economic Research Project, *Study,* 485–510. The sugar controversy in the United States can be followed in *FRUS, 1952–1954,* IV, 900–903, 910–13, and *FRUS, 1955–1957,* VI, 777–89, 791–820, 821–32. See also Thomas J. Heston, *Sweet Subsidy: The Economic and Diplomatic Effects of the U.S. Sugar Acts, 1934–1974* (New York, 1987).

34. Farr & Company, *Manual of Sugar Companies, 1955/56* (New York, 1956), 287; Farr, Whitlock & Co., *Manual of Sugar Companies* (New York, 1960; 35th ed.), 259, 266, 271.

35. "The Sugar Situation in Washington," Sherlock Davis to Laurence A. Crosby, June 4, 1957, Series 50, Braga Brothers Collection.

36. March 9, 1955, *FRUS, 1955–1957,* VI, 101.

37. Memorandum, Gorrín, Mañas, Macía y Alamilla, Abogados, Havana, to Cuban Trading Company, May 17, 1956, Reed Clark Files, Series 61, Braga Brothers Collection.

38. Harold A. Wolf, "The United States Sugar Policy and Its Impact upon Cuba: A Re-appraisal" (Ph.D. dissertation, University of Michigan, 1958), 74–75.

39. Vladimir P. Timoshenko and Boris C. Swerling, *The World's Sugar: Progress and Policy* (Stanford, CA, 1957), 325. See also Brunner, *Cuban Sugar Policy,* 19–20.

40. Bonachea and Valdés, *Revolutionary,* 186, 269.

41. Franqui, *Diary,* 98. Although Castro later, in early 1958, seemed to change his mind when he said nationalization would be "cumbersome," impede industrialization and private enterprise, and scare off foreign investment, he might have been uttering what he knew a U.S. audience wanted to hear. Bonachea and Valdés, *Revolutionary,* 366, 370.

42. A 1960 figure later raised, after a 1962 audit, to $301,136,610. Cuban Electric Company, "Information on Claims for Tax Deduction," Nov. 27, 1968, Box 0142087, Boise Cascade Records. The story of Cuban Electric is told in *HAR* IX (Jan. 1957), 582; X (June 1957), 247; X (Oct. 1957), 469; XI (March 1958), 147; R. Hart Phillips, *Cuba: Island of Paradise* (New York, 1959), 126–27; Newman, *Cuba,* 76–77; Cuban Economic Research Project, *Study,* 580–81; Commerce, *Investment,* 104–5; Thomas, *Cuban Revolution,* 385; Morley, *Imperial,* 52, 99, 112.

43. Bonachea and Valdés, *Revolutionary,* 295.

44. Staff Meeting, Oct. 22, 1958, Box 0143908, Boise Cascade Records; S.G. Menocal (Havana) to Division Managers and Department Heads, Dec. 15, 1958, Box 0144400, *ibid.;* Franqui, *Diary,* 185.

45. MemoConv, Feb. 6, 1959, Box 25, Lot File 61 D 411, DSR, NA; "Scared Money," *Time* LXXIV (Sept. 7, 1959), 30.

46. The story of the telephone company is reported in Cuban Economic Research Project, *Study,* 582–83; Erasmo Dumpierre, "El monopolio de la Cuban Telephone Company," *Bohemia* LXVII (Sept. 12, 1975), 88–92; Newman, *Cuba,* 106–11; Commerce, *Investment,* 114–16; *HAR* X (April 1957), 129; *FRUS, 1952–1954,* IV, 906; López-Fresquet, *My 14 Months,* 26; Thomas, *Cuban Revolution,* 381; Padula, "Fall," 380–82, 388; Morley, *Imperial,* 54, 76.

47. Dec. 6, 1956, *FRUS, 1955–1957,* VI, 839.

48. See Chapters 14, 15, and 17 below.

49. W. H. Gallienne to Sir Anthony Eden, "Cuba: Annual Review for 1954," Feb. 19, 1955, AK 1011/1, File 10463-AH-40, Vol 2494, RG 25, NAC.

4. Curve Balls, Casinos, and Culture

1. "General Background and Importance of Anglo-Cuban Relations," Aug. 30, 1957, AK1052/2, Fdr 126470, FO 371, PRO.

2. Pérez, *Cuba and the United States,* xvi.

3. Pablo Medina, *Exiled Memories: A Cuban Childhood* (Austin, TX, 1990), 59.

4. Oscar Zanetti, "American History: A View from Cuba," *Journal of American History* LXXIX (Sept. 1992), 530.

5. Harry A. Scott to SS for External Affairs, April 3, 1953, File 513-40, Vol 2750, RG 25, NAC.

6. Ronald Radosh, *American Labor and United States Foreign Policy: The Cold War in the Unions from Gompers to Lovestone* (New York, 1970), 375–82; Morley, *Imperial,* 92.

7. Serafino Romualdi, *Presidents and Peons: Recollections of a Labor Ambassador in Latin America* (New York, 1967), 182. See also p. 198.

8. USIS activities are detailed in HaEmb to DS, Desp 429, "Report on OCB Operations Plan for Latin America (CUBA)," 1958, DSR, FOIA; HaEmb to DS, Desp 661, "Progress Report on OCB 'Outline Plan of Operations for Latin America,' " Feb. 19, 1958, Box 2472, DSR, NA.

9. Richard G. Cushing OH, Foreign Affairs Oral History Program, p. 4.

10. For various cultural contacts, see Louis A. Pérez, Jr., "Cuba-United States Relations: A Century of Conflict," paper delivered at University of Connecticut Foreign Policy Seminar, Storrs, CT, March 13, 1990; Pérez, *Cuba and the United States,* 130–34, 207–13; *BW,* No. 1459 (Aug. 17, 1957), 36; *NYT,* Dec. 1, 1956; John H. Parker, *Yankee You Can't Go Home Again* (Miami, 1986); Thomas, *Cuban Revolution,* 320; Padula, "Fall," 26–31, 504, 586; Bourne, *Fidel,* 55; Phillips, *Cuba,* 357–58.

11. Smith to SS, Tel 657, Dec. 31, 1958, 237.1122/12-3158, DSR, NA.

12. Marshall W. Stearns, "Breaking the Rhythm Barrier," *Saturday Review* XL (Dec. 12, 1957), 35.

13. Speech in New York City, Nov. 20, 1958, in HaEmb to DS, Desp 513, Nov. 17, 1958, 737.00/11-1758, DSR, FOIA.

14. Eric A. Wagner, "Baseball in Cuba," *Journal of Popular Culture* XVIII (Summer 1984), 115–17; Pérez, *Cuba and the United States,* 71.

15. Quoted in Bruce Brown, "Cuban Baseball," *The Atlantic* CCLIII (June 1984), 112.

16. Jules Tygiel, *Baseball's Great Experiment: Jackie Robinson and His Legacy* (New York, 1983), 164–66; Leo Durocher with Ed Linn, *Nice Guys Finish Last* (New York, 1975), 203–12.

17. *NYT,* June 7, 1992; Angel Torres, *La historia del beisbol cubano, 1878–1976* (Los Angeles, 1976), 17, 45–47, 59.

18. Bill O'Neal, *The International League: A Baseball History, 1884–1991* (Austin, TX, 1992), 154–64, 275–78.

19. Howie Haak quoted in J. David Truby, "Now Pitching for the Giants . . . Fidel Castro," *Sports History* II (March 1989), 12.

20. Alex Pompez quoted in *ibid.,* 17.

21. Don Hoak with Myron Cope, "The Day I Batted Against Castro," in Charles Einstein, ed., *The Baseball Reader* (New York, 1980), 176–79.

22. Quoted in Miller, *90 Miles,* 32. Also see p. 39.

23. González Pedrero in González, *Cuba Under Castro,* 65.

24. The Economist Intelligence Unit (London), "Economic Review of Cuba, Dominican Republic, Haiti, and Puerto Rico," March 7, 1958, Braga Brothers Collection.

25. For tourism and related industries, see *NYT,* May 13, Dec. 16, 1956; Lyman B. Kirkpatrick, Jr., *The Real CIA* (New York, 1968), 167; Rosalie Schwartz, "Cuban Tourism: A History Lesson," *Cuba Update* XII (Winter/Spring 1991), 24–27; Pérez, "Cuba-United States Relations"; Pérez, *Cuba and the United States,* 207–12; Pérez, *Cuba,* 305.

26. Quoted in Miller, *90 Miles,* 244.

27. JW 4, Jan. 23, 1957, DSR, FOIA.

28. George A. Smathers OH, JFKL.

29. Michael R. Beschloss, *The Crisis Years: Kennedy and Khrushchev, 1960–1963* (New York, 1991), 98–100 (quotations, p. 99).

30. Jack Paar, *I Kid You Not* (Boston, 1960), 207–10.

31. "Memorandum by Lord Reading on His Goodwill Visit to Cuba, March 16–21, 1957," April 4, 1957, AK1051/2, Fdr 126469, FO 371, PRO.

32. G. Cabrera Infante, *Infante's Inferno,* trans. by Suzanne Jill Levine (New York, 1984), 7.

33. Carlos Baker, ed., *Ernest Hemingway: Selected Letters, 1917–1961* (New York, 1981), 882.

34. Miller, *90 Miles,* 22.

35. Peter Valenti, *Errol Flynn: A Bio-Bibliography* (Westport, CT, 1984), 50, 54; *New York Journal-American,* Jan. 7, 1959.

36. For the hotel industry and gambling, see Edwin A. Lahey, "Strange Partners Off on a Gambling Honeymoon: U.S. Underworld and Batista Get 'Married,'" *Miami Herald,* Jan. 8, 1958; Ernest Havemann, "Old Familiar Faces from Las Vegas Show Up in Plush Casinos with Plenty of Fast 'Action' to Take Tourist Dollars," *Life* XLIV (March 10, 1958), 32–37; *HAR* X (Jan. 1958), 663; *HAR* XI (April 1958), 205; *NYT,* Nov. 15, 1957; *New York World Telegram,* Jan. 9, 1958; HaEmb to DS, Desp 190,

Aug. 27, 1958, 837.45/8-2758, Box 4375, DSR, NA; R. Hart Phillips, *The Cuban Dilemma* (New York, 1962), 47–51; Susan Schroeder, *Cuba: A Handbook of Historical Statistics* (Boston, 1982), 459; Franco, *Morning,* 13; Ed Reid, *The Grim Reapers: The Anatomy of Organized Crime in America* (Chicago, 1969), 92, 104, 111, 124–25, 288, 296; Dennis Eisenberg, Uri Dan, and Eli Landau, *Meyer Lansky: Mogul of the Mob* (New York, 1979), 253–56; Newman, *Cuba,* 78, 120–23; Cuban Economic Research Project, *Study,* 569; Virgil Peterson, *The Mob: 200 Years of Organized Crime in New York* (Aurora, IL, 1983), 248, 310, 318–19, 332; Hank Messick, *Lansky* (New York, 1971). The fullest coverage of Meyer Lansky's activities and of Havana gambling in general is Robert Lacey, *Little Man: Meyer Lansky and the Gangster Life* (Boston, 1991), especially chs. 13–14.

37. Quoted in Havenmann, "Old," 33.

38. Quoted in Lacey, *Little Man,* 236.

39. James Robert Parish with Steven Whitney, *The George Raft File: The Unauthorized Biography* (New York, 1973), 154–56.

40. Arthur Gardner to Henry F. Holland, Jan. 13, 1956, Box 2, Lot File 57 D 295, DSR, NA.

41. Arthur Gardner to Henry Holland, April 13, 1956, *ibid.* For similar revulsion at North American behavior in Havana, see Arthur M. Schlesinger, Jr., *A Thousand Days: John F. Kennedy in the White House* (Boston, 1965), 173.

42. Manuel A. de Varona to Senator Estes Kefauver, Feb. 4, 1958, Estes Kefauver Papers, University of Tennessee Library, Knoxville.

43. *New York Daily News,* April 18, 1958.

44. *Ibid.,* April 7, 1958; *New York Mirror,* April 9, 1958.

45. Quoted In Havenmann, "Old," 34.

46. FBIS *Daily Report,* Dec. 12, 1957, p. u2; *BW,* No. 1491 (March 29, 1958), 45.

47. *Los Angeles Times,* March 20, 1958.

48. *BW,* No. 1490 (March 22, 1958), 120; *ibid.,* No. 1491 (March 29, 1958), 102; *WSJ,* Feb. 3, 1958; "Cuba—Notes Week Ending 3/14/58," Box 0143920, Boise Cascade Records; *HAR* XI (March 1958), 146; "Highlights in Latin America," *A Vision Report* (newsletter), April 11, 1958, Box 133, IHALBSP.

5. Supplying Repression: Military, CIA, and FBI Links

1. NSC, "U.S. Policy Toward Latin America," NSC 5902/1, Feb. 16, 1959, Box 26, Policy Papers Subseries, NSC Series, OSANSA, WHO Records, DDEL.

2. "U.S. Policy Toward Latin America," NSC 5613/1, Sept. 25, 1956, Box 18, *ibid.;* "Cuba—U.S. Foreign Assistance," Box 551, John Sherman Cooper Papers, University of Kentucky Library, Lexington; "Mutual Security Program, Fiscal Year 1958 Estimates, Latin America," Box 42, Confidential File, WHCF, DDEL. Cuban requests for military equipment are detailed in Box 3377, 737.5-MSP, DSR, NA.

3. At the very end of 1958, the following U.S. military personnel were stationed in Cuba: U.S. Army, 15; U.S. Navy, 13; U.S. Air Force, 17. Smith to SS, Dec. 31, 1958, Tel 657, 237.1122/12-3158, DSR, NA.

4. SS to HaEmb, June 20, 1955, 737.5-MSP/6-2055, DSR, FOIA; Carlos C. Hall to DS, Desp 181, Sept. 2, 1955, 737.5-MSP/9-255, *ibid;* "Historical Record of the U.S. Naval Mission to Cuba," F.W. Zigler to Chief of Naval Operations, Dec. 31, 1955, Naval Historical Center, Washington Navy Yard, Washington, DC; Louis A. Pérez, Jr., *Army Politics in Cuba, 1898–1958* (Pittsburgh, 1976), 148; Captain Wright H. Ellis (Naval War College) to author, July 23, 1985; Operations Coordinating Board, Washington, DC, "Special Report on Military Training in the U.S. of Foreign Nationals from Selected Countries," March 18, 1959, Box 16, Administration Series, AWF, DDEP, DDEL.

5. Jan. 9, 1953, *FRUS, 1952–1954,* IV, 882; July 14, 1953, *ibid.,* 895.

6. JW 50, Dec. 13, 1956, DSR, FOIA. In October 1962, the USS *Canberra* participated in the U.S. quarantine against Cuba during the missile crisis.

7. Colonel Harold S. Isaacson to ACSI, DA, "Development of Force Bases . . . ," May 27, 1955, 737.5-MSP/6-1755, DSR, NA.

8. JW 34, Aug. 21, 1953, Box 3373, *ibid.*

9. HaEmb to DS, Tel 99, July 14, 1953, Box 3377, 737.5-MSP/7-1453, *ibid.*

10. Carlos C. Hall to Colonel Leonard S. Dysinger, Aug. 29, 1955, 737.5-MSP/8-3155, *ibid.*

11. Col. Leonard S. Dysinger, "MAP Force Objectives," July 30, 1958, Box 57, Carribean Command Records.

12. Henry F. Holland in "Role of the Military in Latin America," 4th Meeting, Feb. 24, 1958, Study Group Report, vol. LXX, Records of Groups, Council on Foreign Relations Archives.

13. Domínguez, *Cuba,* 346–47.

14. Author's interview with Kelley V. Holbert, May 28, 1985, Grants Pass, OR. Holbert was the supply officer in the U.S. Naval Mission to Cuba, Sept. 1955-Aug. 1958.

15. Smith to R. R. Rubottom, Jr., April 22, 1958, 611.37/4-2258, Box 2472, DSR, NA.

16. R. H. Bentley to author, April 30, 1985. Lieutenant Commander Bentley joined the U.S. Naval Mission to Cuba as Sub Chief in September 1957.

17. Harold R. Aaron, "The Seizure of Political Power in Cuba, 1956–1959" (Ph.D. dissertation, Georgetown University, 1964), 178.

18. "Mutual Security Program, Fiscal Year 1958 Estimates, Latin America," Box 42, Confidential File, WHCF, DDEL.

19. Feb. 7, 1958, *FRUS, 1958–1960,* VI, 19.

20. See, for example, Castro's August 1956 letter in Bonachea and Valdés, *Revolutionary,* 336.

21. Harold R. Aaron, "Guerrilla War in Cuba," *Military Review* XLV (May 1965), 40–46; Pérez, *Army Politics,* 156.

22. Aug. 8, 1958, *FRUS, 1958–1960,* VI, 188.

23. *FRUS, 1903,* 350–53.

24. DS, *Treaty Series,* No. 866, *Relations: Treaty Between the United States of America and Cuba* (Washington, DC, 1934).

25. For the Guantánamo Naval Base, see Earl E. T. Smith, *The Fourth Floor: An Account of the Castro Communist Revolution* (New York, 1962), 109–14; U.S. Department of Defense Memo, "Importance of the U.S. Naval Base, Guantánamo, Cuba," *Inter-American Economic Affairs* XIV (Autumn 1960), 101–4; LC, Foreign Affairs Division, to Charles O. Porter, April 1, 1960, "The Status of the Guantánamo Naval Base in Cuba," Box 15, Charles O. Porter Papers, University of Oregon Library, Eugene, OR; U.S. Navy Commander E.A. Grunwald, "An Assessment of Current U.S. Policy Towards Cuba," Research Paper, March 1970, National War College, Washington, DC, pp. 16–21; FBIS *Daily Report,* May 9, 1957, p. u1; "Guantánamo Naval Base," Briefing Paper for Castro Visit, nd, but probably 1959, Lot 61 D 248, DSR, NA; Jules B. Billard, "Guantánamo: Keystone in the Caribbean," *National Geographic* CXIX (March 1961), 420–36; Robert D. Heinl, Jr., "How We Got Guantánamo," *American Heritage* XIII (Feb. 1962), 18–21, 94–97; E. J. Kahn, Jr., "A Reporter at Large: Here Come the Marines," *The New Yorker* LV (Nov. 1979), 190ff; *FRUS, 1958–1960,* VI, 178, 273.

26. *NYT,* July 20, Sept. 2, Dec. 21, 1956; Aug. 11, 1957; Sept. 13, 1959; Smith to SS, Dec. 31, 1958, Tel 657, 237.1122/12-3158, DSR, NA.

27. A letter from CIA Director Allen Dulles to President Batista, July 15, 1955, expressing U.S. interest in having Cuba create BRAC, is reprinted in Oscar Pino-Santos, *Cuba, historia y economía: Ensayos* (La Habana, 1983), 549–51.

28. Quoted in Hugh Thomas, "Cuba: The United States and Batista, 1952–58," *World Affairs* CXLIX (Spring 1987), 175.

29. "Caribbean Command Intelligence Review of Latin America," Feb. 10, 1956, Box 2, Caribbean Command Records; *ibid.,* March 2, 1956; Smith, *Fourth,* 124–25.

30. Phillips, *Cuba,* 351.

31. Kirkpatrick, *Real CIA,* 166, 170.

32. James Reston remembered that the CIA under Allen Dulles tried repeatedly to recruit *New York Times* reporters as informers. James Reston, *Deadline: A Memoir* (New York, 1991), 209.

33. Paul Bethel, an Embassy officer himself, identifies both Caldwell and Noel as CIA in *The Losers* (New Rochelle, NY, 1969), 59, 64. The U.S. Department of State's *Foreign Service List* (Washington, DC, annual) confirms their years in the Havana embassy. Among other CIA personnel in Cuba were Jack Stewart (Havana Embassy political officer, 1955–59) and Basil A. Beardsley (Havana Embassy political officer, 1958–59). For their and Wiecha's identities, see John Dorschner and Roberto Fabricio, *The Winds of December: The Cuban Revolution 1958* (New York, 1980), 93, 94, 147, 504; Szulc, *Fidel,* 428. As Joseph B. Smith, a former CIA officer, has revealed, those embassy officials designated in the Department of State's *Foreign Service List* with an "(R)" after their names were most likely CIA agents. All of the people mentioned here were so listed. Apparently this was one way the Department of State itself kept track of undercover CIA personnel. Joseph B. Smith, *Portrait of a Cold Warrior* (New York, 1976), 436.

34. Charles R. Clark, Jr., in U.S. Senate, Committee on the Judiciary, *State Department Security: The William Wieland Case,* Part 1 (Washington, DC, 1962), 83–84.

35. Robert A. Stevenson OH, Foreign Affairs Oral History Program, pp. 16–17. Stevenson became the Cuban Desk Officer in December 1958.

36. Robert A. Stevenson to Thomas G. Paterson, March 22, 1993.

37. Park F. Wollam to Thomas G. Paterson, Sept. 13, 1992.

38. Lucas Morán Arce, *La revolución cubana (1953–1959): Una versión rebelde* (Ponce, PR, 1980), 203; Andrés Suárez, "The Cuban Revolution: The Road to Power," *Latin American Research Review* VII (Fall 1972), 20.

39. George Thayer, *The War Business: The International Trade in Armaments* (New York, 1969), 99–105, 151.

40. See, for example, the comments of the U.S. Ambassador to El Salvador in the 1950s. Robert C. Hill OH, Columbia University Oral History Project, New York, NY.

41. Szulc, *Fidel,* 427–30, 670–71. Quotations from pp. 427, 428.

42. For an example of the monitoring of telephone calls, see the reference to a December 5, 1957, call by René Rayneri from Miami to Tegucigalpa, Honduras, in DS Instruction A-88 to American Embassy, Tegucigalpa, "Cuban Revolutionary Activities," March 3, 1958, 737.00/3-358, DSR, FOIA.

43. When the author applied under the Freedom of Information Act for documents on the relationship between the FBI and SIM, the FBI answered that the "central records systems files at FBI Headquarters revealed no record responsive" to the request. J. Kevin O'Brien to the author, Dec. 13, 1991. But the FBI was represented in Cuba: see Kirkpatrick, *Real CIA,* 157–66, 170; Szulc, *Fidel,* 320; "Progress Report on OCB," Feb. 19, 1958; Dorschner and Fabricio, *Winds,* 147; Bethel, *Losers,* 94.

44. Park F. Wollam to Terrence G. Leonhardy, April 11, 1958, 737.00/4-1158, Box 3078, DSR, NA.

45. J. Anthony Lukas, *Nightmare: The Underside of the Nixon Years* (New York, 1976), 194–95.

6. Granma *Rebels and the Matthews Interview*

1. For the *Granma* expedition and aftermath, see Franqui, *Diary,* 121–32; Bonachea and Valdés, *Revolutionary,* 87–90; Bonachea and San Martín, *Cuban Insurrection,* 85–89; Szulc, *Fidel,* 372–93.

2. Faustino Pérez, "El Granma Era Invencible Como el Espiritu de los Combatientes," in Luis Pavón, ed., *Días de combate* (La Habana, 1970), 8.

3. Franqui, *Diary,* 125.

4. For debate on the number of *Granma* survivors, see Bonachea and San Martín, *Cuban Insurrection,* 366, n. 49; Szulc, *Fidel,* 381.

5. G. A. Browne to SS for External Affairs, "Current Events in Cuba, November 1–December 8, 1956," Dec. 9, 1956, Box 9, File 7590-N-40, Vol 1986–89/360, RG 25, NAC.

6. Karol, *Guerrillas,* 3.

7. *NYT,* Dec. 4, 11, 1956.

8. "Annual Report of the Work of the Havana Office, 1956," File 18-1-13, Vol 1459, Records of the Department of Industry, Trade and Commerce (RG 20), NAC.

9. Summary of views in Department of State 737.00 file, Dec. 1956–Feb. 1957, DSR, NA. British diplomats were also skeptical that Castro could survive. "Fidel Castro has now faded out of the picture," read one British comment. Minutes by Pease, Jan. 17, 1957, AK1015/1, Fdr 126467, FO 371, PRO.

10. MemoPhone, Dec. 17, 1956, 737.00/12-1756, *ibid.*

11. A. S. Fordham to FO, No. 21, Feb. 22, 1957, AK1015/8, Fdr 126467, FO 371, PRO.

12. Wayne S. Smith, *The Closest of Enemies: A Personal and Diplomatic History of the Castro Years* (New York, 1987), 15.

13. Tel, J. Edgar Hoover, Jan. 18, 1957, FBIR, FOIA; *NYT,* Jan. 13, 1957.

14. "Grand Jury Investigation of Cuban Revolutionary Groups in the U.S.," Jan. 17, 1957, 737.00/1-1657, DSR, NA.

15. *El Diario de Nueva York,* Feb. 11, 1957, identified him as Díaz González.

16. *NYT,* Jan. 18, 31, 1957.

17. Quoted in Szulc, *Fidel,* 395.

18. JW 4, Jan. 23, 1957, DSR, FOIA.

19. Feb. 15, 1957, *FRUS, 1955–1957,* VI, 840; Gardner to Roy R. Rubottom, Jr., Feb. 15, 1957, Box 2, Lot File 59 D 573, DSR, NA.

20. Quoted in Taber, *M-26,* 202.

21. MemoConv, "Conditions in Cuba," Jan. 7, 1957, 737.00/1-757, DSR, NA. See also JW 4, Jan. 23, 1957, *ibid.*

22. JW 7, Feb. 13, 1957, DSR, FOIA; Szulc, *Fidel,* 394.

23. Notes, Feb. 12, 1957, Box 2, Matthews Papers.

24. Matthews's own accounts appear in *The Cuban Story* (New York, 1961) and *Revolution in Cuba* (New York, 1975). Those of his wife and others are cited below. See also Richard E. Welch, Jr., *Response to Revolution: The United States and the Cuban Revolution, 1959–1961* (Chapel Hill, NC, 1985), 141–42; Welch, "Herbert L. Matthews and the Cuban Revolution," *The Historian* XLVII (Nov. 1984), 1–18.

25. Felipe Pazos to Herbert L. Matthews, June 10, 1960, Box 1, Matthews Papers.

26. Ruby Phillips to Herbert L. Matthews, March 5, 1961, Box 1, *ibid.;* Phillips, *Cuba,* 298.

27. Pazos to Matthews, June 10, 1960.

28. Nancie Matthews, "Matthews' Journey to Sierra Maestra: Wife's Version," *Times Talk* X (March 1957), 8.

29. *Ibid.*

30. Matthews, *Cuban Story,* 36–37.

31. Raúl Chibás, "La entrevista de Herbert Matthews con Fidel Castro," from "Memorias de la revolución cubana," Raúl Chibás Collection, Hoover Institution Archives; Carlos Franqui, *The Twelve* (New York, 1968), 94–95, 102.

32. Matthews, "Matthews' Journey."

33. Matthews, *Cuban Story,* 40.

34. *Ibid.,* 43.

35. Llerena, *Unsuspected,* 93–94.

36. *NYT,* Feb. 26, 1957.

37. *NYT,* Feb. 24, 1957.

38. Ted Scott to Herbert Matthews, Feb. 28, 1957, Box 1, Matthews Papers.

39. *NYT,* Feb. 28, March 1, 1957.

40. Arthur Gardner to R. Roy Rubottom, Jr., Feb. 27, 1957, Box 2, Lot File 59 D 573, *ibid.*

41. JW 9, Feb. 27, 1957, DSR, FOIA; HabEmb to DS, Feb. 28, 1957, 737.00/2-2857, Box 3076, DSR, NA.

42. MemoPhone, Feb. 28, 1957, 737.00/2-2857, DSR, NA.

43. DS, *Biographic Register, 1958* (Washington, DC, 1958), 3–4.

44. Cushing to Matthews, Feb. 26, 1957, Box 1, Matthews Papers.

45. Cushing OH, pp. 1, 3.

46. *HAR* X (March 1957), 70; Szulc, *Fidel,* 401.

47. Herbert Matthews, Memo for the Publisher, June 16, 1957, Foreign Desk, Cuba, New York Times Archives, New York, NY.

48. Memo, Emanuel R. Freedman to Mr. Catledge, nd, but probably 1961, *ibid.*

49. Memo, Emanuel R. Freedman to Turner Catledge, Sept. 5, 1962, *ibid.* See also a similar Freedman memo dated Dec. 11, 1963, in *ibid.,* and the critique of the *Times* in "The Penance of Matthews," *The Nation* CCIII (Oct. 31, 1966), 437–38.

50. Senator Frank J. Lausche to Carl M. Meeker, Dec. 28, 1961, Box 131, Frank J. Lausche Papers, Ohio Historical Society, Columbus.

51. U.S. Congress, Senate, Committee on the Judiciary, *Communist Threat to the United States Through the Caribbean,* Part 12 (June 12, 1961), 821.

52. Welch, *Response,* 111.

53. José Bosch to Herbert L. Matthews, March 15, 1961, Box 1, Matthews Papers.

54. Fulgencio Batista, *Cuba Betrayed* (New York, 1962), 52.

55. Jay Mallin, "Reporters' Contacts Sped Castro's Victory," *Editor & Publisher* XCII (Feb. 7, 1959), 16, 62.

56. Another work that exaggerates Matthews's influence, especially on U.S. public opinion, is Van E. Gosse, "History Missing: Cuba, the New Left, and the Origins of Latin American Solidarity in the United States, 1955–1963" (Ph.D. dissertation, Rutgers Univ., 1992). For Matthews's reaction to the witch-hunt-like criticism of his Cuban role, see Herbert L. Matthews, *A World in Revolution: A Newspaperman's Memoir* (New York, 1971), 293–336.

57. Quoted in Herbert L. Matthews, "Dissent Over Cuba," *Encounter* XXIII (July 1964), 83.

7. *Ambassador Gardner and the Propaganda War*

1. Dubois, *Fidel,* 159; *NYT,* May 23, 1957; Thomas, *Cuban Revolution,* 150.

2. FBIS *Daily Report,* March 13, 1957, p. t2; *ibid.,* April 15, 1957, p. u3; *HAR,* X (April 1957), 126–27.

3. *NYT,* March 24, April 22, 1957.

4. JW 11, March 13, 1957, FBIR, FOIA. Under a Freedom of Information Act request, I also received a copy of this document from the Department of State, which, in declassifying it, deleted a good portion of it.

5. JW 11; *HAR* X (April 1957), 126–27.

6. Faure Chomón quoted in Franqui, *Diary,* 147.

7. Carlos Franqui quoted in *ibid.,* 176; Taber, *M-26,* 133.

8. Faure Chomón quoted in Franqui, *Twelve,* 112. See his full account in Faure Chomón, *El asalto al Palacio Presidencial* (La Habana, 1969).

9. *NYT,* March 14, 15, 1957; Szulc, *Fidel,* 417.

10. Enrique Rodríguez-Loeches, "The Crime at 7 Humboldt Street," in McManus, *From the Palm Tree,* 189.

11. Szulc, *Fidel,* 417.

12. MemoPhone, March 13, 1957, 737.00/3-1357, DSR, FOIA; JW 12, March 20, 1957, *ibid.*

13. See Chapter 3 above and Dubois, *Fidel,* 156.

14. JW 13, March 27, 1957, DSR, FOIA; JW 18, April 30, 1957, *ibid.*

15. HaEmb to DS, Desp 609, March 21, 1957, 737.00/3-2157, *ibid.;* MemoConv, "Activities in United States of ex-President Carlos Prío Socarrás," April 2, 1957, 737.00/4-257, *ibid.;* MemoConv, "Activities of Ex-President Prío," April 17, 1957, 737.00/4-1757, *ibid.*

16. Stewart to Rubottom, "Continued Stay of Ex-President Prío and Associates in U.S.," April 26, 1957, 737.00/4-2657, *ibid.*

17. HaEmb to DS, "Political Aspects: Briefing Book on Cuba," by John L. Topping, April 1, 1957, 737.00/4-157, *ibid.*

18. National Intelligence Estimate 80-57, April 23, 1957, *FRUS, 1955–1957,* VI, 629.

19. JW 10, March 5, 1957, DSR, FOIA.

20. DS, Public Studies Division, "American Opinion Series Report on U.S. Relations with Latin America," Report 33, Jan.–March 1957, Box 40, DSR, NA.

21. JW 10; *NYT,* March 6, 7, 1957; *HAR* X (April 1957), 127–28; FBIS *Daily Report,* March 6, 1957, p. g1.

22. "Memo Re Castro," March 6, 1957, Box 84, Lawrence E. Spivak Papers, LC.

23. Carleton Beals, "The New Crime of Cuba," *The Nation* CLXXXIV (June 29, 1957), 560–68; Carleton Beals, letter, *ibid.* CLXXXV (Sept. 21, 1957), inside cover; John A. Britton, *Carleton Beals: A Radical Journalist in Latin America* (Albuquerque, NM, 1989), 211. See also Carleton Beals, "Blackjacking Cuban Labor," *Christian Century* LXXIV (Aug. 21, 1957), 988–89; Llerena, *Unsuspected,* 125.

24. JW 19, May 8, 1957, DSR, FOIA.

25. Quoted in Franqui, *Twelve,* 79–80.

26. Ernesto "Che" Guevara, *Obra Revolucionaria* (Mexico, D. F. Mexico, 1967), 97–99.

27. JW 19.

28. Guevara, *Obra,* 167.

29. For the Taber episode and the end of the story of the three American volunteers, see News Release from CBS Radio, New York, May 10, 1957, Box 94,

IHALBSP; Llerena, *Unsuspected,* 103–8, 144–46; Dubois, *Fidel,* 159; *NYT,* May 10, 20, Oct. 28, 1957; *HAR* X (June 1957), 244, and X (Nov. 1957), 529; JW 21, May 22, 1957, DSR, FOIA; "In Man's War U.S. Boys Quit," *Life* XLII (May 27, 1957), 43; *Bohemia,* May 26, 1957, pp. 72–74, 96–97; JW 42, Oct. 15, 1957, DSR, FOIA. Robert Taber went on to write a book, *M-26: Biography of a Revolution* (New York, 1961), and help organize the Fair Play for Cuba Committee, an organization formed in early 1960 in the United States to gain a favorable American reception for the Castro Revolution. For his pro-Castro sympathies, Taber was called before a Senate investigating committee. His letters for 1959–60 to Congressman Charles O. Porter detail some of Taber's activities. Box 15, Porter Papers.

30. U.S. Army Attaché (Havana) to SS, C-8, May 24, 1957, 737.00/5-2457, Box 3076, DSR, NA.

31. MemoConv, "Revolutionary Activities Against Batista Regime . . . ," two documents with this title dated April 8, 1957, 737.00/4-857, DSR, FOIA; Leonhardy to Rubottom, "Revolutionary Activities in Cuba," April 17, 1957, 737.00/4-757, *ibid.*

32. "Notes on a talk with Rufo López-Fresquet, April 9, 1957," Box 2, Matthews Papers.

33. These M-26-7 club and anti-Batista activities in the United States are detailed in several FBI reports, FBIR, FOIA, as well as in *NYT* and *HAR* for 1957.

34. *Miami Herald,* May 25, 1957; *HAR* X (June 1957), 244; Szulc, *Fidel,* 420; Phillips, *Cuba,* 313–16.

35. Leonhardy to Rubottom, May 24, 1957, 737.00/5-2457, *ibid.*

36. Wieland to Rubottom, with RRR handwritten notes, "Cuban Revolutionary Activities in the United States," May 24, 1957, DSR, FOIA.

37. MemoConv, "Cuban Revolutionary Activities and Other Matters," May 29, 1957, 737.00/5-2957, *ibid.*

38. The bombings are detailed in the U.S. Embassy's Joint Weeka reports to Washington.

39. Gardner to Rubottom, May 31, 1957, Box 2, Lot File 59 D 573, DSR, NA.

40. Franqui, *Diary,* 180–81; Szulc, *Fidel,* 420–21.

41. JW 23, 24, and 25 for June 1957, DSR, FOIA.

42. *NYT,* June 9, 10, 12, 14, 16, 17, 1957.

43. Dwight D. Eisenhower to Arthur Gardner, Jan. 11, 1957, Box 21, DDE Diary Series, AWF, DDEP, DDEL.

44. John Foster Dulles to Arthur Gardner, Feb. 26, 1957, Box 14, Chronological File, John Foster Dulles Papers, DDEL; Memo for the Record, March 6, 1957, Box 18, Memorandums for the Record, Christian Herter Papers, DDEL; Dubois, *Fidel,* 165. Campa quoted in "Informal Comments on the Situation in Cuba," March 20, 1957, DSR, FOIA.

45. MemoPhone, April 20, 1957, 737.00/4-2057, DSR, FOIA.

46. Memo by Herbert Matthews, May 20, 1957, Box 2, Matthews Papers; Matthews Notes, "Havana, arrived June 3, 1957," *ibid.*

47. Senate, *Communist Threat* (Aug. 1960), Part 9, pp. 669–77, and *ibid.* (June 1961), Part 12, p. 807.

48. Mario Lazo, *Dagger in the Heart: American Policy Failures in Cuba* (New York, 1968), 126.

49. Quoted in Phillips, *Cuba,* 325.

50. Quoted in Dubois, *Fidel,* 166.

8. Ambassador Smith Meets the Rebellion

1. T. Michael Ruddy, *The Cautious Diplomat: Charles E. Bohlen and the Soviet Union, 1929–1969* (Kent, OH, 1986), 141; "Matters to Discuss with the President," Jan. 18, 1957, Box 6, Dulles-Herter Series, DDEP; Memo for the President, by John Foster Dulles, March 26, 1957, Box 14, JFD Chronological Series, Dulles Papers, DDEL.

2. Telephone Call, March 27, 1957, Box 22, DDE Diary Series, AWF, DDEP.

3. *NYT,* Aug. 3, 1957.

4. Senate, *Communist Threat,* Part 9 (Aug. 30, 1960), 700; Smith, *Fourth,* 23.

5. Philip W. Bonsal, *Cuba, Castro, and the United States* (Pittsburgh, 1971), 19.

6. "Nomination of Earl E. T. Smith to be Ambassador to Cuba," May 29, 1957, Executive Session, Senate Foreign Relations Committee Records, NA; Smith, *Fourth,* 222.

7. U.S. Congress, Senate, Committee on the Judiciary, *State Department Security: Testimony of William Wieland* (Feb. 8, 1961), Part 5 (Washington, DC, 1962), 552.

8. Smith, *Fourth,* 6–7.

9. Memo, Rubottom to the Secretary, "Cuban Political Situation," June 12, 1957, 737.00/6-1257, DSR, FOIA.

10. U.S. Senate, Committee on Foreign Relations, *Executive Sessions, 1958* (*Historical Series*), X (Washington, DC, 1980), 773.

11. Memo, May 20, 1957, Box 2, Matthews Papers.

12. Quoted in Rabe, *Eisenhower,* 120–21.

13. *HAR* X (Aug. 1957), 353.

14. JW 31, July 31, 1957, DSR, FOIA; FBIS *Daily Report,* July 26, 1957, p. u2; *HAR* X (Aug. 1957), 352–353.

15. JW 30, July 24, 1957, DSR, FOIA.

16. For the País story, see Bonachea and San Martín, *Cuban Insurrection,* 138–46; Szulc, *Fidel,* 422–25; Franqui, *Diary,* 216–17.

17. Vilma Espín, "Deborah," in McManus, *From the Palm Tree,* 62.

18. Smith, *Fourth,* 21.

19. Smith's account of events in Santiago is detailed in Memo, William P. Snow to SS, "Ambassador Smith's Remarks at Santiago de Cuba," Aug. 5, 1957, 737.00/8-557, DSR, FOIA. The rendering of the incident in Smith's memoirs is far more critical of the demonstrators (*Fourth,* 17–26). See also Carlos M. Castañeda, "El Embajador Smith en Santiago," *Bohemia,* Feb. 2, 1958, pp. 64–66, 98; *NYT,* Aug. 1, 1957; *HAR* X (Sept. 1957), 413; Taber, *M-26,* 163–65; Richard G. Cushing to Herbert Matthews, Aug. 6, 1957, Box 3, Matthews Papers.

20. Guerra to SS, Tel 1264, Aug. 2, 1957, 237.1122/8-157, NA, DSR.

21. *DSB* XXXVII (Aug. 26, 1957), 349–50 (Dulles quotation). For the turmoil throughout Cuba following the events in Santiago, see Snow, "Ambassador Smith's Remarks"; *NYT,* Aug. 1–4, 7–8, 17, 22, 30; *HAR* X (Sept. 1957), 413; *FRUS, 1955–1957,* VI, 843; JW 32, 33, 34, 35, 36, Aug. 7, 1957–Sept. 4, 1957, DSR, FOIA; HaEmb to DS, Desp 609, Aug. 23, 1957, 737.00/8-2357, *ibid.;* FBI, "Cuban Revolutionary Activities," Washington, DC, Aug. 12, 1957, FBIR, FOIA; "Telephone Calls, Friday, August 2, 1957," Box 10, CAH Telephone Calls, Herter Papers; Szulc, *Fidel,* 429.

22. MemoPhone, Aug. 13, 1957, with HaEmb Desp 130, Aug. 14, 1957, 611.37/8-1457, DSR, FOIA.

23. Mr. Fordham to FO, No. 58, Aug. 1, 1957, AK1015/32, Fdr 126467, FO 371, PRO; Minutes by Pease, Aug. 7, 1958, *ibid.;* Minutes by Pease, Aug. 14, 1957, AK1015/36. *ibid.*

24. HaEmb to SS, Tel 117, Aug. 23, 1957, 611.37/8-2357, Box 2472, DSR, NA; HaEmb to SS, Tel 118, Aug. 24, 1957, 611.37/8-2457, DSR, FOIA.

25. Christian Herter to Sinclair Weeks, Oct. 9, 1957, Box 20, Herter Papers.

26. DS, "Draft White Paper on Cuba," 43.

27. JW 37, Sept. 11, 1957, DSR, FOIA; JW 38, Sept. 18, 1957, *ibid.;* Bonachea and San Martín, *Cuban Insurrection,* 147–52; Franqui, *Diary,* 225–29.

28. JW 37.

29. Sept. 13, 1957, *FRUS, 1955–1957,* VI, 845–46.

30. *NYT,* Sept. 22, 1957. Press views are summarized in DS, Public Studies Division, "American Opinion Series on U.S. Relations with Latin America," Report 35, July–Sept. 1957, Box 40, DSR, NA.

31. Gardner to SS, Tel 459, March 28, 1957, 737.5 MSP/3-2757, DSR, FOIA.

32. Leonhardy to Rubottom, April 4, 1957, FW 737.5-MSP/4-357, DSR, FOIA; Secretary Dulles to HaEmb, May 16, 1957, 737.5-MSP/5-857, *ibid.*

33. Rubottom, Sept. 23, 1957, *FRUS, 1955–1957,* VI, 853. The *Congressional Record* for 1957 carries just a few pages on Cuban matters.

34. *FRUS, 1955–1957,* VI, 852–55.

35. Senate, *State Department Security,* 541.

36. "Cuban Government Seeks Tanks from Canadians," Oct. 31, 1957, 737.00/10-3157, Box 3077, DSR, NA.

37. Mr. Oliver to FO, No. 79, Oct. 11, 1957, AK1192/1, Fdr 126489, FO 371, PRO. Cuba had earlier approached the British for helicopters. Minutes, July 8, 1957, AK1015/31, Fdr 126467, *ibid.*

38. Minutes by Pease, Nov. 11, 1957, AK1192/3, *ibid.*

39. *FRUS, 1955–1957,* VI, 858.

40. Memo from Clifton Daniel to Mr. O'Neill, Aug. 27, 1957, Fdr E.C. Daniel 1959–70, Cuba, New York Times Archives.

41. *FRUS, 1955–57,* VI, 851n.

42. MemoConv, "Cuban Political Scene: Prío's Status in the United States," Sept. 20, 1957, 737.00/9-2057, DSR, FOIA. See also, *FRUS, 1955–1957,* VI, 850–52.

43. *FRUS, 1955–1957,* VI, 858–62.

44. *Ibid.,* 862–64; MemoPhone, "Cuban Elections and Disposal of Nicaro," Oct. 29, 1957, 737.00/10-2957, DSR, FOIA; MemoConv, "Prosecution of Cuban Revolutionaries . . . ," Nov. 13, 1957, 737.00/11-1357, *ibid.*

9. Expanding Contact with the Rebels

1. An excellent firsthand view of rebel life and conditions in the Sierra Maestra is found in a large, annotated set of 1957 photographs assembled by the journalist Andrew St. George. Box 1, Cuban Collection (Andrew St. George), Yale University Library.

2. Morán Arce, *Revolución,* 238.

3. American Consulate, Santiago de Cuba, to DS, Desp 14, Oct. 22, 1957, 737.00/10-2257, DSR, FOIA.

4. MemoConv, "Threat to Lives and Property of American Owned Sugar Central in Cuba," Nov. 29, 1957, 237.1122/11-2957, DSR, NA.

5. American Consulate, Santiago de Cuba, to DS, Desp 18, 19, 20, Dec. 17, 1957, 737.00/12-1757, *ibid.*

6. Franqui, *Diary,* 262 (quotation); American Consulate, Santiago de Cuba, to DS, Desp 27, Dec. 4, 1957, 737.00/12-457, DSR, FOIA.

7. Miami Office, "July 26 Revolutionary Movement of Miami," Jan. 21, 1958, FBIR, FOIA.

8. MemoConv, "Cuban Political Scene," Nov. 25, 1957, 737.00/11-2557, DSR, NA.

9. Franqui, *Diary,* 258.

10. *NYT,* Nov. 27, 1957.

11. *HAR* XI (Jan. 1958), 662; *NYT,* Nov. 20, 1957.

12. See, for example, Alice-Leone Moats, "The Strange Past of Fidel Castro," *National Review* IV (Aug. 24, 1957), 155–57, who argues that Castro "is playing the Communist game."

13. Fidel Castro, "What Cuba's Rebels Want," *The Nation* CLXXXV (Nov. 30, 1957), 399–401.

14. Llerena tells his story in *Unsuspected.*

15. MemoPhone, "Plans of Prominent Junta Libertadora Cubana Officials to Visit Department," Nov. 1, 1957, 737.00/11-157, DSR, FOIA.

16. Franqui, *Diary,* 271. Also see p. 277.

17. *Ibid.,* 265. For other M-26-7 critiques of the Miami Pact, see pp. 245–50, 263.

18. Quoted in Bonachea and San Martín, *Cuban Insurrection,* 171.

19. MemoConv, "Castro Renunciation of Unity Pact," Jan. 7, 1958, 737.00/1-758, DSR, FOIA.

20. *HAR* X (Nov. 1957), 533; *DSB* XXXVII (Dec. 23, 1957), 1018; HaEmb to DS, JW 48, Nov. 27, 1957, DSR, FOIA.

21. John Hickey, "The Role of Congress in Foreign Policy: Case: The Cuban Disaster," *Inter-American Economic Affairs* XIV (Spring 1961), 74.

22. *Report on Cuba,* Dec. 1957, Box 146, IHALBSP.

23. *HAR* X (Dec. 1957), 597; Dubois, *Fidel,* 181, 242.

24. JW 46, Nov. 13, 1957, DSR, FOIA.

25. MemoConv, "Cuban Political Situation," Nov. 14, 1957, 737.00/11-1457, *ibid.*

26. C. Allan Stewart (DS) to Herbert L. Matthews, March 3, 1958, Box 1, Matthews Papers.

27. See, for example, *NYT,* Oct. 18, Dec. 2, 1957.

28. DS, Public Studies Division, "American Series Report on U.S. Relations with Latin America No. 36," Oct.–Dec. 1957, Box 40, DSR, NA; *HAR* X (Dec. 1957), 595–97.

29. Ruby Phillips to Emanuel R. Freedman, Feb. 5, 1958, Foreign Desk, Ruby Hart Phillips, 1958, Fdr, New York Times Archives.

30. Franqui, *Diary,* 286. Also Matthews, "Matthews' Journey," 8.

31. Herbert L. Matthews, "Matthews, Under Gunfire, Covered Cuban Revolt," *Times Talk* XII (Feb. 1959), 2.

32. The story is told in *New York Herald Tribune,* Dec. 22, 1957; *NYT,* Dec. 21, 22, 27, 1957; *HAR* XI (Jan. 1958), 661–62.

33. Dec. 22, 1957.

34. Letter to the Editor from Joseph Dwork, *NYT,* Dec. 28, 1957.

35. Reports of the meetings are found in memoranda in the State Department's 737.00 file in the National Archives.

36. Dubois, *Fidel,* 284; Franqui, *Diary,* 281; MemoConv, "Cuban Political Situation and Other Matters," Jan. 30, 1958, 737.00/1-3058, DSR, FOIA.

37. Cushing OH, p. 3.

38. López-Fresquet, *My 14 Months,* 33 (quotation), 34.

39. Senate, *Communist Threat,* Part 9 (Aug. 30, 1960), 694. See also Dorschner and Fabricio, *Winds,* 72–74.

40. Dorschner and Fabricio, *Winds,* 141–42.

41. Suárez, "Cuban Revolution," 19–20; Dorschner and Fabricio, *Winds,* 147–48.

42. Davis went to Cuba in June 1956. Holbert Interview.

43. Franqui, *Diary,* 242.

44. Taber, *M-26,* 200–201.

45. Smith, *Fourth,* 35.

46. Franqui, *Diary,* 195. Also pp. 205–6.

47. *Ibid.,* 198.

48. See Chapter 5 above and Szulc, *Fidel,* 427–28.

49. See the discussion of this question in Chapter 5 above.

50. JW 44, Oct. 30, 1957, DSR, FOIA.

51. Smith, *Fourth,* 33–34.

52. HaEmb to SS, Tel 6211, Nov. 12, 1957, 737.00/11-1257, Box 3077, DSR, NA (quotation); Smith, *Fourth,* 33.

53. Senate, *State Department Security* (statement of Feb. 8, 1961), Part 5, pp. 578–79.

54. Smith, *Closest,* 15. Smith arrived in Havana in July 1958.

55. Franqui, *Diary,* 269 (quotation), 274.

56. The numerous memoranda of conversation on these State Department meetings are located in the 737.00 series, DSR, NA.

57. MemoConv, "Cuban Political Situation," Nov. 14, 1957, 737.00/11-1457, DSR, FOIA.

10. Taking Sides: Arms, Arrests, and Elections

1. DS, *Biographic Register* (Washington, DC, 1958), 726.

2. "Possible United States Courses of Action in Restoring Normalcy to Cuba," Nov. 21, 1957, 737.00/11-2157, DSR, FOIA. The document is summarized in *FRUS, 1955–1957,* VI, 865n.

3. William Wieland to C. Allan Stewart, Dec. 3, 1957, 737.00/12-357, Box 3077, DSR, NA. Smith gives a different account of the Wieland visit in *Fourth,* ch. X.

4. *FRUS, 1955–1957,* VI, 865–70.

5. Dec. 19, 1957, *ibid.,* 870–76.

6. Little OH, p. 6.

7. Rubottom, Jan. 17, 1958, *FRUS, 1958–1960,* VI, 9.

8. Wieland, Feb. 3, 1958, *ibid.,* 17.

9. Smith to C. Allan Stewart, Dec. 28, 1957, 611.37/12-2857, DSR, FOIA.

10. Jan. 10, 1958, *FRUS, 1958–1960,* VI, 5–6.

11. FO to British Embassy, Havana, No. 76, Dec. 20, 1957, AK1192/12, Fdr 126480, FO 371, PRO.

12. Smith to SS, Tel 367, 368, 369, Jan. 11, 1958, 737.00/1-1158, Box 3077, DSR, NA.

13. Jan. 17, 1958, *FRUS, 1958–1960,* VI, 8–11; Smith, *Fourth,* 59–60.

14. Press conference statement, *FRUS, 1958–1960,* VI, 12.

15. *Ibid.,* 12, 17; Smith, *Fourth,* 60; Taber, *M-26,* 201–2.

16. Smith, *Fourth,* 61.

17. Wieland to Rubottom, "Arms Shipments to Cuba," Feb. 4, 1958, Box 5, Lot File 60 D 553, DSR, NA. The Defense Department was notified a month later. See Robert G. Barnes to Director, Office of Programming and Control, International Security Affairs, Department of Defense, March 4, 1958, Box 25, Lot File 61 D 411, DSR, NA; *ibid.,* March 5, 1958.

18. HaEmb to DS, Desp 660, Feb. 18, 1958, 611.37/2-1858, DSR, FOIA.

19. *HAR* XI (Feb. 1958), 87–88.

20. Daniel M. Braddock to DS, Desp 600, Feb. 4, 1958, 737.00/2-458, DSR, FOIA.

21. Theodore M. Bernstein to Herbert L. Matthews, Jan. 22, 1958, Box 2, Matthews Papers.

22. MemoConv, April 22, 1958, 737.00/4-2258, DSR, FOIA.

23. Ameringer, *Democratic Left,* 193; MemoConv, "Castro Renunciation of Unity Pact," Jan. 7, 1958, 737.00/1-758, DSR, FOIA; MemoConv, "Cuban Political Situation and Other Matters," Jan. 20, 1958, 737.00/1-2058, Box 3077, DSR, NA; MemoConv, "Cuban Political Situation and Other Matters," Jan. 30, 1958, 737.00/1-

3058, DSR, FOIA; Bonachea and San Martín, *Cuban Insurrection,* 173–87; Matthews, *Revolution,* 100–101; FBIS *Daily Report,* Feb. 24, 1958, p. g1; Herbert L. Matthews to Ernest Hemingway, March 15, 1961, Box 1, Matthews Papers.

24. Sugar interests met with Batista on December 10, 1957, to map a coordinated plan to prevent economic sabotage. Memo to Cuban Trading Company from Gorrín, Mañas, Macía y Alamilla, Dec. 12, 1957, Series 61, Braga Brothers Collection.

25. Alejandro García and Oscar Zanetti, *United Fruit Company: Un caso del dominio imperialista en Cuba* (La Habana, 1976), 357.

26. Quoted in William Wieland, "Attitudes of Politically Influential Groups in Cuba Toward the June One Elections," March 6, 1958, Box 15, Porter Papers.

27. Jan. 19, 1958, letter in Franqui, *Diary,* 282.

28. Franqui, *Diary,* 279.

29. Jan. 21, 1958, *FRUS, 1958–1960,* VI, 30n.

30. Feb. 3, 1958, *ibid.,* 17.

31. MemoConv, "Cuban Political Scene and Other Matters," Feb. 20, 1958, 737.00/2-2058, DSR, FOIA.

32. Feb. 10, 1958, *FRUS, 1958–1960,* VI, 21.

33. Ernesto F. Betancourt to Charles Porter, March 4, 1958, Box 15, Porter Papers; *NYT,* Feb. 26, 1958.

34. *NYT,* Feb. 14, 1958.

35. *Ibid.,* Feb. 15, 1958.

36. José Pedro Llada quoted in Smith to SS, Tel 457, Feb. 17, 1958, 737.00/2-1758, Box 3077, DSR, NA. Other criticism in *Bohemia,* Feb. 23, 1958; *HAR* XI (Feb. 1958), 87; JW 8, Feb. 19, 1958, DSR, FOIA; Taber, *M-26,* 203.

37. *FRUS, 1958–1960,* VI, 28–29.

38. Feb. 18, 1958, *ibid.,* 26–27.

39. MemoPhone, "Cuban Revolutionaries Request Funds from American Sugar Company," Jan. 22, 1958, 737.00/1-2258, Box 3077, DSR, NA.

40. JW 6, Feb. 5, 1958, DSR, FOIA.

41. *FRUS, 1958–1960,* VI, 33; *HAR* XI (Feb. 1958), 85; *New York Herald Tribune,* Feb. 21, 1958.

42. JW 9, Feb. 26, 1958, DSR, FOIA. See also "Unhappy Cuba's Cockeyed Week," *Life* XLIV (March 10, 1958), 28–31.

43. JW 9.

44. Llerena, *Unsuspected,* 191.

45. Oscar Guerra to DS, Feb. 21, 1958, *FRUS, 1958–1960,* VI, 31.

46. The *Coronet* piece is reprinted in Bonachea and Valdés, *Revolutionary,* 364–67.

47. "Chronological Resume," nd, Fdr Foreign Desk, Andrew St. George, 1958–59, New York Times Archives.

48. Guevara, *Obra,* 161, 207. Dorschner and Fabricio have written that the CIA station chief in Havana, James Noel, named St. George as a CIA collaborator. Some CIA agents apparently did pose as journalists. *Winds,* 94n. See also Franqui, *Diary,* 179, 236.

49. Andrew St. George, "Interview with Fidel Castro," *Look* XXII (Feb. 4, 1958), 30.

50. Obituary, *NYT,* April 17, 1991.

51. *Ibid.,* Feb. 26, 27, March 1, 1958.

52. Feb. 28, 1958, *FRUS, 1958–1960,* VI, 38–42 (quotation p. 40).

53. *Ibid.,* 46–48.

54. *Ibid.,* 22, 28, 38.

55. Enrique Meneses, *Fidel Castro,* trans. by J. Halcro Ferguson (New York, 1966), 58–64 (quotation p. 62).

56. U.S. Senate, Committee on Foreign Relations, *Review of Foreign Policy, 1958,* Part I (March 5, 1958), 359–60, 361–66.

57. FBIS *Daily Reports,* March 10, 1958, p. u1.

58. Smith, *Fourth,* 53.

11. Batista's Self-destruction and Arms Suspension

1. *FRUS, 1958–1960,* VI, 42.

2. *Ibid.,* VI, 42–43, 48, 49–50 (quotation p. 50).

3. See, for example, March 4, 1958, *ibid.,* 50–51.

4. MemoPhone, "Cuban Situation," March 8, 1958, 737.00/3-858, DSR, FOIA.

5. Legat, Havana, to FBI Director, March 6, 1958, FBIR, FOIA.

6. U.S. Army Caribbean, Assistant Chief of Staff G-2 (Fort Amador, Canal Zone), "Weekly Intelligence Summary," March 7, 1958, ACSI, Army Staff Records, RG 319, NA.

7. U.S. Navy, Office of the Chief of Naval Operations, Office of Naval Intelligence, "Communist Cold War Reports in Mexico, Central America, and the Caribbean," March 3, 1958, No. 1981 (318B), DDRS.

8. *FRUS, 1958–1960,* VI, 36–38.

9. Cushing OH, p. 3.

10. Bonachea and Valdés, *Revolutionary,* 373–78.

11. March events can be followed especially in the U.S. Embassy's detailed weekly Joint Weeka (JW) reports and the *NYT.*

12. Quoted in Dubois, *Fidel,* 224.

13. Taber, *M-26,* 209.

14. Notes, Box 2, Matthews Papers.

15. MemoPhone, March 13, 1958, 737.00/3-1358, DSR, FOIA; Smith, *Fourth,* 90.

16. *Foreign Commerce Weekly,* Feb. 3, 1958; *Journal of Commerce,* Jan. 24, 1958; *BW,* No. 1481 (Jan. 18, 1958), 47; "Cuba—Notes Week Ending 3/28/58," Box 0143920, Boise Cascade Records.

17. The First National City Bank of New York, Foreign Information Service, "Customers Confidential Bulletin," March 1958, Box 129, IHALBSP.

18. Memo by Steve Kann (Editor, "American Exporter Industrial Reports from Cuba"), March 21, 1958, *ibid.*

19. March 12, 1958, *ibid.,* 56.

20. British Embassy, Havana, to FO, Jan. 23, 1961, AK1015/13, Fdr 156138, FO 371, PRO.

21. Lockwood, *Castro's Cuba,* 161.

22. Rabe, *Eisenhower,* 87.

23. Smith, *Fourth,* 81.

24. *FRUS, 1958–1960,* VI, 57.

25. *Ibid.,* 59.

26. March 14, 1958, *ibid.,* 60.

27. John Foster Dulles Desk Calender, 1958, Box 382, John Foster Dulles Papers, Princeton University Library.

28. Unidentified author, Chapter IV, "American Political Strategy and the Castro Revolution," nd, Box 25, Theodore Draper Papers, Hoover Institution; Confidential letter from former State Department official to Thomas G. Paterson, Dec. 12, 1991.

29. Senate, *State Department Security* (Feb. 8, 1961), 591–94.

30. Snow to Acting Secretary, "Order to Stop Shipment of Rifles to Cuban Army," March 14, 1958, 737.00/3-1458, DSR, FOIA. The records of telephone calls for Herter, Dulles, and Eisenhower at DDEL reveal no conversations on Cuba for this period.

31. March 24, 1958, *FRUS, 1958–1960,* VI, 67, 67n.

32. Tel, March 14, 1958, *ibid.,* 60.

33. Smith, *Fourth,* 91.

34. March 26, 1958, telegram approved by Snow, *FRUS, 1958–1960,* VI, 71.

35. Unnamed quoted in *NYT,* March 2, 1958.

36. Smith, *Fourth,* 88.

37. Snow, "Order to Stop Shipment," March 14, 1958.

38. DS to HaEmb, Tel 494, March 15, 1958, 737.00/3-1558, Vox 3077, DSR, NA; *NYT,* March 15, 1958.

39. Taber, *M-26,* 203–4; Smith, *Fourth,* 84–85.

40. Wieland to HaEmb, March 15, 1958, 737.00/3-1458, Box 3077, DSR, NA; March 14, 1958, *FRUS, 1958–1960,* VI, 60.

41. *FRUS, 1958–1960,* VI, 65.

42. March 16, 1958, *ibid.,* 61.

43. See meeting with Ambassador Campa in MemoConv, "Arms for Cuba," March 18, 1958, 737.00/3-1856, DSR, FOIA.

44. *NYT,* March 26, 1958; Taber, *M-26,* 216.

45. No press release on the subject, for example, appeared in the *Department of State Bulletin.*

46. *FRUS, 1958–1960,* VI, 62n.

47. March 17, 1958, *ibid.,* 63.

48. *NYT,* March 28, May 15, 23, 25, 1958; *HAR* XI (March 1958), 146; Taber, *M-26,* 207; SAC, New York, to FBI Director, May 29, 1958, 97-3243-155, FBIR, FOIA.

49. Hoyt to Snow, "Communism in the Cuban Situation," March 28, 1958, Box 7, Lot File 60 D 513, DSR, NA.

50. *CR* CIV (March 31, 1958), 5815.

51. U.S. Army Caribbean, Assistant Chief of Staff G-2 (Fort Amador, Canal Zone), "Weekly Intelligence Summary," April 11, 1958, p. 5, ACSI, Army Staff Records.

52. Senate, *State Department Security* (Feb. 1961), 592, 650–59.

53. Box 804 of the White House Central Files at the Eisenhower Library is full of critical letters and reports from Cubans.

54. *PPP, DDE, 1958,* 294–304; Pre-Press Conference, April 9, 1958, Box 32, DDE Diary Series, AWF, DDEP.

55. Rabe, *Eisenhower,* 96–99.

56. Llerena, *Unsuspected,* 170–79.

57. *CR* CIV (March 20, 1958), 4948–49, and (March 26, 1958), 5496–98. See Powell's account in his *Adam by Adam: The Autobiography of Adam Clayton Powell, Jr.* (New York, 1971), 185–98.

58. U.S. Senate, Committee on Foreign Relations, *Mutual Security Act of 1958* (hearings, March 31, 1958), 448–49.

59. April 1, 1958, *FRUS, 1958–1960,* VI, 75–76.

60. Smith to SS, Tel 642, April 3, 1958, 737.00/4-358, Box 3078, DSR, NA.

61. Daniel M. Braddock to Henry A. Hoyt, March 17, 1958, Box 7, Lot File 60 D 513, *ibid.;* DS to HaEmb, Tel 552, April 3, 1958, 737.5-MSP/4-358, DSR, FOIA; Smith to SS, Tel 648, April 4, 1958, *ibid.*

62. Memo by Chief of Naval Operations for JCS, "Suspension of Delivery of Military Equipment and Military Sales to Cuba," April 19, 1958, CCS 381, JCS Records, NA. Burke pressed his protest in a May 2, 1958, meeting with State Department officials: *FRUS, 1958–1960,* VI, 89–91.

63. See, for example, Chester J. Pach, Jr., *Arming the Free World: The Origins of the United States Military Assistance Program, 1945–1950* (Chapel Hill, NC, 1991), ch. 2.

64. M. M. Hammond to William P. Snow, "Outstanding Valid Arms Licenses and Pending Applications for the Cuban Government," April 7, 1958, with W.P.S. memo of April 9, 1958, Box 25, Lot File 61 D 411, DSR, NA; Dulles to HaEmb, Tel 557, April 7, 1958, 737.00/4-358, Box 3078, *ibid.;* Herter to HaEmb, "Export Licenses . . . ," April 18, 1958, 737.00/4-1658, *ibid.*

65. April 22, 1958, *FRUS, 1958–1960,* VI, 87.

66. May 6, 1958, *ibid.,* 93.

67. FBIS *Daily Report,* April 16, 1958, p. y2–3.

68. Rubottom to Smith, April 14, 1958, Box 5, Lot File 60 D 553, DSR, NA.

69. Smith to Rubottom, April 22, 1958, Box 2472, *ibid.*

70. Manuel Antonio de Varona to J. William Fulbright, April 18, 1958, BCN 121, J. William Fulbright Papers, University of Arkansas Library, Fayetteville.

71. HaEmb to SS, Tel 603, March 27, 1958, 237.1122/3-2758, DSR, NA.

12. Terrible Mood: Castro and the General Strike

1. The Havana Embassy's Joint Weeka reports detail events for the time.

2. Smith, *Fourth,* 102–3.

3. Franqui, *Diary,* 293 (March 25, 1958).

4. Park F. Wollam to DS, Desp 66, "Political Developments—March 26 to April 1, 1958," April 2, 1958, 737.00/4-258, DSR, FOIA. See also "Castro on Eve of His Big Bid," *Life* XLIV (April 14, 1958), 26–27, which includes photographs of Castro and followers by Andrew St. George.

5. Llerena, *Unsuspected,* 202.

6. Franqui, *Diary,* 312–13 (April 30, 1958).

7. Ray Brennan, *Castro, Cuba, and Justice* (New York, 1959), 208.

8. Park F. Wollam to DS, Desp 66, "Political Developments—March 26 to April 1, 1958," April 2, 1958, 737.00/4-258, DSR, FOIA.

9. Gordon Arneson, April 1, 1958, *FRUS, 1958–1960,* VI, 77–78.

10. *Foreign Commerce Weekly,* May 12, 1958, pp. 6, 16; *WSJ,* April 7, 14, 1958; Smith to SS, Tel 597, March 26, 1958, 737.00/3-2658, DSR, FOIA; JW 13, April 2, 1958; *ibid.;* HaEmb to DS, Desp 781, "Inspection Trip to Oriente Province," April 1, 1958, 737.00/4-158, *ibid.;* CIA, Information Report, March 26, 1958, attached to Amery Adams to Donaldson, April 1, 1958, 737.00/4-158, *ibid;* CIA, Information Report, March 27, 1958, attached to Emery Adams to Donaldson, April 2, 1958, 737.00/4-258, *ibid.;* Wollam to DS, Desp 63, March 25, 1958, 237.1122/3-2558, DSR, NA; Edward S. Little to George H. W. Bush, April 15, 1958, 737.00/4-758, Box 3078, *ibid.; FRUS, 1958–1960,* VI, 78–80; Dubois, *Fidel,* 240–41.

11. Brennan, *Castro, Cuba, and Justice,* 211.

12. MemoPhone, "Cuban Situation," April 9, 1958, 737.00/4-958, DSR, FOIA; MemoPhone, "Situation in Cuba," April 9, 1959, *ibid.;* MemoPhone, "Cuba," April 10, 1958, 737.00/4-1058, *ibid.;* MemoConv, "Political Developments in Cuba and Detention of American Newsman," *ibid.; NYT,* April 9, 1958.

13. SAC, New York, to FBI Director, March 19, 1958, 97-3243-123, FBIR, FOIA; SAC, Chicago, to FBI Director, April 4, 1958, 97-3382-32, *ibid.;* SAC, New York, to FBI Director, April 9, 1958, 97-3243-141, *ibid.*

14. *WSJ,* April 4, 1958.

15. *The Post-Times* (West Palm Beach, FL), April 5, 1958.

16. Letter of March 23, 1958, quoted in Régis Debray, *Revolution in the Revolution?: Armed Struggle and Political Struggle in Latin America* (New York, 1967), 77.

17. Assistant Chief of Staff G-2, U.S. Army Caribbean (Fort Amador, Canal Zone), Weekly Intelligence Summary, April 4, 1958, ACSI, Army Staff Records.

18. Franqui, *Diary,* 301.

19. Quoted in Meneses, *Fidel Castro,* 77.

20. Quoted in Szulc, *Fidel,* 441.

21. Quoted in Dubois, *Fidel,* 253; Taber, *M-26,* 238.

22. Quoted in Phillips, *Cuba,* 352. See also Szulc, *Fidel,* 440–41; Bonachea and San Martín, *Cuban Insurrection,* 220–21.

23. Quoted in Meneses, *Fidel Castro,* 78.

24. "Strongman's Round," *Time* LXXI (April 21, 1958), 27.

25. Quoted in Dubois, *Fidel,* 254.

26. Bonachea and San Martín, *Cuban Insurrection,* which provides an otherwise

compelling account of the general strike and its aftermath, come very close to charging Castro with launching a general strike he did not think would work in order to disarm rivals. Pp. 214–15.

27. A point made by many, including Szulc, *Fidel,* 440.

28. Franqui, *Diary,* 300–301 (April 16, 1958).

29. Quoted in "Strongman's Round."

30. Franqui, *Diary,* 310 (April 25, 1958).

31. *Ibid.,* 312 (April 28, 1958).

32. FBIS *Daily Report,* April 16, 1958, p. y2–3.

33. Quoted in Jorge Ricardo Masetti, *Los que luchan y los que lloran: El Fidel Castro que yo ví* (Buenos Aires, 1958), 97.

34. MemoPhone, April 9, 1958, 737.00/4-958, Box 3078, DSR, NA.

35. *New York Daily News,* April 10, 1958; MemoPhone, "Situation in Cuba," April 9, 1958, 737.00/4-958, DSR, FOIA.

36. Smith quotes himself from the *Miami Herald* in *Fourth,* 137.

37. *NYT,* April 14–21, 1958; *Hartford Courant,* April 16–21, 1958.

38. JW 16, April 16, 1958, DSR, FOIA; *FRUS, 1958–1960,* VI, 81–83. In his flawed if not deliberately distorted memoirs, Ambassador Smith has written that the Embassy had "evidence that the Communists were actively supporting the 26th of July Movement in their plans to make the general strike effective." *Fourth,* 105.

39. *FRUS, 1958–1960,* VI, 84–85.

40. See, for example, "Fuse That Fizzled," *Newsweek* LI (April 21, 1958), 65; Eric Sevareid, "CBS Radio News Analysis, April 11, 1958," Box 133, IHALBSP.

41. April 13, 1958, *FRUS, 1958–1960,* VI, 83–84.

42. Henry A. Hoyt, April 24, 1958, *ibid.,* 88–89.

43. See, for example, JW 17, April 23, 1958, DSR, FOIA.

44. JW 17; JW 18, April 30, 1958, *ibid.*

45. Wollam to DS, Desp 86, "Political Events—Oriente—April 23–29, 1958," April 29, 1958, 737.00/4-2958, *ibid.*

46. FBIS *Daily Report,* April 21, 1958, p. g1; *ibid.,* April 29, 1958, p. u2; JW 17, 18.

47. Guevara, *Obra,* 237–41.

48. Bonachea and San Martín, *Cuban Insurrection,* 218–20; "Agonizing Reappraisal," *Time* LXXI (April 28, 1958), 37.

49. Szulc, *Fidel,* 444.

50. Dubois, *Fidel,* 279.

51. Llerena, *Unsuspected,* 231–35 (quotations, p. 232).

52. Deletion in the original. Franqui, *Diary,* 317 (May 8, 1958).

13. Operation Fin de Fidel *and U.S. Weapons*

1. Smith, *Fourth,* 56.

2. *FRUS, 1958–1960,* VI, 93–94, 96.

3. For the British case, see British Embassy, Washington, to FO, No. 1065, May

2, 1958, AK1191/14, Fdr 132175, FO 371, PRO; H.B.C. Keeble (British Embassy, Washington) to J. Doyle (FO), May 20, 1958, AK1191/21, *ibid.*

4. May 21, 1958, report by Operations Coordinating Board, *FRUS, 1958–1960,* V, 4. For events surrounding the Nixon trip, see *ibid.,* 222–48; Marvin R. Zahniser and W. Michael Weis, "A Diplomatic Pearl Harbor? Richard Nixon's Goodwill Mission to Latin America in 1958," *Diplomatic History* XIII (Spring 1989), 163–90.

5. Quoted in John Hickey, "Blackmail, Mendicancy and Intervention: Latin America's Conception of the Good Neighbor Policy," *Inter-American Economic Affairs* XII (Summer 1958), 69.

6. Manuel Antonio de Varona to Senator Bourke Hickenlooper, May 3, 1958, Box 123, Bourke Hickenlooper Papers, Herbert Hoover Presidential Library, West Branch, IA.

7. *FRUS, 1958–1960,* VI, 95–96; JW 20, May 14, 1958, DSR, FOIA; JW 21, May 21, 1958, *ibid.;* JW 22, May 27, 1958, *ibid.;* Daniel M. Braddock to DS, May 28, 1958, "President Batista's Interview with NBC Correspondent," 737.00/5-2858, *ibid.;* Franqui, *Diary,* 340–41. Nixon reported to the cabinet on his trip: "Minutes of Cabinet Meeting," May 16, 1958, Box 32, DDE Diary Series, AWF, DDEP.

8. JW 19, May 5, 1958, DSR, FOIA.

9. Robert H. Shields, "Confidential Memorandum for Company Heads, Re: 1958 Cuban Sugar Crop," May 5, 1958, United States Beet Sugar Association Records, Washington, DC; Robert H. Shields, Letter No. 19-1958 to Members, May 9, 1958, *ibid.*

10. JW 23, June 4, 1958, DSR, FOIA.

11. Park F. Wollam to DS, Desp 90, "Political Developments—April 30–May 9, 1958," May 9, 1958, 737.00/5-958, DSR, FOIA; JW 20. For the summer offensive, see Bonachea and San Martín, *Cuban Insurrection,* 226–65; Szulc, *Fidel,* 445–48; Franqui, *Diary,* 316–63.

12. Ruby Phillips to Freedman, June 27, 1958, Fdr Foreign Desk, Ruby Hart Phillips, 1958, New York Times Archives.

13. Franqui, *Diary,* 318.

14. *Ibid.,* 318.

15. *Ibid.,* 320.

16. Dubois, *Fidel,* 261–65.

17. *Report on Cuba,* July 11, 1958, Box 146, IHALBSP.

18. Dulles to HaEmb, Tel 67, July 24, 1958, 737.00/7-2358, Box 3079, DSR, NA; Tel G-2 to HaEmb, July 25, 1958, *ibid.*

19. *HAR* XI (May 1958), 256–57; MemoConv, "Assault Against Cuban Pilots by Cuban Exiles at Miami Airport," June 2, 1958, 737.00/6-258, DSR, FOIA.

20. Constantine N. Kangles to author, March 28, 1984; Chicago Office report, CG 109-12, June 9, 1958, FBIR, FOIA.

21. *Sunday Star Ledger* (Newark, NJ), July 27, 1958; *Newark Sunday News,* July 27, 1958.

22. *NYT,* June 19, 1958.

23. Armando Lora quoted in *Washington Star,* May 26, 1958; *Los Angeles Examiner,* May 25, 1958; *Los Angeles Times,* May 25, 1958.

24. Franqui, *Diary,* 326–28 (quotation p. 328).

25. Quoted in Phillips, *Cuba,* 367.

26. Llerena, *Unsuspected,* 242.

27. Michael J. P. Malone to George A. Braga, July 11, 1958, Series 63, Braga Brothers Papers; H. M. Cross, Jr., to George A. Braga, "Visit to Havana and Manati, May 30–June 6, 1958," June 9, 1958, *ibid.;* "Cuba—Notes Week ending 5/23/58," Box 0143920, Boise Cascade Records; *Business International,* May 30, 1958; *Guaranty Survey,* June 1958, Box 142, IHALBSP; First National City Bank of New York, Foreign Information Service, Confidential for Customers, June 1958, Box 137, *ibid.; Foreign Commerce Weekly,* July 21, 1958, p. 6; *BW,* No. 1507 (July 19, 1958), 98.

28. Franqui, *Diary,* 335.

29. Bonachea and Valdés, *Revolutionary,* 379.

30. Quoted in Dubois, *Fidel,* 268–69.

31. Franqui, *Diary,* 351.

32. July 30, 1958, *FRUS, 1958–1960,* VI, 175.

33. Quoted in Franqui, *Diary,* 384.

34. Dulles through Rubottom to HaEmb, April 1, 1958, 737.00/4-158, Box 3078, DSR, NA; MemoPhone, "Kidnapping of American Citizens," July 1, 1958, 737.00/7-158, DSR, FOIA; Smith to SS, Tel 4, July 1, 1958, *ibid.;* Memo for the Chief of Naval Operations, "Situation Report No. 3," July 2, 1958, DDRS 1981 (171B); "Cuban Sitrep," July 2, 1958, CNO/Admiral Burke Files, Naval Historical Center, Washington Navy Yard, Washington, DC; *FRUS, 1958–1960,* VI, 174–79, 181–85; Christian A. Herter to Arleigh Burke, July 31, 1958, Box 5, Chronological File, Herter Papers; *HAR* XI (July 1958), 377, and XI (Aug. 1958), 433–34; *DSB* XXXIX (Aug. 18, 1958), 282; Smith, *Fourth,* 111–13.

35. *FRUS, 1958–1960,* VI, 99–102, 106–7.

36. *Ibid.,* 108–16.

37. *FRUS, 1958–1960,* VI, 193–94; JW 40, Oct. 1, 1958, DSR, FOIA.

38. DS, "Considerations Affecting United States Policy Toward President Batista's Regime . . . ," p. 39, nd but probably 1961, Box WH-5, Arthur M. Schlesinger, Jr., Papers, JFKL.

39. DS, "Draft White Paper on Cuba," 67.

40. *FRUS, 1958–1960,* VI, 109–10, 116.

41. *Ibid.,* 117–18. See also pp. 113–16.

42. Franqui, *Diary,* 316.

43. They are reproduced in Llerena, *Unsuspected,* 244–45.

44. Wollam to DS, Desp 102, May 27, 1958, 737.00/5-2758, DSR, FOIA; Franqui, *Diary,* 325.

45. Snow to Murphy, "Exchange of Rocket Heads with Cuban Navy," May 23, 1958, Box 25, Lot File 61 D 411, DSR, NA (with note by Frank J. Devine).

46. *DSB* XXXIX (July 28, 1958), 153; Llerena, *Unsuspected,* 243; Dubois, *Fidel,* 265, 270–71.

14. Rocket Heads, Kidnappers, and the Castros

1. This chapter's account of the kidnapping episode draws upon Wollam's extensive diary of events from June 27 through July 19: Park F. Wollam to DS, "Kidnapping of Americans in Oriente," Desp 3, July 31, 1958, 737.00/7-3158, Box 3079, DSR, NA. Also useful were Department of State records in the National Archives and *FRUS, 1958–1960,* VI, cited throughout; *HAR* for June and July 1958; Dubois, *Fidel,* 269–79; *NYT* for late June and July 1958; and Peter Lisagor and Marguerite Higgins, *Overtime in Heaven: Adventures in the Foreign Service* (Garden City, NY, 1964), ch. 9. A factually inaccurate article with a heavy anti-Castro bias and an unpersuasive case that the United States should and could have used military force to gain the hostages' release is Roberta Wohlstetter, "Kidnapping to Win Friends and Influence People," *Survey* (Great Britain) XX, No. 4 (1974), 1–40.

2. Quoted in *Time* LXXII (July 7, 1958), 33.

3. The early handling of the crisis is detailed in several MemoConv, most titled "Rebel Kidnapping of U.S. Engineers at Moa Bay," June 27, 1958, 737.00/6-2758, DSR, FOIA, and Box 3078, NA; Memo for the Record, "Naval Base Guantánamo—Cuban Insurrectos," Robert J. Stroh, Office of the Chief of Naval Operations, U.S. Navy, CNO/Admiral Burke Files; *FRUS, 1958–1960,* VI, 117.

4. MemoConv, "Rebel Kidnapping of U.S. Engineers at Moa Bay," June 27, 1958, 737.00/6-2758, DSR, FOIA.

5. Jules Dubois called the town Puriales in Jules Dubois, *Freedom Is My Beat* (Indianapolis, 1959), 248.

6. Manuel Fajardo quoted in Franqui, *Twelve,* 104–5.

7. See Wollam's biography in U.S. Department of State, *The Biographic Register, 1960* (Washington, DC, 1960), 836.

8. Quoted in Lisagor and Higgins, *Overtime,* 215.

9. Wollam to Bal, Sunday, 10:30 a.m., June 29, 1958, 737.00/7-458, Box 3078, DSR, NA.

10. *Ibid.*

11. *Ibid.*

12. Smith to SS, Tel 883, June 29, 1958, 737.00/6-2958, *ibid.; FRUS, 1958–1960,* VI, 121.

13. "The Rebels Were 'Swell,'" *Newsweek* LII (July 14, 1958), 47. See also "Caught in a War," *Time* LXXII (July 14, 1958), 29.

14. Quoted in "Castro's 'Anti-Aircraft,'" *Newsweek* LII (July 21, 1958), 49.

15. Quoted in Lee Hall, "Inside Rebel Cuba with Raul Castro," *Life* XLV (July 21, 1958), 30.

16. *HAR* XI (June 1958), 315, and (July 1958), 375.

17. *FRUS, 1958–1960,* VI, 119–20.

18. July 2, 1958, *ibid.,* 124–25.

19. Rufo López-Fresquet to Herbert L. Matthews, July 6, 1958, Box 1, Matthews Papers.

21. Raúl Castro explained his reasons for the kidnapping to Jules Dubois of the *Chicago Tribune* (Dubois, *Fidel,* 274–75) and in articles in *Verde Olivo* (Havana), Sept. 15, 22, 1963, translated by Beatrice M. Ash in "Operation Antiaircraft," Mallin, *Strategy,* 286–314.

21. *NYT,* June 20, 1958.

22. Quoted in DS, Public Studies Division, "American Opinion Series Report on U.S. Relations with Latin America," April–June 1958, Box 40, DSR, NA. For other press opinion, see this DS report and *HAR* XI (July 1958), 375.

23. Charles L. Blume to the President, July 2, 1958, Box 804, General File 122 (Cuba), WHCF, DDEP.

24. Quoted in *NYT,* July 2, 1958. See also July 1. Republican Senators Homer E. Caphart of Indiana and William Knowland of California and Democratic Representative Victor Anfuso of New York also demanded stern U.S. action. Senate Republican Memo No. 26, July 3, 1958, Box 131, Hickenlooper Papers; *HAR* XI (July 1958), 375; *NYT,* July 14, 1958.

25. July 2, 1958, *PPP, DDE, 1958,* 512.

26. Baker, *Hemingway,* 884.

27. Mrs. William H. Koster to the President, June 30, 1958, Box 804, General File 122 (Cuba), WHCF, DDEP.

28. MemoConv, "Whereabouts and Welfare of William H. Koster," July 2, 1958, 737.00/7-258, DSR, FOIA.

29. Quoted in Hall, "Inside," 29.

30. *DSB* XXXIX (July 21, 1958), 104, 109, 110, 111.

31. Franqui, *Diary,* 356.

32. MemoPhone, "United States Victims in Cuba . . . ," July 2, 1958, 737.00/7-258, DSR, FOIA.

33. Raúl Castro Ruz to Park F. Wollam, 2 de Julio de 1958, enclosure in Wollam, "Kidnapping of Americans in Oriente," July 31, 1958.

34. July 3, 1958, *FRUS, 1958–1960,* VI, 126.

35. July 3, 1958, *ibid.,* 125–26.

36. Smith to SS, Tel 18, July 2, 1958, 737.00/7-358, Box 3078, DSR, NA.

37. MemoPhone, "Kidnapped Americans," July 4, 1958, 737.00/7-458, DSR, FOIA.

38. Brennan, *Castro, Cuba and Justice,* 238.

39. Karl Meyer of *Washington Post* quoting Fidel Castro's officers after a visit to the Sierra Maestra, in Park F. Wollam to DS, Desp 19, Sept. 5, 1958, 737.00/9-558, Box 3079, DSR, NA.

40. FBIS *Daily Report,* June 30, 1958, p. g1.

41. Bonachea and Valdés, *Revolutionary,* 383–84.

42. Dubois, *Freedom,* 239–59.

43. MemoPhone, "Possible Release of More Kidnapped Americans," July 6, 1958, 737.00/7-658, Box 7, Lot File 60 D 513, DSR, NA.

44. Wollam to HaEmb for DS, July 7, 1958, in E. A. Gilmore, Jr., to DS, "Documents Pertaining to Kidnapping of Americans by Rebel Forces in Oriente Province," Desp 40, July 10, 1958, Box 3078, *ibid.*

45. Wollam, "For the Embassy," July 7, 1958, enclosure in *ibid.*

46. July 9, 10, 1958, *FRUS, 1958–1960,* VI, 134–36, 141–43.

47. Smith, *Fourth,* 145.

46. Memo for the Files by William Wieland, July 3, 1958, Box 25, Lot File 61 D 411, DSR, NA.

49. July 2, 1958, *FRUS, 1958–1960,* VI, 123–24.

50. Memo for the Record, "Discussion at Admiral Burke's Residence Regarding the Situation in Cuba as of 10 July," July 12, 1958, CNO/Admiral Burke Files.

51. Robert Murphy, *Diplomat Among Warriors* (Garden City, NY, 1964), 369.

52. Burke to JCS, July 10, 1958, *FRUS, 1958–1960,* VI, 140; N. F. Twining to the Secretary of Defense, July 11, 1958, *ibid.,* 145; Burke to Herter, July 11, 1958, CCS 381, JCS Records.

53. July 10, *FRUS, 1958–1960,* VI, 140n.

54. JW 29, July 16, 1958, DSR, FOIA.

55. Raúl Castro Ruz to Park F. Wollam, 14 de Julio de 1958, enclosure in Wollam, "Kidnapping of Americans in Oriente," July 31, 1958.

56. Dubois, *Fidel,* 278–79.

57. JW 30, July 23, 2958, DSR, FOIA; *FRUS, 1958–1960,* VI, 156–57.

58. Dubois, *Freedom,* 256–57.

59. Taber, *M-26,* 259.

60. Park F. Wollam to Thomas G. Paterson, May 17, 1993.

61. Carlos C. Hall to Hughes, Aug. 9, 1961, Box 56, Countries-Cuba, NSF, JFKP, JFKL.

15. Frankenstein and a Toughened Attitide

1. Quoted in Lisagor and Higgins, *Overtime,* 239.

2. Ramón M. Barquín, *Las luchas guerrilleras en Cuba: De la colonia a la Sierra Maestra,* 2 vols. (Madrid, Spain, 1975), II, 642; Bonachea and San Martín, *Cuban Insurrection,* 245.

3. Bonachea and Valdés, *Revolutionary,* 388.

4. DS, Bureau of Intelligence and Research, Division of Research and Analysis for American Republics, Aug. 15, 1958, *FRUS, 1958–1960,* VI, 197.

5. July 20, 1958, *ibid.,* 159.

6. Office of Secretary of Defense to CINCARIB, DEF 945297, July 23, 1958, Box 57, 092 Cuba, Caribbean Command Records; Robert G. Barnes to Charles H. Shuff, Sept. 19, 1958, 737.5-MSP/8-2158, DSR, FOIA; *FRUS, 1958–1960,* VI, 198. The Caribbean Command Records document U.S. decisions on non-combat equipment.

7. July 24, 1958, *FRUS, 1958–1960,* VI, 162–68.

8. *Ibid.,* 168–74, 186–92.

9. MemoConv, "Cuban Political Situation," Aug. 7, 1958, 737.00/8-758, DSR, FOIA; Nov. 24, 1958, *FRUS, 1958–1960,* VI, 266n (quotation).

10. Aug. 8, 1958, *FRUS, 1958–1960,* VI, 189, 190.

11. "Discussion at the 375th Meeting of the National Security Council, Thursday, August 7, 1958," Aug. 8, 1958, Box 10, NSC Summaries of Discussion, AWF, DDEP.

12. Michael J. Francis, "The U.S. Press and Castro: A Study in Declining Relations," *Journalism Quarterly* XLIV (Summer 1967), 257–66.

13. Jon D. Cozean, "The U.S. Elite Press and Foreign Policy: The Case of Cuba" (Ph.D. dissertation, American University, 1979), 1046–54.

14. U.S. Department of Commerce, World Trade Information Service, "Economic Developments in Cuba 1958," Part 1, No. 59–42 (Washington, DC, April 1959); First National City Bank of New York, Foreign Information Service, Confidential for Customers, Oct. 1958, Box 162, and Nov. 1958, Box 167, IHALBSP.

15. New York Office of FBI, Report of Nov. 20, 1958, 97-3400-72, FBIR, FOIA.

16. Batista, *Cuba Betrayed,* 95.

17. Quoted in Franqui, *Diary,* 417. The rebel advance is followed in *ibid.,* ch. 16; Szulc, *Fidel,* 455–56; and Bonachea and San Martín, *Cuban Insurrection,* ch. 12.

18. Franqui, *Diary,* 391, 396–99, 428, 436.

19. See, for example, MemoConv, "Sabotage of American Properties . . . ," Sept. 30, 1958, 737.00/9-3058, DSR, FOIA.

20. MemoPhone, "Cuban Rebel Threats Against U.S. Companies," Oct. 8, 1958, 737.00/10-858, *ibid.*

21. Oct. 10, 1958, *FRUS, 1958–1960,* VI, 240–41.

22. MemoConv, "Rebel Demands Against American Business Firms in Cuba," Oct. 10, 1958, 737.00/10-1058, DSR, FOIA; Dulles to HaEmb, Tel 201, Oct. 14, 1958, 737.00/10-1458, Box 3079, DSR, NA.

23. MemoConv, "Activities of Cuban Rebels," Oct. 25, 1958, 737.00/10-2558, DSR, FOIA.

24. Smith to SS, Tel 303, Sept. 22, 1958, 737.00/9-2258, Box 3079, DSR, NA.

25. Wollam to DS, "Revolutionary Actions—October 2–6, 1958," Oct. 6, 1958, 737.00/10-658, DSR, FOIA.

26. Dulles to HaEmb, Tel 175, Oct. 7, 1958, 737.00/10-758, *ibid.;* JW 41, Oct. 8, 1958, *ibid.* At an Oct. 28, 1958, meeting in New York, executives of the following companies were represented: United Fruit, Lone Star Cement, Francisco Sugar, Manatí Sugar, Freeport Sulpher, Czarnikow-Rionda, Texaco, Punta Alegre Sugar, Chase Manhattan Bank, West Indies Sugar, King Ranch, First National City Bank of New York, Compañía Cubana Primadera, Cuban-American Sugar, First National City Bank of Boston, and American Sugar Refining. (List in Box 5, Lot File 60 D 553, DSR, NA.) Three days later several of them and others met with Rubottom in Washington. *FRUS, 1958–1960,* VI, 245–49.

27. MemoConv, "Cuban Political Situation and Possible Effects on American Interests Operating in Eastern Cuba," Oct. 31, 1958, 737.00/10-3158, DSR, FOIA.

28. FBIS *Daily Report,* Aug. 7, 1958, p. u1; Smith to SS, Tel 190, Aug. 10, 1958,

737.00/8-1058, Box 3079, DSR, NA; Park F. Wollam to DS, Desp 6, Aug. 12, 1958, 737.00/8-1258, DSR, FOIA; JW 35, Aug. 27, 1958, *ibid.;* "Considerations Affecting United States Policy Toward President Batista's Regime . . . ," nd (probably 1958 because it covers 1958 events even thought it is tagged 611.37/1-157), Box 2472, DSR, NA.

29. General Services Administration representative quoted in Park F. Wollam to DS, Desp 23, Sept. 24, 1958, 737.00/9-2458, DRS, FOIA. Also for September events: JW 37, Sept. 10, 1958, *ibid.; HAR* XI (Sept. 1958), 496; Smith to DS, Desp 263, Sept. 15, 1958, 737.00/9-1558, Box 3079, DSR, NA; JW 38, Sept. 17, 1958, DSR, FOIA; Staff Meeting, Oct. 22, 1958, Box 0143908, Boise Cascade Records.

30. MemoConv in Smith to DS, Desp 263, Sept. 15, 1958, 737.00/9-1558, Box 3079, DSR, NA; *FRUS, 1958–1960,* VI, 223.

31. *HAR* XI (Oct. 1958), 551; FBIS *Daily Report,* Oct. 23, 1958, p. g3; Smith to SS, Tel 339, Oct. 3, 1958, 737.00/10-358, Box 3079, DSR, NA; Park F. Wollam to DS, Desp 35, Oct. 11, 1958, 737.00/10-1158, DSR, FOIA; *FRUS, 1958–1960,* VI, 244.

32. Brennan, *Castro, Cuba, and Justice,* 245. Brennan was with Castro in the Sierra Maestra at the time.

33. Texaco's troubles and the kidnapping are detailed in several documents in the 837.3932 series for Oct.–Nov. 1958, Box 4375, and the 737.00 series, Box 3079, DSR, NA; JW 44, Oct. 28, 1958, DSR, FOIA; *FRUS, 1958–1960,* VI, 243–44; Dubois, *Fidel,* 320; *HAR* XI (Oct. 1958), 551; *NYT,* Oct. 22, 1958; Franqui, *Diary,* 429–30.

24. Wieland to Snow, "Rebel Raid at Nicaro," July 23, 1958, Box 25, Lot File 61 D 411, DSR, NA.

35. Aug. 1, 1958, *FRUS, 1958–1960,* VI, 180.

36. Little to Rubottom, "Rebel Raid on Nicaro," Sept. 30, 1958, Box 25, Lot File 61 D 411, DSR, NA.

37. For Nicaro's troubles and the evacuation, see Wollam to DS, Desp 30, Oct. 1, 1958, 737.00/10-158, DSR, FOIA; Wollam to DS, Desp 38, Oct. 20, 1958, 737.00/10-2058, *ibid.;* JW 43, Oct. 21, 1958, *ibid.;* JW 44, Oct. 28, 1958, *ibid.;* HaEmb to SS, Tel 414, Oct. 24, 1958, 237.1122/10-2458, DSR, NA; several documents dated Oct. 24, 1958, in 737.00/10-2458, Box 3079, *ibid.;* "Synopsis of Intelligence and State Material Reported to the President," by John Eisenhower, Oct. 27, 1958, Box 37, DDE Diary Series, AWF, DDEP; *HAR* XI (Oct. 1958), 551; Dubois, *Fidel Castro,* 320–24; Franqui, *Diary,* 436.

38. Franqui, *Diary,* 429.

39. MemoConv, "Activities of Cuban Rebels," Oct. 25, 1958, 737.00/10-2558, DSR, FOIA.

16. The Quest for Communists and Arms

1. Roy R. Rubottom, Jr., OH, Foreign Affairs Oral History Program.

2. Smith, *Fourth,* 146.

3. See, for example, Smith to DS, Desp 263, Sept. 15, 1958, 737.00/9-1558, Box 3079, DSR, NA.

4. Braddock to DS, Desp 285, Sept. 18, 1958, 737.00/9-1858, DSR, FOIA.

5. July 24, 1958, *FRUS, 1958–1960,* VI, 161. See also pp. 187–90.

6. Kirkpatrick, *Real,* 172; *FRUS, 1958–1960,* VI, 160–61.

7. The scholar Geoffrey S. Smith has demonstrated that during the Cold War "little distance separated smashing Communists from bashing gays and lesbians." See his "National Security and Personal Isolation: Sex, Gender, and Disease in the Cold-War United States," *International History Review* XIV (May 1992), 327. Ambassador Smith in *FRUS, 1958–1960,* VI, 241 (Oct. 22, 1958).

8. July 30, 1958, *FRUS, 1958–1960,* VI, 161n.

9. S. B. Donahoe to A. H. Belmont, March 14, 1958, 109-551-53, FBIR, FOIA.

10. Wollam to DS, Desp 5, Aug. 4, 1958, 737.00/8-458, Box 3079, DSR, NA.

11. Sept. 23, 1958, *FRUS, 1958–1960,* VI, 212; MemoConv, "Fidel Castro and his '26 of July Movement'," Oct. 14, 1958, 737.00/10-1458, DSR, FOIA. See also Javier Pazos's recollection of the Fidel he knew in the mountains: He was "definitely not a Marxist" and "not even particularly progressive." Quoted in Hugh Thomas, "Middle-Class Politics and the Cuban Revolution," in Claudio Véliz, ed., *The Politics of Conformity in Latin America* (London, 1967), 266.

12. Dorschner and Fabricio, *Winds,* 93.

13. Bonsal, *Cuba, Castro,* 18.

14. Wollam to DS, Desp 12, Aug. 22, 1958, 737.00/8-458, DSR, FOIA.

15. DS, Division of Research and Analysis for American Republics, Bureau of Intelligence and Research, "The 26th of July Movement Since the Abortive General Strike of April 9, 1958," Intelligence Report No. 7780, Aug. 15, 1958, DDRS, 71-B (1979). An abstract of this document is printed in *FRUS, 1958–1960,* VI, 196–97.

16. Guevara, *Obra,* 250.

17. Quoted in Bethel, *Losers,* 64–65.

18. On the relationship between the PSP and Castro, see the discussion in Chapter 2 above.

19. MemoConv, "Cuban Political Situation and Possible Effects on American Interests Operating in Eastern Cuba," Oct. 31, 1958, 737.00/10-3158, DSR, FOIA.

20. Nov. 4, 1958, *FRUS, 1958–1960,* VI, 251.

21. Cuban Minister of State in Smith to DS, Aug. 1, 1958, 737.00/8-158, Box 3079, DSR, NA. See also MemoConv, Aug. 19, 1958, Box 2472, *ibid.*

22. Aug. 8, 1958, *FRUS, 1958–1960,* VI, 190. Also p. 198.

23. *Ibid.,* 201–3.

24. Quoted in Dubois, *Fidel,* 318.

25. The British sale is documented in various folders in FO 371, PRO.

26. JW 39, Sept. 24, 1958, DSR, FOIA.

27. P. R. Oliver to C. J. A. Whitehouse, Aug. 29, 1958, AK1192/11, Fdr 132175, FO 371, PRO.

28. A. S. Fordham to Selwyn Lloyd, No. 114, Nov. 18, 1958, AK1015/62, Fdr 132165, *ibid.*

29. Minutes by D. J. Shillcock, Oct. 1, 1958, AK1192/20, Fdr 132175, *ibid.*

30. FBIS *Daily Report,* Oct. 21, 1958, p. gl; DS to HaEmb, Desp 425, Oct. 22,

1958, 737.00/10-2258, Box 3079, DSR, NA; A. S. Fordham to Mr. Selwyn Lloyd, "Cuba: Annual Review for 1958," Feb. 12, 1959, File 10463-AH-40, Vol 2494, RG 25, NAC: Canadian Ambassador (Havana) to SS for External Affairs, "Annual Review of Events in Cuba—1958," Jan. 20, 1959, *ibid.; NYT,* Nov. 28, 1958; Phillips, *Cuba,* 338–39; MemoConv, "British Policy Toward Shipment of Arms to Cuba," Dec. 23, 1958, 737.56/12-2358, DSR, FOIA.

31. Comité del Exilio to Neil McElroy (Defense Department), Aug. 11, 1958, Box 93, Legislation Files, Pre-Presidential Papers, JFKP, JFKL; Charles O. Porter to John Foster Dulles, Aug. 25, 1958, Box 15, Porter Papers; William B. Macomber, Jr., to Charles O. Porter, Oct. 7, 1958, *ibid.;* JW 33, Aug. 13, 1958, DSR, FOIA.

32. José Miró Cardona quoted in Dubois, *Fidel,* 301.

33. MemoConv, "Suggestion of Military Attaché, Habana . . . ," Aug. 29, 1958, Box 7, Lot File 60 D 513, DSR, NA.

34. JW 36, Sept. 3, 1958, DSR, FOIA; Dubois, *Fidel,* 285.

35. U.S. Army Caribbean, Assistant Chief of Staff G-2 (Fort Amador, Canal Zone), "Weekly Intelligence Summary," Aug. 22, 1958, ACSI, Army Staff Records.

36. *DSB* XXXIX (Sept. 15, 1958), 440, and (Sept. 29, 1958), 505.

37. Dubois, *Fidel,* 317.

38. JW 42, Oct. 14, 1958, DSR, FOIA; Park F. Wollam to DS, Desp 37, Oct. 15, 1958, 737.00/10-1558, *ibid.*

39. Dubois, *Fidel,* 327; Brennan, *Castro, Cuba, and Justice,* 249–50.

40. *FRUS, 1958–1960,* VI, 154.

41. Daniel M. Braddock to DS, Desp 180, Aug. 22, 1958, 737.00/8-2258, which includes a MemoConv with Hormel, DSR, FOIA; Nicolás Arroyo y Márquez to John Foster Dulles, Aug. 27, 1958, 737.00/8-2758, *ibid.; FRUS, 1958–1960,* VI, 203–4; *Miami News,* Oct. 5, 1958.

42. *HAR* XI (Sept. 1958), 496, and (Oct. 1958), 552; *NYT,* Sept. 11, 1958.

43. Andrew St. George to Emanuel R. Freedman (*New York Times*), Oct. 22, 1958, Fdr Foreign Desk, Andrew St. George, 1958–59, New York Times Archives.

44. Braddock to DS, Desp 409, Oct. 20, 1958, 737.00/10-2058, DSR, FOIA.

45. Herter to HaEmb, Tel 127, Aug. 19, 1958, 737.00/8-1958, Box 3079, DSR, NA; Braddock to SS, Tel 231, Aug. 20, 1958, 737.00/8-2058, *ibid.*

46. Herter to HaEmb, Tel 121, Aug. 18, 1958, 737.00/8-1858, DSR, FOIA.

47. Aug. 8. 1958, *FRUS, 1958–1960,* VI, 188.

48. HaEmb to DS, Desp 429, Oct. 22, 1958, 611.37/10-2258, Box 2472, DSR, NA.

17. A Pox on Both Their Houses

1. Sept. 26, 1958, *FRUS, 1958–1960,* VI, 222.

2. British Embassy (Havana) to Selwyn Lloyd, No. 110, Oct. 23, 1958, AK 1015/52, Fdr 132165, FO 371, PRO.

3. The U.S. Embassy followed the elections in its Joint Weeka reports for November and December 1958 in 737.00 (W), DSR, FOIA. The embassy's overall

summary is HaEmb to DS, Desp 596, Dec. 15, 1958, 737.00/12-1558, *ibid.* See also *FRUS, 1958–1960,* VI, 249–50; Phillips, *Cuba,* 380–81.

4. Quoted in Phillips, *Cuba,* 381.

5. *PPP, DDE, 1958,* 831.

6. Smith, *Fourth,* 155.

7. Nov. 6, 1958, *FRUS, 1958–1960,* VI, 251 (quotation), 252n.

8. HaEmb to DS, Desp 556, Nov. 26, 1958, 737.00/11-2658, DSR, FOIA.

9. *FRUS, 1958–1960,* VI, 252–56 (quotation, p. 253).

10. *NYT,* Nov. 5, 11, 19, 22, 1958.

11. A. N. Noble (British Ambassador to Mexico) to Selwyn Lloyd, Dec. 17, 1958, AK10326/1, Fdr 132166, FO 371, PRO.

12. HaEmb to DS, Desp 513, Nov. 17, 1958, 737.00/11-1758, DSR, FOIA.

13. Unless otherwise noted, the quotations from this meeting are drawn from *FRUS, 1958–1960,* VI, 262–64.

14. Wieland and Smith quoted in Dorschner and Fabricio, *Winds,* 57–58.

15. CIA, "The Situation in Cuba," Special National Intelligence Estimate 85–58, Nov. 24, 1958, CIA Files, FOIA; *FRUS, 1958–1960,* VI, 265–66.

16. *FRUS, 1958–1960,* VI, 266–69.

17. For the events of November and early December, see the Joint Weeka reports in 737.00 (W) and Consul Park Wollam's periodic reports from Santiago de Cuba in 737.00, DSR, FOIA; Thomas, *Cuban Revolution,* 234–36; Bonachea and San Martín, *Cuban Insurrection,* 290–91; Bonachea and Valdés, *Revolutionary,* 432–42; Franqui, *Diary,* 443–61.

18. Bonachea and Valdés, *Revolutionary,* 435, 437.

19. A former North American M-26-7 fighter who became an academic historian at the University of Florida has calculated that the rebel army numbered 7,250 by December 20, 1958. Neill Macaulay, "The Cuban Rebel Army: A Numerical Survey," *Hispanic American Historical Review* LVIII (May 1978), 284–95.

20. Franqui, *Diary,* 445.

21. Dec. 2, 1958, *FRUS, 1958–1960,* VI, 276–77. Smith gives his account of this meeting in *Fourth,* 161–63. For other examples of talk about a military coup or junta, see *FRUS, 1958–1960,* VI, 259–60; Dubois, *Fidel Castro,* 329.

22. Wollam to DS, Desp 53, Nov. 21, 1958, 737.00/11-2158, DSR, FOIA.

23. Quoted in *NYT,* Nov. 4, 1958.

24. The saga of flight 495 is told in JW 45, Nov. 4, 1958, DSR, FOIA; JW 46, Nov. 11, 1958, *ibid.; HAR* XI (Nov. 1958), 608; Franqui, *Diary,* 445.

25. Wollam to DS, Desp 45, Nov. 12, 1958, 737.00/11-1258, DSR, FOIA; *HAR* XI (Nov. 1958), 608–9; *NYT,* Nov. 12, 1958.

26. Wollam Desp 53, Nov. 21, 1958.

27. *NYT,* Dec. 4, 1958.

28. Quoted in *WSJ,* Nov. 12, 1958.

29. For the travails of Americans and U.S. companies: HaEmb to SS, Tel 519, 521, and 522, Nov. 18, 1958, 737.00/11-1858, Box 3079, DSR, NA; Wollam to DS, Desp 54, Nov. 18, 1958, 737.00/11-1858, DSR, FOIA; MemoConv, "Activities of

Cuban Rebels," Nov. 26, 1958, 737.00/11-2658, *ibid.;* MemoConv, "Texaco Company's Problems in Operating Refinery in Santiago," Nov. 28, 1958, 837.2553/11-2858, *ibid.;* MemoConv, "Situation in Cuba," Dec. 8, 1958, 737.00/12-858, *ibid.;* MemoConv, "Bombing of American Sugar Property and General Cuban Situation," Dec. 10, 1958, 737.00/12-1058, *ibid.;* Wollam to DS, Desp 64, Dec. 15, 1958, 737.00/12-1558, *ibid.; HAR* XI (Nov. 1958), 609, and (Dec. 1958), 669; S. G. Menocal (Cuban Electric, Havana) to Division Managers and Department Heads, Dec. 15, 1958, Box 0144400, Boise Cascade Records.

30. Unidentified, quoted in *WSJ,* Dec. 23, 1958.

31. *FRUS, 1958–1960,* VI, 257.

32. *Ibid.,* 273.

33. JW 48, Nov. 25, 1958, DSR, FOIA; *FRUS, 1958–1960,* VI, 273.

34. FBIS *Daily Report,* Dec. 1, 1958, p. g1.

35. Wollam to DS, Desp 59, Dec. 3, 1958, 737.00/12-358, DSR, FOIA.

36. *NYT,* Dec. 11, 1958.

37. *FRUS, 1958–1960,* VI, 274–76.

38. Canadian Ambassador to SS for External Affairs, "Annual Review of Events in Cuba—1958," Jan. 20, 1959, File 10463-AH-40, Vol 2494, RG 25, NAC.

39. Howard J. Wiarda, "Friendly Tyrants and American Interests," in Daniel Pipes and Adam Garfinkle, eds., *Friendly Tyrants: An American Dilemma* (London, 1991), 10–11.

18. Pawley's Plot and Smith's Blow

1. Eisenhower comments in "Memorandum of Meeting with the President (Tuesday, November 29, 1960 at 11:00 a.m.)," Dec. 5, 1960, Box 5, OSANSA, WHO Records, DDEL.

2. Pawley's personal account of his mission to push Batista out appears in several sources: "William D. Pawley's Book," an undated, unpublished memoir which Anita Pawley allowed the author to read; Senate, *Communist Threat: Testimony of William D. Pawley,* Part 10 (Sept. 2, 8, 1960), 711–69; and William D. Pawley OH, Hoover Library, IA. Using interviews with Cubans and Americans, Dorschner and Fabricio in *Winds,* ch. 15, add detail to Pawley's trip.

3. Pawley's career is reported in *NYT,* Jan. 7, 1977; "Wizard at Work," *Time* LV (March 20, 1950), 36; various reports and memoranda titled "Subject: William D. Pawley," 1945–72, in File 12-79985, FBIR, FOIA.

4. See, for example, the following memoranda in FBIR, FOIA: Horton R. Telford (Legal Attaché, Paris) to FBI Director, Sept. 7, 1948; A. H. Belmont to D. M. Ladd, March 2, 1953; A. H. Belmont to L. V. Boardman, July 14, 1955; S. B. Donahoe to A. H. Belmont, Sept. 16, 1960. Braden counterattacked in Spruille Braden, *Diplomats and Demagogues: The Memoirs of Spruille Braden* (New Rochelle, NY, 1971), 379–81.

5. CIA, "Situation," Nov. 24, 1958; Memo, R. R. Roach to Mr. Belmont, Dec. 16, 1958, 97-3243, DBIR, FOIA.

6. Senate, *Communist Threat,* Part 10 (Sept. 1960), 746.

7. "Pawley's Book," ch. 19, pp. 9, 11.

8. Note, William A. Wieland to Charles Porter, nd but probably 1958, Box 15, Porter Papers.

9. Senate, *State Department Security: Wieland.*

10. Senate, *Communist Threat,* Part 10, p. 738.

11. "Pawley's Book," ch. 19, p. 13.

12. *Ibid.*

13. *Ibid.,* 14.

14. Gardner and Nixon knowledge suggested in Beschloss, *Crisis Years,* 94.

15. Lazo, *Dagger,* 157–58.

16. Dorschner and Fabricio, *Winds,* 93.

17. Bethel, *Losers,* 88–89.

18. FBIS *Daily Report,* Dec. 12, 1958, p. u1.

19. An undated biographical profile of Barquín, probably prepared by the Department of State, is located in Box 36, Porter Papers.

20. "Pawley's Book," ch. 19, pp. 17, 18.

21. *Ibid.,* 22.

22. Dorschner and Fabricio interviewed Güell on this point. *Winds,* 158–59.

23. "Pawley's Book," ch. 19, p. 15.

24. *Ibid.,* 18.

25. Smith, *Fourth,* 164–68.

26. "Pawley's Book," ch. 20.

27. S. B. Donahoe to A. H. Belmont, Jan. 20, 1960, File 62-79985, FBIR, FOIA.

28. "Pawley's book," ch. 22, pp. 11, 21.

29. *NYT,* Jan. 7, 1977.

30. *FRUS, 1958–1960,* VI, 282.

31. Allen J. Ellender's personal notes/diary of his trip to Cuba are located in his papers at the Ellender Library, Nicholls State University, Thibodaux, LA. The author received from the State Department under an FOIA request a slightly different version, "Diary of Senator Allen J. Ellender Written on His 1958 Foreign Aid Investigation Trip to Latin America." Some of the diary is published in *FRUS, 1958–1960,* VI, 285–89, and Senate, *A Review . . . Latin America,* 514–19.

32. Thomas A. Becnel, "Fulbright of Arkansas v. Ellender of Louisiana: The Politics of Sugar and Rice, 1937–1974," *Arkansas Historical Quarterly* XLIV (Winter 1984), 289–303.

33. For the press interview and other Ellender activities, see HaEmb to DS, Desp 633, Dec. 16, 1958, 033.1100-EL/12-1658, DSR, FOIA; HaEmb to DS, Desp 636, Dec. 17, 033.1100-EL/12-1758, *ibid.*

34. Dec. 15, 1958, *FRUS, 1958–1960,* VI, 291.

35. *The Times of Havana,* Dec. 13, 1958.

36. Smith, *Fourth,* 170–74 (quotations, p. 174); Batista, *Cuba Betrayed,* 96, 138–39.

37. Quoted in Dorschner and Fabricio, *Winds,* 200.

38. Roy M. Melbourne (Operations Coordinating Board) to James S. Lay, Jr. (NSC), Dec. 22, 1958, Box 18, Policy Planning Subseries, NSC Series, OSANSA,

WHOF, DDEL. The State Department's circular telegram went out December 8 and is reprinted in *FRUS, 1958–1960,* VI, 279–81. The DSR 737.00 file for December 1958 includes many responses, as well as other documents related to this subject. See also *FRUS, 1958–1960,* VI, 289, 307, 313, 323, 332; "Positions of American Republics on Cuban Situation," Dec. 30, 1958, Box 25, Lot File 61 D 411, DSR, NA.

39. DS to HaEmb, Tel 364, Dec. 23, 1958, 737.00/12-1958, DSR, FOIA; "Synopsis of Intelligence and State Material Reported to the President, December 30 material reported January 2, 1959," Box 37, DDE Diary Series, AWF, DDEP, DDEL; *FRUS, 1958–1960,* VI, 301.

40. For an approach by a pro-rebel priest in Santiago de Cuba that may have represented a rebel attempt to open talks with U.S. officials, see Wollam to SS, Tel 37, Dec. 15, 1958, 737.00/12-1558, DSR, FOIA; Wollam to DS, Desp 65, Dec. 16, 1958, 737.00/12-1658, *ibid.* This contact is summarized in *FRUS, 1958–1960,* VI, 311–12, 373. See also another contact involving Haydée Santamaría in *ibid.,* 181.

41. Pérez, *Army Politics,* 153–158.

42. Smith, *Fourth,* 177–78.

43. Dorschner and Fabricio, *Winds,* 232–33.

44. Wollam to DS, Desp 68, Dec. 22, 1958, 737.00/12-2258, DSR, FOIA.

45. Quoted in Dorschner and Fabricio, *Winds,* 200.

46. Franqui, *Diary,* 477. For the battles of December, see Bonachea and San Martín, *Cuban Insurrection,* ch. 13.

47. Franqui, *Diary,* 471.

19. U.S. Third-Force Conspiracies

1. Franqui, *Diary,* 469–70.

2. *Ibid.,* 469–70.

3. Javier Figueroa's interview with Raúl Chibás, San Juan, Puerto Rico, Jan. 7, 1985. Notes of the interview shared with the author.

4. Suárez, "Cuban Revolution," 20; Dorschner and Fabricio, *Winds,* 231.

5. The Cantillo-Castro meeting and Cantillo-Tabernilla maneuverings are described in Szulc, *Fidel,* 457–58; Bonachea and San Martín, *Cuban Insurrection,* 305–11; and Dorschner and Fabricio, *Winds, passim.* These works rely upon interviews with Cantillo and others.

6. Dec. 29, 1958, *FRUS, 1958–1960,* VI, 315n.

7. Quoted in Dorschner and Fabricio, *Winds,* 336.

8. Bonachea and San Martín, *Cuban Insurrection,* 164; Draper, *Castro's Cuba,* 34–35.

9. Justo Carrillo, "Información Histórica para Teodoro Draper," p. 43, nd, Box 23, Draper Papers.

10. Carrillo, in *ibid.,* gives December as the month, but a November CIA contact with Carrillo is mentioned in Memo for the Record, "First Meeting of General Maxwell Taylor's Board of Inquiry on Cuban Operations Conducted by CIA," April 23, 1961, Box 61A, Paramilitary Study Group, Countries-Cuba, NSF, JFKP.

11. Dorschner and Fabricio, *Winds,* 286, 291, 357; *FRUS, 1958–1960,* VI, 325.

12. For the Varona scheme, see Dorschner and Fabricio, *Winds,* 215–16, 251–52, 286, 311–13, 324, 338–39; *FRUS, 1958–1960,* VI, 325.

13. Quoted in Dorschner and Fabricio, *Winds,* 216.

14. Dec. 18, 1958, *FRUS, 1958–1960,* VI, 300.

15. *Ibid.,* 296.

16. Dec. 26, 1958, *ibid.,* 311.

17. Dec. 23, 1958, *ibid.,* 302–3. Eisenhower's displeasure with not having been told earlier about the desire to block Castro is also expressed in Dwight D. Eisenhower, *Waging Peace, 1956–1961* (Garden City, NY, 1965), 521.

18. Dec. 23, 1958, *FRUS, 1958–1960,* VI, 304–7.

19. The President said more, but part of this exchange with Gray remains security classified. Gordon Gray, "Memorandum of Conversation with the President (Wednesday, 24 December 1958, 9:45 a.m.)," Box 3, Presidential Subseries, Special Assistant Series, OSANSA, WHO Records, DDEL.

20. Dec. 26, 1958, *FRUS, 1958–1960,* VI, 311.

21. Irving P. Pflaum, *Tragic Island: How Communism Came to Cuba* (Englewood Cliffs, 1961), 1.

22. For this picture of December 30: John D. Eisenhower, "Synopsis of Intelligence and State material reported to the President," Dec. 29, 1958, Box 37, Goodpaster Briefings, DDE Diary Series, DDEP, DDEL; *BW,* 1532 (Jan. 10, 1959), 30; *FRUS, 1958–1960,* VI, 328; Dorschner and Fabricio, *Winds,* 336, 339; JW 53, Dec. 30, 1958, DSR, FOIA; *NYT,* Dec. 31, 1958.

23. Dec. 30, 1958, *FRUS, 1958–1960,* VI, 321.

24. *Ibid.*

25. Senate, *Executive Sessions, 1958,* X, 787. For the full hearing, pp. 767–800.

26. Dec. 31, 1958, *FRUS, 1958–1960,* VI, 323–29; Herter's handwritten notes in Box 4, Subject Subseries, NSC Series, OSANSA, WHO Records, DDEL.

27. Quoted in Neill Macaulay, *A Rebel in Cuba: An American Memoir* (Chicago, 1970), 126.

28. Quoted in Howard L. Lewis, "The Cuban Revolt Story: AP, UPI and 3 Papers," *Journalism Quarterly* XXXVII (Autumn 1961), 577. See also "Cuban Revolt Story Was Underplayed," *Editor and Publisher* XCII (Feb. 21, 1959), 56.

29. For the last hours of the Batista regime, see Dorschner and Fabricio, *Winds,* chs. 36–40; Matthews, *Revolution,* 119–20; Dubois, *Fidel,* 342–45; Bonachea and San Martín, *Cuban Insurrection,* 311–14.

30. Dec. 31, 1958, *FRUS, 1958–1960,* VI, 330–31.

31. Quoted in *ibid.,* 360.

32. Nancie Matthews (in Havana) to Eric Matthews, Jan. 1, 1959, Box 27, Matthews Papers.

33. Quoted in Dorschner and Fabricio, *Winds,* 385.

34. Phillips, *Cuba,* 396.

35. David A. Phillips, *The Night Watch* (New York, 1977), 77.

36. Batista, *Cuba Betrayed,* 64.

37. Quoted in Dorschner and Fabricio, *Winds,* 393.

38. Jan. 8, 1959, *FRUS, 1958–1960,* VI, 348.

39. The Carrillo-CIA story is told in Justo Carrillo, "Información complementaria para Teodoro Draper," 43–45, nd, Box 23, Draper Papers.

40. Dorschner and Fabricio, *Winds,* 338.

41. Justo Carrillo, "Vision and Revision: U.S.-Cuban Relations, 1902 to 1959," in Jaime Suchlicki et al., eds., *Cuba: Continuity and Change* (Miami, 1985), 166–67.

42. Dorschner and Fabricio, *Winds,* 369, 404, 427–28, 505; *NYT,* July 2, 1959; Thomas, *Cuban Revolution,* 445, 518.

43. *NYT,* Feb. 11, June 18, July 2, Sept. 3, Dec. 22, 1959, Jan. 26, June 12, Oct. 30, 1961, April 6, 1977; Jaime Suchlicki, *Historical Dictionary of Cuba* (Metuchen, NJ, 1988), 232–33.

44. Quoted in Dubois, *Fidel,* 345.

45. Szulc, *Fidel,* 458.

46. Stevenson OH, 29.

20. Madhouse: Castro's Victory, Smith's Defeat

1. For the events of early January, see Phillips, *Cuba,* 396–400; Dorschner and Fabricio, *Winds,* chs. 41–55; Matthews, "Matthews, Under Gunfire"; Thomas, *Cuban Revolution,* 245–52; Bonachea and Valdés, *Revolutionary,* 446–49; Bonachea and San Martín, *Cuban Insurrection,* 313–28; Daniel M. Braddock to DS, Desp 736, "The First Two Weeks of the Revolutionary Government," Jan. 15, 1959, 737.00/1-1559, DSR, FOIA; *FRUS, 1958–1960,* VI, 334–46; Franqui, *Diary,* ch. 18; Nicolás Rivero, *Castro's Cuba: An American Dilemma* (Washington, DC, 1962), 197; *NYT* (daily).

2. Phillips, *Cuba,* 396–97.

3. Quoted in Smith, *Closest,* 40.

4. Nancie Matthews to Eric Matthews, letter of Jan. 1–8, 1959, Box 27, Matthews Papers.

5. A. S. Fordham to Selwyn Lloyd, Jan. 9, 1959, AK1015/20, Fdr 139398, FO 371, PRO.

6. Franqui, *Diary,* 501.

7. Smith to SS, Tel 682, Jan. 1, 1959, 737.00/1-159, DSR, NA.

8. Casuso, *Cuba and Castro,* 143–47.

9. Smith, *Closest,* 39.

10. Lacey, *Little Man,* 250–51.

11. Bethel, *Losers,* 80–81; MemoConv, "The Cuban Situation," Jan. 2, 1959, 737.00/1-259, DRS, FOIA; "Various Telephone Conversations Regarding Cuba," Jan. 2, 1959, *ibid.;* U.S. Army Staff Communications Office to CINCLANT, Jan. 5, 1959, CCS 9123/9440, JCS Records; Smith, *Closest,* 40–41; House Representative Thomas B. Curtis to author, Jan. 6, 1984.

12. Rubottom in "Various," Jan. 2, 1959.

13. CINCLANT to JCS, Jan. 2, 1959, CCS 9123/3204, JCS Records, NA; "The Sailing of Ships . . . ," memo by Arleigh Burke, Jan. 3, 1959, *ibid.;* Tel, CINCLANT to DS, Jan. 4, 1959, No. 1980-62C, DDRS; Memo for the President, Jan. 6, 1959, Box

25, Administration Series, AWF, DDEP; "Evacuation of American Citizens from Cuba," Gerald G. Jones to Files, Jan. 13, 1959, Box 25, Lot 61 D 411, DSR, NA.

14. Franqui, *Diary,* 487; Dorschner and Fabricio, *Winds,* 400.

15. Quoted in Dubois, *Fidel,* 347.

16. USARMA Havana to ACSI DA, Jan. 1, 1959, C-3-59, Box 319, ACSI, Army Records.

17. "Synopsis of Intelligence and State material reported to the President," Jan. 2, 1959 material reported Jan. 3, Box 14, Intelligence Briefings Notes, Alphabetical Subseries, Subject Series, Office of the Staff Secretary, WHO Records, DDEL. See also Jan. 1, 1959, *FRUS, 1958–1960,* VI, 338.

18. Carlos Franqui, *Family Portrait with Fidel* (New York, 1984), 6.

19. Barquín gives a detailed account of his brief ascension to power in Ramón M. Barquín, *El día que Fidel Castro se apoderó de Cuba: 72 horas trágicas para la libertad en las Américas* (San Juan, PR, 1978).

20. Quoted in Dorschner and Fabricio, *Winds,* 446. The Cantillo-Barquín-Carrillo episode is told in *ibid.,* 430–31, 444–49, 459–61; Carrillo, "Información complementaria."

21. Wollam to SS, Jan. 1, 1959, 737.00/1-159, DSR, FOIA.

22. Franqui, *Diary,* 489–91.

23. Barquín, *El día,* 70–73.

24. JW 1 (mismarked as 54), Jan. 7, 1959, DSR, FOIA.

25. Quoted in Dorschner and Fabricio, *Winds,* p. 461.

26. Barquín, *El día,* p. 73.

27. Quoted in Rivero, *Castro's Cuba,* 207.

28. Neill Macaulay, "I Fought for Fidel," *American Heritage* XLII (Nov. 1991), 88.

29. HaEmb to SS, Tel 862, Jan. 20, 1959, 737.00/1-2059, Box 3081, DSR, NA (quotation); Braddock to SS, Tel 879, Jan. 23, 1959, 737.00/1-2359, *ibid.*

30. "Synopsis of Intelligence and State material reported to the President," Jan. 6, 1959 material reported Jan. 6, Box 14, Intelligence Briefing Notes, Alphabetical Subseries, Subject Series, WHO Records, DDEL.

31. Braddock, "First Two Weeks," 5. See also *FRUS, 1958–1960,* VI, 370–72.

32. Quoted in Dorschner and Fabricio, *Winds,* 474. For a brief depiction of his entrance into Havana, see Manuel Urrutia Lleó, *Fidel Castro & Company, Inc.: Communist Tyranny in Cuba* (New York, 1964), 31.

33. Smith, *Fourth,* 202.

34. On the BRAC case, see Bethel, *Losers,* 93–95; Dorschner and Fabricio, *Winds,* 479–80.

35. Jan. 6, 1959, *FRUS, 1958–1960,* VI, 345.

36. *Revolución,* Jan. 5, 10, 20, 1959.

37. Braddock to SS, Tel 756, Jan. 6, 1959, 837.2614/1-659, DSR, FOIA.

38. A. H. Belmont to the Director, Jan. 5, 1959, FBIR, FOIA.

39. Quoted in *WSJ,* Jan. 8, 1959.

40. Jan. 7, 1959, *FRUS, 1958–1960,* VI, 347.

41. Senate, *State Department Security* (Feb. 2, 1962), Part 5, pp. 666–67.

42. For Castro's march to Havana, see Dubois, *Fidel,* 352–63; Nancie Matthews to Eric Matthews, letter of Jan. 1–8, 1959; Szulc, *Fidel,* 465–69; Dorschner and Fabricio, *Winds,* ch. 55.

43. Franqui, *Diary,* 495, 501.

44. Quoted in Dubois, *Fidel,* 354.

45. Sullivan also wrote a newspaper column which criticized Ambassador Smith for having "swallowed Batista's propaganda hook, line and sinker." *New York Daily News,* Jan. 12, 1959.

46. Matthews to Matthews, Jan. 1–8, 1959.

47. A. S. Fordham to Selwyn Lloyd, Jan. 10, 1959, AK1015/21, Fdr 139398, FO 371, PRO.

48. Paar, *I Kid,* 212–13.

49. Quoted in Casuso, *Cuba and Castro,* 156.

50. HaEmb to SS, Tel 840, Jan. 16, 1959, 611.37/1-1659, DSR, FOIA; Dubois, *Fidel,* 367.

51. Braddock, "First Two Weeks."

52. MemoComv, "Cuba," Jan. 8, 1959, 737.00/1-859, DSR, FOIA.

53. Franqui, *Family,* 5.

54. Jan. 30, 1959, *FRUS, 1958–1960,* VI, 388.

55. *NYT,* Jan. 26, 1959.

56. Quoted in Brown, "Cuban Baseball," 112. In 1960, however, after Cuban-American relations had deteriorated, the Sugar Kings departed Havana and became the home club for Jersey City.

57. Quoted in Lacey, *Little Man,* 252.

58. HaEmb to DS, Desp 889, Feb. 12, 1959, 837.45/2-1259, Box 4375, DSR, NA; Phillips, *Cuban Dilemma,* 49–53.

59. Legal Attaché, Havana, to FBI Director, April 29, 1959, Errol Flynn File 92 2781-95, FBIR, J. Edgar Hoover Building, Washington, DC.

60. Smith, *Fourth,* 196–98.

61. Constantine N. Kangles to John Foster Dulles, Jan. 5, 1959, Box A789, File 57, Series BCN 140, Fulbright Papers.

62. A. S. Fordham to Selwyn Lloyd, June 25, 1958, AK1902/1, Fdr 132189, FO 371, PRO.

63. Quoted in G. Bernard Noble, *Christian A. Herter* (New York, 1970), 178.

64. *FRUS, 1958–1960,* VI, 344.

65. "Telephone Call from the President," Jan. 7, 1959, 11:17 a.m, Box 13, Telephone Conversations Series, Dulles Papers, DDEL.

66. Jan. 8, 1959, *FRUS, 1958–1960,* VI, 348.

67. "Telephone Call to the President," Jan. 9, 1959, 2:56 p.m., Box 13, Telephone Conversations Series, Dulles Papers, DDEL.

68. *FRUS, 1958–1960,* VI, 354–55.

69. Smith, *Fourth,* 206.

70. MemoPhone, Sept. 12, 1960, Box 551, Senate Files, Pre-Presidential Papers, JFKP.

71. Chester Bowles OH, JFKL.

72. *NYT,* Feb. 15, 1991; Beschloss, *Crisis Years,* 100.

21. A Complete Break

1. JW 2, Jan. 13, 1959, DSR, FOIA.

2. Feb. 12, 1959, *FRUS, 1958–1960,* VI, 398.

3. "Fidel Castro, Cuban Revolutionary Leader," Jan. 9, 1959, Staff Summary Biographic Supplement, Box 19, Staff Research Group, WHO, DDEL.

4. Braddock, "First Two Weeks." Also see Jan. 20, March 17, 1959, *FRUS, 1958–1960,* VI, 378, 431; "Guevara, Ernesto ('Che')," Jan. 1959, DS, Division of Biographic Information, Box 36, Porter Papers.

6. "Dynamic Boss Takes Over a U.S. Neighbor," *Life* XLVI (Jan. 12, 1959), 14 ("austere"); S. J. Rundt, "The Last Plane from Habana," *Export Trade* LXXVIII (Jan. 12, 1959), 27 ("U.S.-type"); Braddock, "First Two Weeks" ("danger").

6. "Synopsis of State and Intelligence material reported to the President," Jan. 12, 1959, Box 38, Goodpaster Briefings, DDE Diary Series, AWF, DDEP.

7. *HAR* XV (May 1960), 412; Ruth Pearson, "Cuba's Double Jeopardy," *Bulletin of the Atomic Scientists* XLVIII (Dec. 1992), 49.

8. Braddock, "First Two Weeks," 1.

9. Quoted by González, *Cuba Under Castro,* 53, from *Revolución,* Jan. 14, 1959. For other Castro attacks on the Platt amendment for stealing the nation's freedom, see *El pensamiento de Fidel Castro: Selección temática, enero 1959–abril 1961* (La Habana, 1983), vol. I, book I, pp. 4–5, 9; book II, pp. 388, 607; *FRUS, 1958–1960,* VI, 383.

10. June 1959 statement quoted in Benjamin, *United States and Origins,* 182.

11. Statement of July 1959 quoted in Padula, "Fall," 548.

13. Rubottom in Senate, *Executive Sessions, 1958,* X, 773.

13. Stevenson OH, 15.

14. Statement of Jan. 26, 1959, in U.S. Senate, Foreign Relations Committee, *Executive Sessions of the Senate Foreign Relations Committee* (*Historical Series*), XI, *1959* (Washington, DC, 1982), 125.

15. Operations Coordinating Board, "Special Report on Latin America (NSC 5613/1)," Nov. 26, 1958, Box 18, NSC Series, OSANA, WHO Records, DDEL.

16. Unidentified quoted in A. S. Fordham to FO, July 4, 1958, AK1015/34, Fdr 132165, FO 371, PRO.

17. Quoted in J. L. Pimsleur, "Civil War in Cuba," *The New Leader* XL (Sept. 30, 1957), 5.

18. For the tenacity of nationalism and its strong anti-American component, see Introduction above. For a contrary view, that nationalistic sentiment had subsided by the early 1950s, see Domínguez, *To Make,* 9.

19. Author's interview with Juan Antonio Blanco, Cuban historian and diplomat, Storrs, CT, April 13, 1993.

20. For "even-handed," see Cole Blasier, *The Hovering Giant: U.S. Responses to Revolutionary Change in Latin America* (Pittsburgh, 1976), 32; for "passive," see Jules R. Benjamin, "Interpreting the U.S. Reaction to the Cuban Revolution, 1959–1960," *Cuban Studies 19* (Pittsburgh, 1989), 146.

21. Quoted in William Attwood, *The Twilight Struggle: Tales of the Cold War* (New York, 1987), 254.

22. Gordon Chase and John Plank, "U.S./Cuban Relations—January 2, 1959 to January 3, 1961," p. 2, Box 5, Plank/Chase Cuban Project, Files of Gordon Chase, NSF, LBJL.

23. Harry S. Truman to Dean Acheson, Oct. 12, 1962, Box 31, Series I, Dean Acheson Papers, Yale University Library.

24. CIA, "Situation in Cuba," Nov. 24, 1958.

25. Aaron, "Seizure."

26. *FRUS, 1958–1960,* VI, 435.

27. Quoted in Pérez, *Army Politics,* 155.

28. Smith, *Closest,* 34.

29. Richard Cushing to the author, April 29, 1985.

30. For the question of U.S. decline, see Paterson, *On Every Front,* chs. 9–10.

31. See, for example, *FRUS, 1958–1960,* VI, 263–64.

32. Richard M. Nixon, *Six Cases* (New York, c. 1962, 1968), 247; Milton S. Eisenhower, *The Wine Is Bitter: The United States and Latin America* (Garden City, NY, 1963), 78.

33. DS, "Draft White Paper on Cuba," 2.

34. A question raised in unidentified author, "American Political Strategy."

35. Quoted in Morley, *Imperial State,* 43.

36. Robert U. Brown, "Shop Talk at Thirty," *Editor and Publisher* XCII (Jan. 31, 1959), 68; "Cuban Revolt Story Was Underplayed," *ibid.,* Feb. 21, 1959), 56.

37. See NSC Summaries of Discussion in NSC Series, AWF, DDEP.

38. Emphasis added. Chase and Plank, "U.S./Cuban Relations," 32.

39. Gleijeses, *Shattered Hope,* 361.

40. U.S. policy as set by the National Security Council permitted unilateral military intervention in the event of the communization of a Latin American state. See Stephen G. Rabe, "The Johnson (Eisenhower?) Doctrine for Latin America," *Diplomatic History* IX (Winter 1985), 95–100.

41. Senate, *Executive Sessions, 1958,* X, 787. For Wieland's similar view that Castro might be subject to the moderates, see Senate, *State Department Security,* Part 5, p. 628.

42. Quoted in Matthews, *Fidel Castro,* 112.

43. Pardo Llada was an Ortodoxo party leader and radio commentator who later broke with the Castro revolution and went into exile in Colombia. Quoted in Luis Báez, *Los que se fueron* (La Habana, 1991), 123–24.

44. "Avowed friend" in Braddock, "First Two Weeks"; Topping quoting López-Fresquet in Dorschner and Fabricio, *Winds,* 73.

45. Urrutia quoted in López-Fresquet, *My 14 Months,* 12.

46. MemoConv, Oct. 14, 1958, 737.00/10-1458, DSR, FOIA.

47. A. F. Montero, "1958 Year End Report on Cuba," Dec. 28, 1958, Bank of America Archives, San Francisco, CA. See also *BW,* No. 1532 (Jan. 10, 1959), 84.

48. Feb. 17, 1959, *FRUS, 1958–1960,* VI, 400.

49. Llerena, *Unsuspected,* 60.

50. Padula, "Fall," 5; MemoConv, March 24, 1959, 737.00/3-2459, Box 3081, DSR, NA.

51. Quoted in Bonsal, *Cuba, Castro,* 6.

52. Llerena, *Unsuspected,* 60.

53. Quoted in *WSJ,* Sept. 6, 1957.

54. Quoted in Lockwood, *Castro's Cuba,* 159.

55. For an example of such thinking, see *FRUS, 1958–1960,* VI, 297.

56. Braddock, Feb. 18, 1959, *ibid.,* 401.

22. The United States and Cuba in the Castro Era

1. Rubottom, *ibid.,* 656; unidentified official quoted in Morley, *Imperial,* 114.

2. Quoted in Dubois, *Fidel,* 372.

3. Braddock to SS, Tel 874, Jan. 22, 1959, 737.00/1-2259, DSR, FOIA.

4. Jan. 19, 1959, *FRUS, 1958–1960,* VI, 373.

5. "Threats to the Stability of the US Military Facilities Position in the Caribbean Area and Brazil," SNIE 100-3-59, March 10, 1959, CIA Records, FOIA.

6. Daniel Braddock, Feb. 18, 1959, *FRUS, 1958–1960,* VI, 401.

7. William Wieland, Feb. 19, 1959, *ibid.,* 405.

8. April 14, 1959, *ibid.,* 455.

9. April 14, 1959, *ibid.,* 457.

10. March 26, 1959, *ibid.,* 441.

11. State Department records do not reveal U.S. zeal for loans for Cuba, and Rubottom indicates U.S. coolness to the unidentified author of "American Political Strategy." Pazos quoted at length in Carrillo, "Vision," 172–73.

12. April 14, 1959, *FRUS, 1958–1960,* VI, 450–51, 469–70 (quotations).

13. March 26, 1959, *ibid.,* 442.

14. Quoted in Morley, *Imperial,* 79. See also Casuso, *Cuba and Castro,* 216; Szulc, *Fidel,* 487.

15. López-Fresquet, *My 14 Months,* 106.

16. Suárez, *Cuba,* 48n.

17. "Meeting with . . . Fidel Castro," April 23, 1959, vol. XXIX, Records of Meetings, Council on Foreign Relations Archives.

18. Jeffrey Safford, "The Nixon-Castro Meeting of 19 April 1959," *Diplomatic History* IV (Fall 1980), 426–31.

19. Quoted in Stevenson OH, 23.

20. April 18, 1959, *FRUS, 1958–1960,* VI, 475.

21. Rivero, *Castro's Cuba,* 18. See also *FRUS, 1958–1960,* VI, 387, 455, 466.

22. Smith, *Closest,* 49–50.

23. JW 4, Jan. 27, 1959, DSR, FOIA; Caracas Embassy, Jan. 30, 1959, *FRUS, 1958–1960,* VI, 387 (quotation).

24. Severo Aguirre (Cuban delegate to the 21st Communist Party of the Soviet Union Congress), Jan. 31, 1959, in Clissold, *Soviet Relations,* 254.

25. Bonsal to SS, Nov. 6, 1959, 611.37/11-659, Box 2473, DSR, NA; Eisenhower quoted in Trumbull Higgins, *The Perfect Failure: Kennedy, Eisenhower, and the CIA at the Bay of Pigs* (New York, 1987), 48. For 1959–61 events, see Rabe, *Eisenhower;* Stephen E. Ambrose with Richard H. Immerman, *Ike's Spies* (Garden City, NY, 1981), 303–16; Welch, *Response;* Blasier, *Hovering Giant.*

26. "Why the Cuban Revolution of 1958 Led to Cuba's Alignment with the USSR," Feb. 21, 1961, CIA Records, FOIA.

27. Nov. 5, 1959, *FRUS, 1958–1960,* VI, 656–57.

28. Nov. 20, 1959, *ibid.,* 676.

29. Quoted in Smith, *Closest,* 64.

30. For the Bay of Pigs, see Higgins, *Perfect Failure; Haynes Johnson, The Bay of Pigs* (New York, 1964); Karl Meyer and Tad Szulc, *The Cuban Invasion* (New York, 1962); Lucien S. Vandenbroucke, "Anatomy of a Failure: The Decision to Land at the Bay of Pigs," *Political Science Quarterly* XCIX (Fall 1984), 471–91; Peter Wyden, *Bay of Pigs* (New York, 1979).

31. The Kennedy campaign against Castro is detailed in Thomas G. Paterson, "Fixation with Cuba: The Bay of Pigs, Missile Crisis, and Covert War Against Fidel Castro," in Thomas G. Paterson, ed., *Kennedy's Quest for Victory: American Foreign Policy, 1961–1963* (New York, 1989), 123–55, 343–52; Thomas G. Paterson, "Commentary: The Defense-of-Cuba Theme and the Missile Crisis," *Diplomatic History* XIV (Spring 1990), 249–56. I have drawn from these works for the next several paragraphs. For searching analysis of the events that led to the missile crisis, see the several essays in James A. Nathan, ed., *The Cuban Missile Crisis Revisited* (New York, 1992).

32. Bonsal, *Cuba, Castro,* 192.

33. Quoted in Pierre Salinger, "Gaps in the Cuban Missile Crisis Story," *NYT,* Feb. 5, 1989.

34. *DSB* XLVII (Oct. 22, 1962), 592–94.

35. "The Cuba Project," Program Review by Brig. Gen. Lansdale, Feb. 20, 1962, National Security Archive, Washington, DC.

36. Quoted in Bill Keller, "Papers Show 1962 U.S. Plan Against Castro," *NYT,* Jan. 27, 1989.

37. "Re: Anti-Fidel Castro Activities; Internal Security—Cuba," Oct. 5, 1962, FBI Miami Office, FBIR, FOIA.

38. Quoted in Frank Mankiewicz and Kirby Jones, *With Fidel: A Portrait of Castro and Cuba* (New York, 1975), 130.

39. Statement of July 26, 1962, FBIS, "Radio Propaganda Report: Castro on the Normalization of U.S.-Cuba Relations," March 2, 1964, Box 21, Cuba Country File, NSF, LBJL.

40. "Review of the Cuban Situation and Policy," Feb. 28, 1963, Box 115, POF, JFKL.

41. "Lisa Howard Interview of Fidel Castro," 1963, Box 23-I-9-10F, Hubert H. Humphrey Papers, Minnesota Historical Society, St. Paul; Lisa Howard, "Castro's Overture," *War/Peace Report,* Sept. 1963, pp. 3–5. Also, Arthur M. Schlesinger, Jr., *Robert F. Kennedy and His Times* (Boston, 1978), 541.

42. Ernest Halperin, *The Rise and Decline of Fidel Castro* (Berkeley, CA, 1972), 210–46.

43. DS to American Embassy, Paris, May 11, 1963, DSR, FOIA; Foy Kohler (Moscow) to SS, July 31, 1963, *ibid.*

44. John A. McCone, Memo for NSC Standing Group Members, May 1, 1963, CIA Records, FOIA.

45. Reports by George Volsky (USIA-Miami), Aug. 16, 23, Sept. 16, and Oct. 4, 25, 1963, Box 26, Draper Papers; D. J. Brennan to W. C. Sullivan, "Anti-Fidel Castro Activities," Nov. 4, 1963, FBIR, FOIA.

46. William Attwood, *The Reds and the Blacks* (New York, 1967), 142–46; Attwood, *Twilight Struggle,* 257–64; Schlesinger, *Robert F. Kennedy,* 551–52, 556.

47. Jean Daniel, "Unofficial Envoy: An Historic Report from Two Capitals," *The New Republic* CXLIX (Dec. 14, 1963), 20; Tad Szulc, "Friendship Is Possible, But . . . ," *Parade Magazine,* April 1, 1984, p. 6.

48. *PPP, Kennedy, 1963,* 876.

49. "Meeting with the President," Dec. 19, 1963, Box 19, Aides Files-Bundy, NSF, LBJL.

50. Jean Daniel, "When Castro Heard the News," *The New Republic* CXLIX (Dec. 7, 1963), 7.

51. U.S. Congress, Senate, Select Committee to Study Governmental Operations with Respect to Intelligence Activities, *Alleged Assassination Plots Involving Foreign Leaders: An Interim Report* (Washington, DC, 1975), 88–89.

52. Quoted in Herbert L. Matthews diary of trip to Cuba, Oct. 24–Nov. 3, 1963, Box 27, Matthews Papers.

53. Verbatim record of telephone conversation, Bundy and George Ball, Dec. 18, 1963, Box 2, George Ball Papers, LBJL.

54. "Meeting with the President," Dec. 19, 1963; Attwood, *Reds and Blacks,* 146.

55. For events since the early 1960s, see Cole Blasier and Carmelo Mesa-Lago, eds., *Cuba in the World* (Pittsburgh, 1979); Philip Brenner, *From Confrontation to Negotiation: U.S. Relations with Cuba* (Boulder, CO, 1988); Lynn Darrell, *The Politics of Hostility: Castro's Revolution and United States Policy* (Hato Rey, PR, 1975); del Aguila, *Cuba;* Domínguez, *Cuba;* Domínguez, *To Make;* Jorge I. Domínguez and Rafael Hernández, eds., *U.S.-Cuban Relations in the 1990s* (Boulder, CO, 1989); H. Michael Erisman, *Cuba's International Relations* (Boulder, CO, 1985); Pamela S. Falk, *Cuban Foreign Policy: Caribbean Tempest* (Lexington, MA, 1986); Howard H. Frederick, *Cuban-American Radio Wars: Ideology in International Telecommunications* (Norwood, NJ, 1986); Gary C. Hufbauer and Jeffrey J. Schott, *Economic Sanctions Reconsidered: History and Current Policy* (Washington, DC, 1990; 2nd ed.); Francisco López Segrera,

"Cuba-EEUU: Percepciones mutuas (1959–1993)," paper, March 9, 1993, Paris, France; Francisco López Segrera, *Cuba: Política exterior y revolución 1959–88* (La Habana, 1989); Morley, *Imperial State;* Pérez, *Cuba and the United States;* John Plank, ed., *Cuba and the United States* (1967); Carla Anne Robbins, *The Cuban Threat* (New York, 1983); Smith, *Closest;* Wayne S. Smith and Esteban Morales Domínguez, eds., *Subject to Solution: Problems in Cuba-U.S. Relations* (Boulder, CO, 1988).

56. Marilyn B. Young, *The Vietnam War, 1945–1990* (New York, 1991), 99.

57. For opposing views on comparative dependency, see William M. LeoGrande, "Cuban Dependency: A Comparison of Pre-Revolutionary and Post-Revolutionary International Economic Relations," *Cuban Studies/Estudios Cubanos* IX (July 1979), 1–28, and Robert A. Packenham, "Capitalist and Socialist Dependency: The Case of Cuba," in Jan F. Triska, ed., *Dominant Powers and Subordinate States: The United States in Latin America and the Soviet Union* (Durham, NC, 1986), 310–41.

58. Wayne S. Smith, "U.S.-Cuba Relations: Twenty-five Years of Hostility," in Sandor Halebsky and John M. Kirk, eds., *Cuba: Twenty-Five Years of Revolution, 1959–1984* (New York, 1985), 333–51.

59. Castro quoted in National Information Agency (Cuba), *Aggression over the Airwaves* (Havana, 1989), 28.

60. *NACLA Report on the Americas* XXIV (Aug. 1990), 12.

61. *WP National Weekly Edition,* Jan. 23–29, 1989 (Jeff MacNelly); Jan. 1–7, 1990 (Herblock); March 12–18, 1990 (Jeff MacNelly).

62. Quoted in *Hartford Courant,* Sept. 23, 1991.

63. Andres Oppenheimer, *Castro's Final Hour* (New York, 1992).

64. Quoted in *NYT,* Oct. 12, 1992.

65. Quoted in *WP National Weekly Edition,* July 5–11, 1993.

66. *Ibid.*

67. *Ibid.,* August 2–8, 1993; *NYT,* Sept. 26, 30, Oct. 1, 6, 7, 1993; *Hartford Courant,* Sept. 19, 1993.

SOURCES

Unpublished Personal Papers and Records

Dean Acheson Papers, Yale University Library, New Haven, CT
George Aiken Papers, University of Vermont Library, Burlington
Hanson Baldwin Papers, Yale University Library
Bangor Punta Corporation Files, Greenwich, CT
Bank of America Archives, San Francisco, CA
Samuel Flagg Bemis Papers, Yale University Library
Boise Cascade Records, Boise Cascade Company, Boise, ID
Chester Bowles Papers, Yale University Library
Braga Brothers Collection, University of Florida Library, Gainesville
Raúl Chibás Collection, Hoover Institution Archives, Stanford, CA
John Sherman Cooper Papers, University of Kentucky Library, Lexington
Council on Foreign Relations Archives, New York, NY
Cuban Collection, Andrew St. George Papers, Yale University Library
Theodore Draper Papers, Hoover Institution Archives
John Foster Dulles Papers, Princeton University Library, Princeton, NJ
Allen J. Ellender Papers, Nicholls State University Library, Thibodaux, LA
Ford Industrial Archives, Redford, MI
J. William Fulbright Papers, University of Arkansas Library, Fayetteville
Bourke Hickenlooper Papers, Herbert Hoover Presidential Library, West Branch, IA
Robert C. Hill Papers, Hoover Institution Archives
Hubert H. Humphrey Papers, Minnesota Historical Society, St. Paul
Institute of Hispanic American and Luso-Brazilian Studies Papers, Hoover Institution Archives
Estes Kefauver Papers, University of Tennessee Library, Knoxville
King Ranch, Inc., Files, Kingsville, Texas

Frank J. Lausche Papers, Ohio Historical Society, Columbus
Herbert L. Matthews Papers, Columbia University Library, New York, NY
Walter Lippmann Papers, Yale University Library
Wayne L. Morse Papers, University of Oregon Library, Eugene
National Security Archive, Washington, DC
New York Times Archives, New York, NY
Charles O. Porter Papers, University of Oregon Library
Walter P. Reuther Collection, Archives of Labor and Urban Affairs, Wayne State University, Detroit, MI
Eric Sevareid Papers, Library of Congress, Washington, DC
George A. Smathers Papers, University of Florida Library
Lawrence E. Spivak Papers, Library of Congress
United States Beet Sugar Association Records, Washington, DC

Papers and Records at U.S. Archives and Presidential Libraries

*(*Indicates some agency records received under Freedom of Information Act)*

Central Intelligence Agency Records, Washington, DC*
Dwight D. Eisenhower Library, Abilene, Kansas
- Robert Cutler Papers
- John Foster Dulles Papers
- Dwight D. Eisenhower Papers, Ann Whitman File
- Milton S. Eisenhower Papers
- Gordon Gray Papers
- James C. Hagerty Papers
- Christian A. Herter Papers
- C. D. Jackson Papers
- Gerald D. Morgan Records
- Don Paarlberg Papers
- White House Central Files
- White House Office Records

Federal Archives and Records Center, Laguna Niguel, CA
- Richard M. Nixon Papers, Pre-Presidential Papers

Federal Bureau of Investigation Records, Washington, DC*
- Jules Dubois File
- Errol Flynn File
- Samuel Giancana File
- Internal Security, Cuba File
- William D. Pawley File

Lyndon B. Johnson Library, Austin, TX
- George Ball Papers
- National Security Files

John F. Kennedy Library, Boston MA
John F. Kennedy Papers
National Security Files*
Pre-Presidential Papers, Sentate Files
President's Office Files
Arthur M. Schlesinger, Jr., Papers
National Archives of the United States, Washington, DC
Department of the Army, Army Staff, Assistant Chief of State, Intelligence, Records, Record Group 319*
Department of State Records, Record Group 59*
Decimal Files
Lot Files
Public Studies Division, Office of Public Opinion Studies
Joint Chiefs of Staff Records, Record Group 218
Joint Commands, Caribbean Command Records, Record Group 349*
Senate Foreign Relations Committee Records, Record Group 46
Naval Historical Center, Washington Navy Yard, Washington, DC
Chief of Naval Operations/Admiral Burke Files
"Historical Record of the U.S. Naval Mission to Cuba"
Franklin D. Roosevelt Library, Hyde Park, NY
Adolf A. Berle Papers

Foreign Archives

National Archives of Canada, Ottawa, Ontario
Department of External Affairs Records, Record Group 25
Department of Industry, Trade and Commerce Records, Record Group 20
Public Record Office, London, Great Britain
Foreign Office Correspondence (FO 371)

Oral Histories and Interviews

William Attwood, Foreign Affairs Oral History Program (FAOHP), Lauinger Library, Georgetown University, Washington, DC
Chester Bowles, Kennedy Library
Richard G. Cushing, FAOHP
Milton S. Eisenhower, Columbia University Oral History Project, New York, NY
J. William Fulbright, Columbia University
Andrew J. Goodpaster, Columbia University
Gordon Gray, Columbia University
Robert C. Hill, Columbia University
Pat M. Holt, Senate Historical Office, Washington, DC

Edward S. Little, FAOHP
Thomas C. Mann, Columbia University
Thomas C. Mann, Eisenhower Library
Thomas C. Mann, Johnson Library
Livingston Merchant, Eisenhower Library
William D. Pawley, Hoover Library
Roy R. Rubottom, Jr., FAOHP
George A. Smathers, Kennedy Library
George A. Smathers, Senate Historical Office
Robert A. Stevenson, FAOHP

Interviews Conducted by Thomas G. Paterson

Juan Antonio Blanco, April 13, 1993, Storrs, CT
John H. Crimmins, February 4, 1985, Washington, DC
Kelley V. Holbert, May 28, 1985, Grants Pass, OR
Francisco López Segrera, April 9, 1985, La Habana, Cuba
Hugo Pons, April 10, 1985, La Habana, Cuba
G. Harvey Summ, May 28, 1985, Washington, DC
Carlos Ciaño Zanetti, April 12, 1985, La Habana, Cuba

Letters to Thomas G. Paterson

R. H. Bentley, April 30, 1985
Thomas B. Curtis, January 6, 1984
Richard Cushing, April 29, 1985
Eugene Desvernine, April 29, 1985
Wright H. Ellis, July 23, 2985
Constantine N. Kangles, March 28, 1984
J. Kevin O'Brien, December 13, 1991
R. Roy Rubottom, Jr., December 12, 1991
Robert A. Stevenson, March 22, 1993
Park F. Wollam, September 13, 1992; May 17, 1993

Published U.S. Government Documents

Congress, Senate

Committee on Appropriations, *A Review of United States Government Operations in Latin America* (by Allen J. Ellender), 1959.

Committee on Foreign Relations, *Events in United States-Cuban Relations: A Chronology, 1957–1963,* 1963.

Committee on Foreign Relations, *Executive Sessions, 1958, Historical Series,* X, 1980.

Committee on Foreign Relations, *Executive Sessions, 1959, Historical Series,* XI, 1982.

Committee on Foreign Relations, *Review of Foreign Policy, 1958,* Part 1, 1958.

Committee on Foreign Relations, *Mutual Security Act of 1958,* 1958.

Committee on Foreign Relations, *Review of Foreign Policy, 1958,* Part 1, 1958.

Committee on Foreign Relations, *Study Mission in Caribbean Area, December 1957. Report of George D. Aiken to Committee on Foreign Relations,* 1958.

Committee on the Judiciary, *Communist Threat to the United States Through the Caribbean,* Part 9, 1960; Part 10, 1960; Part 12, 1961.

Committee on the Judiciary, *State Department Security: The Case of William Wieland,* 1962.

Committee on the Judiciary, *State Department Security: Testimony of William Wieland,* Part 5, 1962.

Committee on the Judiciary, *State Department Security: The William Wieland Case,* Part 1, 1962.

Select Committee to Study Governmental Operations, *Alleged Assassination Plots Involving Foreign Leaders,* 1975.

Congressional Record

Department of Commerce, *Investment in Cuba: Basic Information for the United States Businessman,* 1956.

Department of State

Biographic Register

Department of State Bulletin

Foreign Relations of the United States

Foreign Service List

Treaty Series

Export-Import Bank of Washington, *Reports to Congress*

Foreign Broadcast Information Service, *Daily Reports*

Foreign Commerce Weekly

Public Papers of the Presidents, Dwight D. Eisenhower

Public Papers of the Presidents, John F. Kennedy

Memoirs, Participant Accounts, and Documentary Collections

Attwood, William. *The Reds and the Blacks.* New York: Harper & Row, 1967.

———. *The Twilight Struggle: Tales of the Cold War.* New York: Harper & Row, 1987.

Báez, Luis. *Los que se fueron.* La Habana, Cuba: Editorial "José Martí," 1991.

Baker, Carlos, ed. *Ernest Hemingway: Selected Letters, 1917–1961.* New York: Charles Scribners' Sons, 1981.

Ball, George W. "JFK's Big Moment," *New York Review of Books,* Feb. 13, 1992, pp. 16, 18.

Barquín, Ramón M. *El día que Fidel Castro se apoderó de Cuba: 72 horas trágicas para la libertad en las Américas.* San Juan, PR: Editorial Rambar, 1978.

———. *Las luchas guerrilleras en Cuba: De la colonia a la Sierra Maestra,* 2 vols. Madrid, Spain: Playor, 1975.

Batista, Fulgencio. *Cuba Betrayed.* New York: Vantage, 1962.

Bethel, Paul. *The Losers.* New Rochelle, NY: Arlington House, 1969.

Bonachea, Rolando E., and Nelson P. Valdés, eds. *Revolutionary Struggle, 1947–1958,* Vol. I of *The Selected Works of Fidel Castro.* Cambridge, MA: MIT Press, 1972.

Bonsal, Philip W. *Cuba, Castro, and the United States.* Pittsburgh: Univ. of Pittsburgh Press, 1971.

Braden, Spruille. *Diplomats and Demagogues: The Memoirs of Spruille Braden.* New Rochelle, NY: Arlington House, 1971.

Brennan, Ray. *Castro, Cuba and Justice.* Garden City, NY: Doubleday, 1959.

Carrillo, Justo. "Vision and Revision: U.S.-Cuban Relations, 1902 to 1959," in Jaime Suchlicki et al., eds., *Cuba: Continuity and Change.* Miami: Univ. of Miami North-South Center, 1985. Pp. 163–74.

Casuso, Teresa. *Castro and Cuba.* New York: Random House, 1961.

Chomón, Faure. *El asalto al Palacio Presidencial.* La Habana, Cuba: Editorial de Ciencias Sociales, 1969.

Clissold, Stephen, ed. *Soviet Relations with Latin America, 1918–1968: A Documentary Survey.* London: Oxford Univ. Press, 1970.

Dubois, Jules. *Fidel Castro, Rebel—Liberator or Dictator?* Indianapolis: Bobbs-Merrill, 1959.

———. *Freedom Is My Beat.* Indianapolis: Bobbs-Merrill, 1959.

Durocher, Leo, with Ed Linn. *Nice Guys Finish Last.* New York: Simon and Schuster, 1975.

Eisenhower, Dwight D. *Waging Peace, 1956–1961.* Garden City, NY: Doubleday, 1965.

Eisenhower, Milton S. *The Wine Is Bitter: The United States and Latin America.* Garden City, NY: Doubleday, 1963.

Franco, Victor. *The Morning After: A French Journalist's Impression of Cuba Under Castro,* trans. by Ivan Kats and Philip Pendered. New York: Praeger, 1963.

Franqui, Carlos. *Diary of the Cuban Revolution.* New York: Viking, 1980.

———. *Family Portrait with Fidel.* New York: Random House, 1984.

———. *The Twelve.* New York: Lyle Stuart, 1968.

Guevara, Ernesto "Che." *Obra revolucionaria.* México, D.F.: Ediciones ERA, 1967.

Gunther, John. *Inside Latin America.* New York: Harper & Brothers, 1941.

Hoak, Don, with Myron Cope. "The Day I Batted Against Castro," in Charles Einstein, ed., *The Baseball Reader.* New York, 1980. Pp. 176–79.

Howard, Lisa. "Castro's Overture," *War/Peace Report,* Sept. 1963, pp. 3–5.

Kirkpatrick, Lyman B., Jr. *The Real CIA.* New York: Macmillan, 1968.

Lazo, Mario. *Dagger in the Heart: American Policy Failures in Cuba.* New York: Funk & Wagnalls, 1968.

López-Fresquet, Rufo. *My 14 Months with Castro.* Cleveland: World, 1966.

Llerena, Mario. *The Unsuspected Revolution: The Birth and Rise of Castroism.* Ithaca, NY: Cornell Univ. Press, 1978.

Lockwood, Lee. *Castro's Cuba, Cuba's Fidel.* New York: Vintage, 1969 ed.
Macaulay, Neill. *A Rebel in Cuba: An American Memoir.* Chicago: Quadrangle, 1970.
———. "I Fought for Fidel," *American Heritage* XLII (Nov. 1991), 78–92.
Mallin, Jay, ed. *Strategy for Conquest: Communist Documents on Guerrilla Warfare.* Coral Gables, FL: Univ. of Miami Press, 1970.
Mankiewicz, Frank, and Kirby Jones. *With Fidel: A Portrait of Castro and Cuba.* New York: Ballantine, 1975.
Márquez-Sterling, Carlos. *Historia de Cuba desde Colon hasta Castro.* New York: Las Américas, 1963.
Masetti, Jorge Ricardo. *Los que luchan y los que lloran: El Fidel Castro que yo ví.* Buenos Aires, Argentina: Ed. Freeland, 1958.
Matthews, Herbert L. *A World in Revolution: A Newspaperman's Memoir.* New York: Charles Scribner's Sons, 1971.
———. *The Cuban Story.* New York: George Braziller, 1961.
———. "Dissent Over Cuba," *Encounter* XXIII (July 1964), 82–90.
———. *Fidel Castro.* New York: Simon and Schuster, 1969.
———. "Matthews, Under Gunfire, Covered Cuban Revolt," *Times Talk* XII (Feb. 1959), 1–2.
———. *Revolution in Cuba.* New York: Charles Scribner's Sons, 1975.
Matthews, Nancie. "Matthews' Journey to Sierra Maestra: Wife's Version," *Times Talk* X (March 1957), 8.
McManus, Jane, ed. *From the Palm Tree: Voices of the Cuban Revolution.* Secaucus, NJ: Lyle Stuart, 1983.
Medina, Pablo. *Exiled Memories: A Cuban Childhood.* Austin: Univ. of Texas Press, 1990.
Meneses, Enrique. *Fidel Castro,* trans. by J. Halcro Ferguson. New York: Taplinger, 1966.
Miller, Warren. *90 Miles from Home: The Face of Cuba Today.* Boston: Little, Brown, 1961.
Morán Arce, Lucas. *La revolución cubana (1953–1959): Una versión rebelde.* Ponce, PR: Universidad Católica, 1980.
Murphy, Robert. *Diplomat Among Warriors.* Garden City, NY: Doubleday, 1964.
Nixon, Richard M. *Six Crises.* New York: Pyramid, 1968 ed.
Paar, Jack. *I Kid You Not.* Boston: Little, Brown, 1960.
Parker, John H. *Yankee You Can't Go Home Again.* Miami: American Club of Miami, 1986.
Pavón, Luis, ed. *Días de combate.* La Habana, Cuba: Instituto del Libro, 1970.
El pensamiento de Fidel Castro: Selección temática, enero 1959–abril 1961. La Habana, Cuba: Editora Política, 1983.
Pflaum, Irving P. *Tragic Island: How Communism Came to Cuba.* Englewood Cliffs, NJ: Prentice Hall, 1961.
Phillips, David A. *The Night Watch.* New York: Atheneum, 1977.
Phillips, R. Hart. *Cuba: Island of Paradise.* New York: McDowell, Obolensky, 1959.
———. *The Cuban Dilemma.* New York: Ivan Obolensky, 1962.

Powell, Adam Clayton. *Adam by Adam: The Autobiography of Adam Clayton Powell, Jr.* New York: Dial, 1971.

Reston, James. *Deadline: A Memoir.* New York: Random House, 1991.

Rivero, Nicolás. *Castro's Cuba: An American Dilemma.* Washington, DC: Luce, 1962.

Romualdi, Serafino. *Presidents and Peons: Recollections of a Labor Ambassador in Latin America.* New York: Funk & Wagnalls, 1967.

Schlesinger, Arthur M., Jr. *A Thousand Days: John F. Kennedy in the White House.* Boston: Houghton Mifflin, 1965.

———. *Robert F. Kennedy and His Times.* Boston: Houghton Mifflin, 1978.

Smith, Earl E. T. *The Fourth Floor: An Account of the Castro Communist Revolution.* New York: Random House, 1962.

Smith, Joseph B. *Portrait of a Cold Warrior.* New York: G. P. Putnam's Sons, 1976.

Smith, Wayne S. *The Closest of Enemies: A Personal and Diplomatic History of the Castro Years.* New York: W. W. Norton, 1987.

Suárez, Andrés. *Cuba: Castroism and Communism, 1959–1966.* Cambridge, MA: MIT Press, 1967.

———. "The Cuban Revolution: The Road to Power," *Latin American Research Review* VII (Fall 1972), 5–29.

Taber, Robert. *M-26: Biography of a Revolution.* New York: Lyle Stuart, 1961.

Urrutia Lleó, Manuel. *Fidel Castro & Company, Inc.: Communist Tyranny in Cuba.* New York: Praeger, 1964.

Newspapers and Periodicals

AUFS Reports; Bohemia; Business Week; Christian Century; Editor and Publisher; El Diario de Nueva York; Export Trade; Hartford Courant; Hispanic American Report; Inter-American Economic Affairs; Journal of Commerce; Life; Look; Los Angeles Examiner; Los Angeles Times; Miami Herald; Miami News; The Nation; National Review; The New Leader; The New Republic; New York Daily News; New York Herald Tribune; New York Journal-American; New York Mirror; New York Times; New York World Telegram; Newark Sunday News; Newsweek; The Post-Times; Report on Cuba; Revolución; Saturday Review; Sunday Star Ledger (Newark, NJ); *Time; The Times of Havana; Wall Street Journal; Washington Post; Washington Star*

Secondary Books and Articles

Aaron, Harold R. "Guerrilla War in Cuba," *Military Review* XLV (May 1965), 40–46.

Aguilar, Luis E. *Cuba 1933: Prologue to Revolution.* New York: W. W. Norton, 1974 ed.

Ambrose, Stephen E., with Richard H. Immerman. *Ike's Spies.* Garden City, NY: Doubleday, 1981.

Ameringer, Charles D. *The Democratic Left in Exile: The Antidictatorial Struggle in the Caribbean, 1945–1959.* Coral Gables, FL: Univ. of Miami Press, 1974.

———. "The Auténtico Party and the Political Opposition in Cuba, 1952–57," *Hispanic American Historical Review* LXV (May 1985), 327–52.

Andrew, Christopher, and Oleg Gordievsky. *KGB: The Inside Story of Its Foreign Operations from Lenin to Gorbachev.* New York: HarperCollins, 1990.

"Anti-Americanism: Origins and Context," *The Annals* CDXCVII (May 1988), 9–171.

Barnet, Richard J. *Intervention & Revolution: The United States in the Third World.* New York: New American Library, rev. ed., 1972.

Becnel, Thomas A. "Fulbright of Arkansas v. Ellender of Louisiana: The Politics of Sugar and Rice, 1937–1974," *Arkansas Historical Quarterly* XLIV (Winter 1984), 289–303.

Bender, Lynn Darrell. *The Politics of Hostility: Castro's Revolution and United States Policy.* Hato Rey, PR: Inter American Univ. Press, 1975.

Benjamin, Jules R. "Interpreting the U.S. Reaction to the Cuban Revolution, 1959–1960," in *Cuban Studies 19.* Pittsburgh: Univ. of Pittsburgh Press, 1989). Pp. 145–65.

———. *The United States and the Origins of the Cuban Revolution: An Empire for Liberty in an Age of National Liberation.* Princeton, NJ: Princeton Univ. Press, 1990.

Beschloss, Michael R. *The Crisis Years: Kennedy and Khrushchev, 1960–1963.* New York: HarperCollins, 1991.

Bhattacharya, Sauripada. "Cuban-Soviet Relations under Castro, 1959–1964," *Studies on the Soviet Union* IV (No. 3, 1965), 27–36.

Billard, Jules B. "Guantánamo: Keystone in the Caribbean," *National Geographic* CXIX (March 1961), 420–36.

Blasier, Cole. *The Giant's Rival: The USSR and Latin America.* Pittsburgh: Univ. of Pittsburgh Press, 1983.

———. *The Hovering Giant: U.S. Responses to Revolutionary Change in Latin America.* Pittsburgh: Univ. of Pittsburgh Press, 1976.

——— and Carmelo Mesa-Lago, eds. *Cuba in the World.* Pittsburgh: Univ. of Pittsburgh Press, 1979.

Bonachea, Ramón, and Marta San Martín. *The Cuban Insurrection, 1952–1959.* New Brunswick, NJ: Transaction Books 1974.

Boughton, George J. "Soviet-Cuban Relations, 1956–1960," *Journal of Inter-American Studies* XVI (Nov. 1974), 436–53.

Bourne, Peter G. *Fidel: A Biography of Fidel Castro.* New York: Dodd, Mead, 1986.

Brenner, Philip. *From Confrontation to Negotiation: U.S. Relations with Cuba.* Boulder, CO: Westview, 1988.

Britton, John A. *Carleton Beals: A Radical Journalist in Latin America.* Albuquerque: Univ. of New Mexico Press, 1989.

Brown, Bruce. "Cuban Baseball," *The Atlantic* CCLIII (June 1984), 109–14.

Brunner, Heinrich. *Cuban Sugar Policy from 1963 to 1970,* trans. by Marguerite Borchardt and H.F. Broch de Rothermann. Pittsburgh: Univ. of Pittsburgh Press, 1977.

Burton, David H. *Theodore Roosevelt: Confident Imperialist.* Philadelphia: Univ. of Pennsylvania Press, 1968.

Bustamante, Jorge A. "Demystifying the United States-Mexico Border," *Journal of American History* LXXIX (Sept. 1992), 485–90.

Infante, G. Cabrera. *Infante's Inferno,* trans. by Suzanne Jill Levine. New York: Harper & Row, 1984.

Corbitt, Duvon C. "Cuban Revisionist Interpretations of Cuba's Struggle for Independence," *Hispanic American Historical Review* XLIII (Aug. 1963), 395–404.

Cuban Economic Research Project, *A Study of Cuba.* Coral Gables, FL: Univ. of Miami Press 1965.

Debray, Régis. *Revolution in the Revolution?: Armed Struggle and Political Struggle in Latin America.* New York: Monthly Review Press, 1967.

del Aguila, Juan M. *Cuba: Dilemmas of a Revolution.* Boulder, CO: Westview, rev. ed., 1988.

Dinerstein, Herbert S. *The Making of a Missile Crisis: October 1962.* Baltimore, MD: John Hopkins Univ. Press, 1976.

Domínguez, Jorge I. *Cuba: Order and Revolution.* Cambridge, MA: Harvard Univ. Press, 1978.

______. *To Make a World Safe for Revolution: Cuba's Foreign Policy.* Cambridge, MA: Harvard Univ. Press, 1989.

______ and Rafael Hernández, eds. *U.S.-Cuban Relations in the 1990s.* Boulder, CO: Westview, 1989.

Dorschner, John, and Roberto Fabricio. *The Winds of December: The Cuban Revolution 1958.* New York: Coward, McCann & Geoghegan, 1980.

Draper, Theodore. *Castroism: Theory and Practice.* New York: Praeger, 1965.

Dumpierre, Erasmo. "El monopolio de la Cuban Telephone Company," *Bohemia* LXVII (Sept. 12, 1975), 88–92.

Eisenberg, Dennis, Uri Dan, and Eli Landau. *Meyer Lansky: Mogul of the Mob.* New York: Paddington, 1979.

Erisman, H. Michael. *Cuba's International Relations.* Boulder, CO: Westview, 1985.

Falk, Pamela S. *Cuban Foreign Policy: Caribbean Tempest.* Lexington, MA: Lexington Books, 1986.

Farber, Samuel. "The Cuban Communists in the Early Stages of the Cuban Revolution: Revolutionaries or Reformists?" *Latin American Research Review* XVIII, No. 1 (1983), 59–84.

______. *Revolution and Reaction in Cuba, 1933–1960: A Political Sociology from Machado to Castro.* Middletown, CT: Wesleyan Univ. Press, 1976.

Francis, Michael J. "The U.S. Press and Castro: A Study in Declining Relations," *Journalism Quarterly* XLIV (Summer 1967), 257–66.

Frederick, Howard H. *Cuban-American Radio Wars: Ideology in International Telecommunications.* Norwood, NJ: Ablex, 1986.

García, Alejandro, and Oscar Zanetti. *United Fruit Company: Un caso del dominio imperialista en Cuba.* La Habana, Cuba: Editorial de Ciencias Sociales, 1976.

Gleijeses, Piero. *Shattered Hope: The Guatemalan Revolution and the United States, 1944–1954.* Princeton, NJ: Princeton Univ. Press, 1991.

González, Edward. "Castro's Revolution, Cuban Communist Appeals, and the Soviet Response," *World Politics* XXI (Oct. 1968), 39–68.

______. *Cuba Under Castro: The Limits of Charisma.* Boston: Houghton Mifflin, 1974.

Halperin, Ernest. *The Rise and Decline of Fidel Castro.* Berkeley: Univ. of California Press, 1972.

Healy, David. *Drive to Hegemony: The United States in the Caribbean, 1898–1917.* Madison: Univ. of Wisconsin Press, 1988.

Heinl, Robert D., Jr. "How We Got Guantánamo," *American Heritage* XIII (Feb. 1962), 18–21, 94–97.

Hennessy, C.A.M. "The Roots of Cuban Nationalism," *International Affairs* XXXIX (July 1963), 346–58.

Hernández, José M. *Cuba and the United States: Intervention and Militarism, 1868–1933.* Austin: Univ. of Texas Press, 1993.

Heston, Thomas J. *Sweet Subsidy: The Economic and Diplomatic Effects of the U.S. Sugar Acts, 1934–1974.* New York: Garland, 1987.

Higgins, Trumbull. *The Perfect Failure: Kennedy, Eisenhower, and the CIA at the Bay of Pigs.* New York: W. W. Norton, 1987.

Hinckle, Warren, and William W. Turner. *The Fish Is Red: The Story of the Secret War Against Castro.* New York: Harper & Row, 1981.

Hitchman, James H. "The Platt Amendment Revisited: A Bibliographic Survey," *The Americas* XXIII (April 1967), 343–69.

Hufbauer, Gary C., and Jeffrey J. Schott. *Economic Sanctions Reconsidered: History and Current Policy.* Washington, DC: Institute for International Economics, 2nd ed., 1990.

Hunt, Michael H. *Ideology and U.S. Foreign Policy.* New Haven: Yale Univ. Press, 1987.

Immerman, Richard H. *The CIA in Guatemala: The Foreign Policy of Intervention.* Austin: Univ. of Texas Press, 1982.

Jackson, D. Bruce. *Castro, the Kremlin, and Communism in Latin America.* Baltimore, MD: John Hopkins Univ. Press, 1969.

Johnson, Haynes. *The Bay of Pigs.* New York: W.W. Norton, 1964.

Johnson, John J. *Latin America in Caricature.* Austin: Univ. of Texas Press, 1980.

Johnson, Leland L. "U.S. Business Interests in Cuba and the Rise of Castro," *World Politics* XVII (April 1965), 440–59.

Kahn, E.J., Jr. "A Reporter at Large: Here Come the Marines," *The New Yorker* LV (Nov. 1979), 190ff.

Karol, K. S. *Guerrillas in Power: The Course of the Cuban Revolution,* trans. by Arnold Pomerans. New York: Hill and Wang, 1970.

Keohane, Robert O., and Joseph Nye, *Power and Interdependence.* Glenview, IL: Scott, Foresman, 2nd ed., 1989.

Kirk, John M. *José Martí: Mentor of the Cuban Nation.* Gainesville: Univ. Presses of Florida, 1983.

Lacey, Robert. *Little Man: Meyer Lansky and the Gangster Life.* Boston: Little, Brown, 1991.

LaCharité, Norman A. *Case Studies in Insurgency and Revolutionary Warfare: Cuba, 1953–1959.* Washington, DC: Special Operations Research Office, American Univ., 1963.

LaFeber, Walter. *Inevitable Revolutions: The United States in Central America.* New York: W.W. Norton, rev. ed., 1993.

LeoGrande, William M. "Cuban Dependency: A Comparison of Pre-Revolutionary and Post-Revolutionary International Economic Relations," *Cuban Studies/Estudios Cubanos* IX (July 1979), 1–28.

Lévesque, Jacques. *The USSR and the Cuban Revolution: Soviet Ideological and Strategical Perspectives, 1959–77.* New York: Praeger, 1978.

Lewis, Howard L. "The Cuban Revolt Story: AP, UPI and 3 Papers," *Journalism Quarterly* XXXVII (Autumn 1961), 573–78.

Lisago, Peter, and Marguerite Higgins. *Overtime in Heaven: Adventures in the Foreign Service.* Garden City, NY: Doubleday, 1964.

López Segrera, Francisco. *Cuba: Politíca exterior y revolución 1959–88.* La Habana: I.S.R.I., 1989.

Lukas J. Anthony. *Nightmare: The Underside of the Nixon Years.* New York: Viking, 1976.

Luxenberg, Alan H. "Did Eisenhower Push Castro into the Arms of the Soviets?," *Journal of Interamerican Studies and World Affairs* XXX (Spring 1988), 37–71.

Macaulay, Neill. "The Cuban Rebel Army: A Numerical Survey," *Hispanic American Historical Review* LVIII (May 1978), 284–95.

Mazrui, Ali A. "Uncle Sam's Hearing Aid," in Sanford J. Ungar, ed., *Estrangement: America and the World.* New York: Oxford Univ. Press, 1985. Pp. 181–92.

Mesa-Lago, Carmelo. *The Economy of Socialist Cuba: A Two-Decade Appraisal.* Albuquerque: Univ. of New Mexico Press 1981.

———, ed. *Revolutionary Change in Cuba.* Pittsburgh: Univ. of Pittsburgh Press, 1971.

Messick, Hank. *Lansky.* New York: G. P. Putnam's Sons, 1971.

Meyer, Karl, and Tad Szulc. *The Cuban Invasion.* New York: Praeger, 1962.

Miller, Nicola. *Soviet Relations with Latin America, 1959–1987.* Cambridge, UK: Cambridge Univ. Press, 1989.

Montaner, Carlos Alberto. "The Roots of Anti-Americanism in Cuba: Sovereignty in an Age of World Cultural Homogeneity," *Caribbean Review* XIII (Spring 1984), 13–16, 42–46.

Morley, Morris H. *Imperial State and Revolution: The United States and Cuba, 1952–1986.* New York: Cambridge Univ. Press, 1987.

Nathan, James A., ed. *The Cuban Missile Crisis Revisited.* New York: St. Martin's, 1992.

Newman, Philip C. *Cuba Before Castro: An Economic Appraisal.* India: Foreign Studies Institute, 1965.

Noble, G. Bernard. *Christian A. Herter.* New York: Cooper Square, 1970.

O'Connor, James. *The Origins of Socialism in Cuba.* Ithaca, NY: Cornell Univ. Press, 1970.

O'Neal, Bill. *The International League: A Baseball History, 1884–1991.* Austin, TX: Eakin Press, 1992.

Oppenheimer, Andres. *Castro's Final Hour.* New York: Simon & Schuster, 1992.

Pach, Chester J., Jr. *Arming the Free World: The Origins of the United States Military Assistance Program, 1945–1950.* Chapel Hill: Univ. of North Carolina Press, 1991.

Packenham, Robert A., ed. "Capitalist Dependency and Socialist Dependency: The Case of Cuba," in Jan F. Triska, ed., *Dominant Powers and Subordinate States: The United States in Latin America and the Soviet Union in Eastern Europe.* Durham, NC: Duke Univ. Press, 1986. Pp. 310–41.

Padula, Alfred. "Financing Castro's Revolution, 1956–1958," *Revista/Review Interamericana* VIII (Summer 1978), 234–46.

Parish, James Robert, with Steven Whitney. *The George Raft File: The Unauthorized Biography.* New York: Drake, 1973.

Paterson, Thomas G. "Commentary: The Defense-of-Cuba Theme and the Missile Crisis," *Diplomatic History* XIV (Spring 1990), 249–56.

———. "Fixation with Cuba: The Bay of Pigs, Missile Crisis, and Covert War Against Fidel Castro," in Thomas G. Paterson, ed., *Kennedy's Quest for Victory: American Foreign Policy, 1961–1963.* New York: Oxford Univ. Press, 1989. Pp. 123–55, 343–52.

———. *On Every Front: The Making and Unmaking of the Cold War.* New York: W.W. Norton, rev. ed., 1992.

Pearson, Ruth. "Cuba's Double Jeopardy," *Bulletin of the Atomic Scientists* XLVIII (Dec. 1992), 44–50.

Peterson, Virgil. *The Mob: 200 Years of Organized Crime in New York.* Aurora, IL: Green Hill, 1983.

Pérez, Louis A., Jr. *Army Politics in Cuba, 1898–1958.* Pittsburgh: Univ. of Pittsburgh Press, 1976.

———. *Cuba and the United States: Ties of Singular Intimacy.* Athens: Univ. of Georgia Press, 1990.

———. *Cuba: Between Reform and Revolution.* New York: Oxford Univ. Press, 1988.

———. *Cuba Under the Platt Amendment, 1902–1934.* Pittsburgh: Univ. of Pittsburgh Press, 1986.

———. "The Meaning of the *Maine:* Causation and the Historiography of the Spanish-American War," *Pacific Historical Review* LVIII (Aug. 1989), 293–322.

Pérez-López, Jorge F. *The Economics of Cuban Sugar.* Pittsburgh: Univ. of Pittsburgh Press, 1991.

Pike, Frederick B. *The United States and Latin America: Myths and Stereotypes of Civilization and Nature.* Austin: Univ. of Texas Press, 1992.

Pino-Santos, Oscar. *Cuba, historia y economía: Ensayos.* La Habana, Cuba: Editorial de Ciencias Sociales, 1983.

Rabe, Stephen G. *Eisenhower and Latin America: The Foreign Policy of Anticommunism.* Chapel Hill: Univ. of North Carolina Press, 1988.

———. "The Johnson (Eisenhower?) Doctrine for Latin America," *Diplomatic History* IX (Winter 1985), 95–100.

Radosh, Ronald. *American Labor and United States Foreign Policy: The Cold War in the Unions from Gompers to Lovestone.* New York: Vintage, 1970.

Reid, Ed. *The Grim Reapers: The Anatomy of Organized Crime in America.* Chicago: Henry Regnery, 1969.

Robbins, Carla Anne. *The Cuban Threat.* New York: McGraw Hill, 1983.

Ruddy, T. Michael. *The Cautious Diplomat: Charles E. Bohlen and the Soviet Union, 1929–1969.* Kent, OH: Kent State Univ. Press, 1986.

Ruíz, Ramón E. *Cuba: The Making of a Revolution.* New York: W.W. Norton, 1970 ed.

Safford, Jeffrey. "The Nixon-Castro Meeting of 19 April 1959," *Diplomatic History* IV (Fall 1980), 426–31.

Schroeder, Susan. *Cuba: A Handbook of Historical Statistics.* Boston: G. K. Hall, 1982.

Schwartz, Rosalie. "Cuban Tourism: A History Lesson," *Cuba Update* XII (Winter/Spring 1991), 24–27.

Sims, Harold D. "Cuban Labor and the Communist Party, 1937–1958: An Interpretation," *Cuban Studies* XV (Winter 1985), 43–58.

Smith, Geoffrey S. "National Security and Personal Isolation: Sex, Gender, and Disease in the Cold-War United States," *International History Review* XIV (May 1992), 307–37.

Smith, Robert Freeman. "Twentieth-Century Cuban Historiography," *Hispanic American Historical Review* XLIV (Feb. 1964), 44–73.

Smith, Tony. "The Spirit of the Sierra Maestra: Five Observations on Writing about Cuban Foreign Policy," *World Politics* XL (Oct. 1988), 98–119.

Smith, Wayne S. "U.S.-Cuba Relations: Twenty-Five Years of Hostility," in Sandor Halenbsky and John M. Kirk, eds., *Cuba: Twenty-Five Years of Revolution, 1959–1984.* New York: Praeger, 1985. Pp. 333–51.

———, and Esteban Morales Domínguez, eds. *Subject to Solution: Problems in Cuba-U.S. Relations.* Boulder, CO: Lynne Rienner, 1988.

Stiller, Jesse H. *George S. Messersmith: Diplomat of Democracy.* Chapel Hill: Univ. of North Carolina Press, 1987.

Szulc, Tad. "Friendship Is Possible, But . . . ," *Parade Magazine,* April 1, 1984, pp. 4–6.

———. *Fidel: A Critical Portrait.* New York: Morrow, 1986.

Thayer, George. *The War Business: The International Trade in Armaments.* New York: Simon and Schuster, 1969.

Thelen, David. "Of Audiences, Borderlands, and Comparisons: Toward the Internationalization of American History," *Journal of American History* LXXIX (Sept. 1992), 436–44.

Thomas, Hugh. "Cuba: The United States and Batista, 1952–58," *World Affairs* CXLIX (Spring 1987), 169–77.

———. *The Cuban Revolution.* New York: Harper & Row, 1977.

———. "Middle-Class Politics and the Cuban Revolution," in Claudio Véliz, ed., *The Politics of Conformity in Latin America.* London: Oxford Univ. Press, 1967. Pp. 249–77.

Timoshenko, Vladimir P., and Boris C. Swerling. *The World's Sugar: Progress and Policy.* Stanford, CA: Stanford Univ. Press, 1957.

Torres, Angel. *La historia del beisbol cubano, 1878–1976.* Los Angeles: Angel Torres, 1976.

Truby, J. David. "Now Pitching for the Giants . . . Fidel Castro," *Sports History* II (March 1989), 12ff.

Tygiel, Jules. *Baseball's Great Experiment: Jackie Robinson and His Legacy.* New York: Oxford Univ. Press, 1983.

Valenti, Peter. *Errol Flynn: A Bio-Bibliography.* Westport, CT: Greenwood, 1984.

Vandenbroucke, Lucien S. "Anatomy of a Failure: The Decision to Land at the Bay of Pigs," *Political Science Quarterly* XCIX (Fall 1984), 471–91.

Wagner, Eric A. "Baseball in Cuba," *Journal of Popular Culture* XVIII (Summer 1984), 113–20.

Welch, Richard E., Jr. *Response to Revolution: The United States and the Cuban Revolution, 1959–1961.* Chapel Hill: Univ. of North Carolina Press, 1985.

———. "Herbert L. Matthews and the Cuban Revolution," *The Historian* XLVII (Nov. 1984), 1–18.

Weston, Rubin F. *Racism in U.S. Imperialism: The Influence of Racial Assumptions on American Foreign Policy, 1893–1946.* Columbia: Univ. of South Carolina Press, 1972.

Wiarda, Howard J. "Friendly Tyrants and American Interests," in Daniel Pipes and Adam Garfinkle, eds., *Friendly Tyrants: An American Dilemma.* London: Macmillan, 1991. Pp. 3–20.

Wohlstetter, Roberta. "Kidnapping to Win Friends and Influence People," *Survey* (Great Britain) XX, No. 4 (1974), 1–40.

Wyden, Peter. *Bay of Pigs.* New York: Simon and Schuster, 1979.

Young, Marilyn B. *The Vietnam War, 1945–1990.* New York: HarperCollins, 1991.

Zahniser, Marvin R., and W. Michael Weis. "A Diplomatic Pearl Harbor? Richard Nixon's Goodwill Mission to Latin America in 1958," *Diplomatic History* XIII (Spring 1989), 163–90.

Zanetti, Oscar. "American History: A View from Cuba," *Journal of American History* LXXIX (Sept. 1992), 530–31.

Zeitlin, Maurice. *Revolutionary Politics and the Cuban Working Class.* Princeton, NJ: Princeton Univ. Press, 1967.

Zuaznábar, Ismael. *La economía cubana en la década del 50.* La Habana, Cuba: Editorial de Ciencias Sociales, 1986.

Reference Works

Arbena, Joseph L. *An Annotated Bibliography of Latin American Sport: Pre-Conquest to the Present.* Westport, CT: Greenwood, 1989.

Blackey, Robert, ed. *Revolutions and Revolutionists: A Comprehensive Guide to the Literature.* Santa Barbara, CA: ABC-Clio, 1982.

Chilcote, Ronald H., with Sheryl Lutjens, eds. *Cuba, 1953–1978: A Bibliographical Guide to the Literature,* 2 vols. White Plains, NY: Kraus, 1986.

Grow, Michael, ed. *Scholar's Guide to Washington, D.C. for Latin American and Caribbean Studies.* Washington, DC: Smithsonian Institution Press, 1979.

Meyer, Michael C., ed. *Supplement to a Bibliography of United States-Latin American Relations Since 1810.* Lincoln: Univ. of Nebraska Press, 1979.

Pérez, Louis A., Jr. *A Guide to Cuban Collections in the United States.* New York: Greenwood, 1991.

———. *Cuba: An Annotated Bibliography.* New York: Greenwood, 1988.

———. *The Cuban Revolutionary War, 1953–1958: A Bibliography.* Metuchen, NJ: Scarecrow, 1976.

———. *Historiography in the Revolution: A Bibliography of Cuban Scholarship, 1959–1979.* New York: Garland, 1981.

Suchlicki, Jaime. *Historical Dictionary of Cuba.* Metuchen, NJ: Scarecrow, 1988.

Trask, David F., et al., eds. *A Bibliography of United States-Latin American Relations Since 1810.* Lincoln: Univ. of Nebraska Press, 1968.

Valdés, Nelson P., and Edwin Liewen, *The Cuban Revolution: A Research Study Guide, 1959–1969.* Albuquerque: Univ. of New Mexico Press, 1971.

Dissertations and Other Unpublished Materials

Aaron, Harold R. "The Seizure of Political Power in Cuba, 1956–1959." Ph.D. dissertation, Georgetown University, 1964.

Cozean, Jon D. "The U.S. Elite Press and Foreign Policy: The Case of Cuba." Ph.D. dissertation, American University, 1979.

Gosse, Van E. "History Mission: Cuba, the New Left, and the Origins of Latin American Solidarity in the United States, 1955–1963." Ph.D. dissertation, Rutgers University, 1992.

Grunwald, E. A. "An Assessment of Current U.S. Policy Towards Cuba," Research Paper, March 1970, National War College, Washington, DC.

López Segrera, Francisco. "Cuba-EEUU: Percepciones mutuas (1959–1993)," paper, March 3, 1993, Paris, France.

Padula, Alfred L., Jr. "The Fall of the Bourgeoisie: Cuba, 1959–1961." Ph.D. dissertation, University of New Mexico, 1974.

Pawley, William D. "William D. Pawley's Book," undated.

Pérez, Louis A., Jr. "Cuba-United States Relations: A Century of Conflict," paper delivered at University of Connecticut Foreign Policy Seminar, Storrs, March 13, 1990.

Tremayne, Russell M. "The Cuban Electric Company: A Study of Expropriation." Master's thesis, Boise State University, 1982.

Wolf, Harold A. "The United States Sugar Policy and Its Impact Upon Cuba: A Reappraisal." Ph.D. dissertation, University of Michigan, 1958.

Miscellaneous Materials

Declassified Documents Reference Service, Carrollton Press.

Farr & Company, *Manual of Sugar Companies, 1955/56,* 33rd ed. New York: Farr, 1956.

Farr, Whitlock & Co., *Manual of Sugar Companies,* 35th ed. New York: Farr, Whitlock, 1960.

Javier Figueroa, notes of interview of Raúl Chibás, San Juan, Puerto Rico, Jan. 7, 1985.

Jane Franklin, *Cuban Foreign Relations: A Chronology*. New York: Center for Cuban Studies, 1984.

International Bank for Reconstruction and Development, *Report on Cuba*. Baltimore, MD: Johns Hopkins Press, 1951.

Sears, Roebuck and Company, *Annual Reports*.

INDEX

Castro's victory. By the time the defiant revolutionary leader entered Havana in early 1959, the foundation of the long, bitter hostility between Cuba and the United States had been firmly laid.

Since the end of the Cold War, the futures of Communist Cuba and Fidel Castro have become clouded. Paterson's gripping and timely account explores the origins of America's troubled relationship with its island neighbor and explains what went wrong and how the United States "let this one get away." Because the paths to the future necessarily cross those of the past, Paterson's timely study provides the context for understanding changes in U.S.-Cuba relations today.

Denise Carreau

Thomas G. Paterson is Professor of History at the University of Connecticut. He is the winner of many awards, and author or editor of more than a dozen books, including *On Every Front: The Making and Unmaking of the Cold War*, *Kennedy's Quest for Victory*, and the popular textbook *A People and a Nation.*